THE *MAPPAE MUNDI* OF MEDIEVAL ICELAND

Studies in Old Norse Literature

Print ISSN 2514-0701
Online ISSN 2514-071X

Series Editors
Professor Sif Rikhardsdottir
Professor Carolyne Larrington

Studies in Old Norse Literature aims to provide a forum for monographs and collections engaging with the literature produced in medieval Scandinavia, one of the largest surviving bodies of medieval European literature. The series investigates poetry and prose alongside translated, religious and learned material; although the primary focus is on Old Norse-Icelandic literature, studies which make comparison with other medieval literatures or which take a broadly interdisciplinary approach by addressing the historical and archaeological contexts of literary texts are also welcomed. It offers opportunities to publish a wide range of books, whether cutting-edge, theoretically informed writing, provocative revisionist approaches to established conceptualizations, or strong, traditional studies of previously neglected aspects of the field. The series will enable researchers to communicate their findings both beyond and within the academic community of medievalists, highlighting the growing interest in Old Norse-Icelandic literary culture.

Proposals or queries should be sent in the first instance to the editors or to the publisher, at the addresses given below.

Professor Sif Rikhardsdottir, Faculty of Icelandic and Comparative Cultural Studies, University of Iceland, Aðalbygging v/Sæmundargötu, S-101 Reykjavik, Iceland

Professor Carolyne Larrington, Faculty of English Language and Literature, St John's College, Oxford University, Oxford, OX1 3JP, UK

Boydell & Brewer, PO Box 9, Woodbridge, Suffolk, IP12 3DF, UK

Previous volumes in the series are listed at the end of the volume.

The *Mappae Mundi* of Medieval Iceland

Dale Kedwards

D. S. BREWER

First published 2020
D. S. Brewer, Cambridge

ISBN 978 1 84384 569 0

D. S. Brewer is an imprint of Boydell & Brewer Ltd
PO Box 9, Woodbridge, Suffolk IP12 3DF, UK
and of Boydell & Brewer Inc.
668 Mt Hope Avenue, Rochester, NY 14620–2731, USA
website: www.boydellandbrewer.com

A catalogue record for this book is available
from the British Library

This publication is printed on acid-free paper

MIX
Paper from
responsible sources
FSC® C013056

Printed and bound in Great Britain by TJ International Ltd, Padstow, Cornwall

This book is dedicated to Julie Kedwards, Sally Kedwards,
Ida Parker and Max the dog.

Contents

Illustrations ix
Acknowledgements xi
Abbreviations xiii

Introduction 1
1. The Icelandic Hemispherical World Maps 23
2. The Icelandic Zonal Map 63
3. The Two Maps from Viðey 101
4. Iceland in Europe 119
5. Forty Icelandic Priests and a Map of the World 147
Conclusion 179

Map Texts and Translations

The Icelandic Hemispherical World Maps (Copenhagen, Arnamagnæan Collection, AM 736 I 4to, f. 1v, *c.* 1300 and Copenhagen, Arnamagnæan Collection, AM 732b 4to, f. 3r, *c.* 1300–25) 187
The Icelandic Zonal Map (Reykjavík, Stofnun Árna Magnússonar, GkS 1812 4to, f.11v, 1315–*c.*1400) 193
The Larger Viðey Map (Reykjavík, Stofnun Árna Magnússonar, GkS 1812 4to, ff. 5v–6r, *c.* 1225–50) 197
The Smaller Viðey Map (Reykjavík, Stofnun Árna Magnússonar, GkS 1812 4to, f. 6v, *c.* 1225–50) 231

Bibliography 241
Index 257

Illustrations

1. Macrobius's celestial-terrestrial zone diagram (Baltimore, Walters Art Gallery, Walters Ms. W.22, f. 41r, *c.* 1175–1200) 29

2. Hemispherical world map in the Liber Floridus (Ghent, University of Ghent Library, MS 92, f. 227v, 1121) (Ghent University Library, BHSL.HS.0092) 30

3. Zonal map in the Liber Floridus (Ghent, Ghent University Library, MS 92, f. 24v, 1121) (Ghent University Library, BHSL. HS.0092) 32

4. The context of the older version of the Icelandic hemispherical world map (Copenhagen, Arnamagnæan Collection, AM 736 I 4to, f. 1v, *c.* 1300. Photograph Suzanne Reitz. Reproduced with permission from the Arnamagnæan Institute) 41

5. An Icelandic geographical treatise and text about the three Biblical Kings (Copenhagen, Arnamagnæan Collection, AM 736 I 4to, f. 1r, *c.* 1300. Photograph Suzanne Reitz. Reproduced with permission from the Arnamagnæan Institute) 46

6. The Biblical Magi following the star on the Catalan World Atlas (Abraham Cresques?) (Paris, Bibliothequè Nationale de France, *c.* 1375) 49

7. Icelandic map of Jerusalem (Copenhagen, Arnamagnæan Collection, AM 736 I 4to, f. 2r, *c.* 1300. Photograph Suzanne Reitz. Reproduced with permission from the Arnamagnæan Institute) 52

8. The context of the younger version of the Icelandic hemispherical world map (Copenhagen, Arnamagnæan Collection, AM 732b 4to, f. 3r, *c.* 1300–25. Photograph Suzanne Reitz. Reproduced with permission from the Arnamagnæan Institute) 54

9. The nested planetary spheres (Copenhagen, Arnamagnæan Collection, AM 732b 4to f. 3v, *c.* 1300–25. Photograph Suzanne Reitz. Reproduced with permission from the Arnamagnæan Institute) 58

10. The heliocentric orbits of the inner planets, Mercury and Venus (Reykjavík, Stofnun Árna Magnússonar, GkS 1812 I 4to, f. 10v, 1315–*c.*1400) 74

11. The revolutionary motions of the outer planets (Reykjavík, Stofnun Árna Magnússonar, GkS 1812 I 4to, f.11r, 1315–*c.*1400) 83

12. The context of the Icelandic zonal map (Reykjavík, Stofnun Árna Magnússonar, GkS 1812 I 4to, f.11v, 1315–*c.*1400) 86

13. Details from the Icelandic zonal and hemispherical world maps (Reykjavík, Stofnun Árna Magnússonar, GkS 1812 I 4to, f.11v, 1315–*c.*1400 and Copenhagen, Arnamagnæan Collection, AM 736 I 4to, f. 1v, *c.* 1300. Photograph Suzanne Reitz. Reproduced with permission from the Arnamagnæan Institute) 90

14. Three maps in a manuscript of William of Conches's *De philosophia mundi*. Philadelphia, University of Penn, LJS 384, ff. 13r, 13v, and 15r, *c.* 1150 (Lawrence J Schoenberg Collection of Manuscripts, Kislak Centre for Special Collections, Rare Books and Manuscripts, University of Pennsylvania) 93

15. Diagram of the eclipses and tides (Reykjavík, Stofnun Árna Magnússonar, GkS 1812 I 4to, f.12v, 1315–*c.*1400) 95

16. The register of forty highborn Icelandic priests (Reykjavík, Stofnun Árna Magnússonar, GkS 1812 III 4to, f. 5r, *c.* 1225–50) 159

17. Icelandic depiction of Christ in Majesty, with T-O map, in the Icelandic Teiknibókin (Reykjavík, Stofnun Árna Magnússonar, AM 673a III 4to, f. 16r, 1450–75) 173

The author and publisher are grateful to all the institutions and individuals listed for permission to reproduce the materials in which they hold copyright. Every effort has been made to trace the copyright holders; apologies are offered for any omission, and the publisher will be pleased to add any necessary acknowledgement in subsequent editions.

Acknowledgements

Writing these acknowledgements is a long anticipated pleasure. I would like to thank Jürg Glauser and colleagues at the University of Zürich (Nationaler Forschungsschwerpunkt Medienwandel – Medienwechsel – Medienwissen. Historische Perspektiven) for granting me a research fellowship that allowed me to compile the texts and translations. Much of the research and writing was completed in the course of a fellowship awarded by the Centre for Medieval Literature, funded by the Danish National Research Foundation (project DNRF102ID), at Syddansk Universitet. This book's indebtedness to the Centre for Medieval Literature reads, I am sure, on every page, and it is my pleasure to thank Lars Boje Mortensen and Elizabeth M. Tyler for their support. Finishing touches were made early in the course of a fellowship generously awarded by the Carlsberg Foundation, and held in collaboration with Stofnun Vigdísar Finnbogadóttur, Þjóðminjasafnið Íslands, and Det Nationalhistoriske Museum. Each of these places has stimulated its own fundamental rethinking of the book's central argument and aims, and I want to thank all my colleagues for their generosity in sharing their learning.

Thanks are also owed to the very many people who have discussed ideas with me. Most of all, I am grateful to Matthew Townend and Alfred Hiatt, whose comments on an earlier version of this book helped to refine many of its arguments. I also warmly recall conversations with Paul Harvey and Catherine Delano-Smith, who both gave generously of their time and expertise. Erika Graham-Goering and Rosa María Rodríguez Porto took the time to advise me on the cartographic depictions of regions in their expertise. Margaret Cormack kindly read the whole manuscript, and taught me much about Icelandic Church history and much else besides. Victoria Flood has heard much about this book, and deserves special thanks. I must also convey my deep gratitude to Carolyne Larrington, Sif Rikharðsdóttir, and Caroline Palmer for their guidance in bringing this book to publication. I convey my warmest thanks to an anonymous reviewer, whose generous comments on my manuscript has elevated the work in countless ways.

My closing thanks go to friends and colleagues who are, I hope, already aware of their important contribution. Special thanks are owed to Sarah

Baccianti, Kristin Bourassa, Christopher Callow, Jennie England, Jóhanna Katrín Friðriksdóttir, Nik Gunn, Lara Hogg, Judith Jesch, Ragnheiður Maren Hafstað, Kevin Müller, Jane-Héloïse Nancarrow, Heidi Stoner, Paul Trower, Elliði Vatnsfjörð, and Teva Vidal. This book has taken form only from their encouragement and inspiration. This book's dedication to Julie Kedwards, Sally Kedwards, Ida Parker and Max the dog only begins to acknowledge the debt I owe them.

I'm finishing this book in Iceland, where my interest in the Middle Ages took root. I came here in 2004, on an A-Level Geology trip with King Edward VI College, Stourbridge, with the purpose of studying Iceland's volcanic and glacial landscapes. A free afternoon in Reykjavík, however, took me to Safnahúsið, where I saw an exhibition on Iceland's medieval manuscripts. In the museum shop, I picked up my copies of *Njál's saga* and Carolyne Larrington's translation of the *Poetic Edda*. These purchases were fatal to my aspirations as a geologist, but set me on the course that has ended with this book. It's a pleasure to think that, in writing a book about Iceland and maps, my interests have come full circle.

Reykjavík, Iceland
Dale Kedwards

Abbreviations

AÍ I–III	*Alfræði Íslenzk: Islandsk encyklopædisk litteratur*. 3 vols., ed. Kristian Kålund (Copenhagen: Møllers bogtrykkeri, 1908–18)
DI	*Diplomatarium Islandicum, Íslenskt fornbréfasafn*

Introduction

The intellectual work of maps is to make the world, or an aspect of it, understandable. The earliest maps from medieval Iceland, drawn between *c.* 1225 and *c.* 1400, take the entire world as their object, their makers motivated by the intellectual desire to know the world and their place within it. They represent ancient and medieval theories about the order of the world and its position in the cosmos: its places, and the peoples who inhabited them. Their purposes, however, were not simply to instruct their viewers in the cosmographic principles that lay behind them. Maps are tools with which to think, as well as to show, and thirteenth- and fourteenth-century Icelanders used the cartographic medium, as we shall see, to think about Iceland's place in the world, its people, and their history. The contemplated world, and its role in Icelandic literature and culture, is the subject of this book.

The five maps that survive in Icelandic manuscripts are not *sui generis*, but represent cartographic genres in pan-European circulation. They date to the apogee of map production in medieval Europe, a period in which maps begin to appear in unprecedented numbers, and in a greater variety of contexts.[1] There was no native Scandinavian tradition of map-making in the centuries before Iceland's conversion to Christianity and the associated advent of alphabetic literacy and bookish culture in *c.* 1000. The Vikings did not record their extensive geographic knowledge in map form, and the maps Icelanders drew in subsequent centuries do not emanate from the period of European oceanic expansion that began in the Viking Age. The maps drawn in medieval Iceland are of the type referred to by modern and sometimes medieval thinkers as *mappae mundi* (singular *mappa mundi*), a compound of the Latin words *mappa* ('cloth') and *mundus* ('world'). The Icelandic examples are few, but include representatives of all the main *mappa mundi* varieties: the hemispherical world map; the zonal map; the tripartite world map; and the schematic T-O map.

These map types are visually and conceptually diverse, and their discussion must of necessity build up from a fundamental understanding

1 Catherine Delano-Smith and Roger J. P. Kain, *English Maps: A History* (London: The British Library, 1999), p. 13.

of what they show. The Icelandic hemispherical and zonal maps represent ancient theories about the Earth's spherical form – a fact well-known to medieval thinkers – its distribution of land and ocean, and the extent of its habitable areas. The world these maps depict is wider than that known first-hand to medieval Europeans, which, as these maps show, occupies a relatively small part of the Earth's surface. These maps show that the known world was situated on the Earth's northern hemisphere, at temperate midlatitudes between the hot equator and the frozen north pole.

The tripartite and schematic T-O maps limit their perspective to the known world, which medieval thinkers conceptualised as a tricontinental land area on the Earth's northern hemisphere comprising Europe, Mediterranean Africa, and Asia. The three continents (from Latin *terra continens*, 'continuous land') centred on the Mediterranean Sea (at the 'middle', *medius*, of 'lands', *terrae*) and were surrounded by ocean. These maps are so-called because they depict the circular 'O' of the known world divided into its continental thirds by the River Tanais, the River Nile, and the Mediterranean Sea, whose intersection forms a watery 'T'. This space encompassed all human civilization and history, and was called in Latin the *orbis terrarum*, 'circle of lands', and in Old Norse, the *kringla heimsins*. These maps derive most of their information from ancient authoritative texts, canonised for the High Middle Ages by scholars such as Orosius (*c*. 385–*c*. 418) and Isidore of Seville (*c*. 560–636), with some additions in their depiction of Scandinavia and Europe from contemporary medieval sources. Both species of map, those that depict the entire globe and those that depict the known world, present complex visual arguments about the order of the world, its cosmic position, and humanity's place within it.

These maps were drawn in Iceland in the thirteenth and fourteenth centuries, a period that witnessed the rise of Icelandic literature in the vernacular, and the intense social dynamics that marked Iceland's transition from Commonwealth to a possession of the Norwegian Crown. This period, however, has been overlooked by map historians. The dedicated histories of Icelandic cartography, Halldór Hermansson's *Cartography of Iceland* (1931) and Haraldur Sigurðsson's *Kortasaga Íslands* (1971), devote little attention to the Icelandic *mappae mundi* in the conviction that they were mechanical copies of traditional images that moved unchanged through the cultures that copied them. Halldór Hermansson dismissed them as the 'conventional products of monks, or men of the traditional learning [that] give no indication of any real knowledge about the country', while Haraldur Sigurðsson's treatment of them extends to less than a page.[2] These histories focus instead on

2 Halldór Hermansson, *The Cartography of Iceland* (London: Milford, 1931), p. 2; Haraldur Sigurðsson, *Kortasaga Íslands: Frá öndverðu til loka 16. Aldar*, vol. 1 (Reykjavík: Bókaútgáfa Menningarsjóðs og Þjóðvinafélagsins, 1971), p. 46.

the appearance of Iceland on maps belonging to the newer cartographic genres that proliferated in the fifteenth century: the maps we call portolan charts, and maps in the Ptolemaic tradition. The first of these, the portolan charts, are navigational maps of coastlines based on compass directions and estimated distances that originated in Italy, and were compiled for practical purposes by pilots at sea.[3] Icelanders who sailed southern waters in the thirteenth century may have encountered such maps of Mediterranean and Black Sea coastlines, but waters north of the British Isles would not become known to these southern cartographers, and then poorly, until the late fifteenth.

The second of these genres, the Ptolemaic maps, appeared in the decades following the reintroduction to western Europe of Claudius Ptolemy's *Geography,* a cartographic treatise and gazetteer written at Alexandria in *c.* 150 CE, and translated from Greek into Latin by the Italian humanist scholar Jacobus Angelus in *c.* 1406. This treatise, which may have arrived into western Europe with Byzantine refugees, acquainted European thinkers with the means to encode and transmit geographical data digitally, in the mathematical language of paired coordinates. In the decades that followed its translation, the *Geography* was revised and expanded with new maps, *tabulae modernae,* of regions poorly known to ancient geographers, the first to show Iceland and Greenland compiled by the Danish cartographer Claudius Clavus (1388– ?) in the 1430s.[4] In the decades following the publication of early printed editions of the *Geography* at Bologna in 1477 and Ulm in 1482, Iceland became a familiar, if not entirely stable, presence on maps produced across Europe. The *mappae mundi,* however, drawn by Icelanders in earlier centuries, have been characterised as conventional reproductions, and not particularly

3 On the earliest nautical charts see Tony Campbell, 'Portolan Charts from the Late Thirteenth Century to 1500', in *The History of Cartography,* vol. 1: *Cartography in Prehistoric, Ancient, and Medieval Europe and the Mediterranean,* ed. J. B. Harley and David Woodward (London: The University of Chicago Press, 1987), pp. 371–462; and Patrick Gautier Dalché, *Carte marine et portolan au XIIe siècle' Le 'Liber de existencia riveriarum et forma maris nostril Mediterranei' (Pise, circa 1200)* (Rome: École Française de Rome, 1995).

4 See William R. Mead, 'Scandinavian Renaissance Cartography', in *The History of Cartography,* vol. 3, part 2, *Cartography in the European Renaissance,* ed. David Woodward (London: University of Chicago Press, 2007), pp. 1781–1805, esp. pp. 1781–86; Halldór Hermansson, *Cartography of Iceland,* pp. 14–20; Edward Lynam, 'Early Maps of Scandinavia and Iceland (Synopsis of Lecture)', *Saga-Book* 11 (1928–36), 1–4, 3; Haraldur Sigurðsson, *Kortasaga Íslands,* p. 11; and Douglas McNaughton, 'A World in Transition: Early Cartography of the North Atlantic', in *Vikings: The North Atlantic Saga,* ed. William W. Fitzhugh and Elisabeth I. Ward (Washington: Smithsonian Institution Press, 2000), pp. 257–69, p. 258.

Icelandic. Their status as *maps* – or at the very least good maps – has been doubted, and they are subsequently elided from these histories.

Precisely what defines a map in any period is an open question; these Icelandic examples might, in their simplicity or expansive scales, challenge our expectations of what a map should be. Given that 'maps' as we know them did not exist in the Middle Ages, it is not surprising that there developed no word in medieval Latin or Old Norse that explicitly denoted the concept. Broadly speaking, a map could be called by any word meaning 'picture', but the terms medieval thinkers used to describe a map – *descriptio, tabula, imago* – may have applied as easily to written descriptions of the world as cartographic images. [5] The English Hereford Map (*c.* 1300) bears a dedicatory inscription, written in Insular French, that describes the map as an *estoire* ('story' or 'history').[6] The term *mappa mundi* appears on the thirteenth-century Ebstorf Map, which bears an inscription that explains to audiences unfamiliar with the term that 'mappa dicitur forma. Inde mappa mundi id est forma mundi' ('*mappa* means image (*forma*); hence a *mappa mundi* is an image of the world').[7] But even the term *mappa mundi* did not designate a map unambiguously. The English chronicler Gervase of Canterbury (*c.* 1141–*c.* 1210) applied the term to his written topographical survey of ecclesiastical foundations in England, Wales, and parts of Scotland.[8] In the High Middle Ages, map-making was rarely, if ever, an activity practiced with any self-awareness by people who were dedicated cartographers. Medieval maps most commonly survive in manuscript books alongside texts and images covering a diversity of themes and genres, and their contents cannot be

5 The term *figura* was used to designate a map in Classical Latin, and, by the medieval period, had come to signify a smaller illustration or diagram in the service of an accompanying scientific text. *Mappa* was not used in Classical Latin, where the preference was for *forma, figura, orbis pictus, orbis terrarum descripto,* or *formula picturarum.* See David Woodward, 'Medieval *Mappaemundi*', in *The History of Cartography, vol. 1: Cartography in Prehistoric, Ancient, and Medieval Europe and the Mediterranean,* ed. J. B. Harley and David Woodward (London: The University of Chicago Press, 1987), pp. 286–370, p. 287; P. D. A. Harvey, *Mappa Mundi: The Hereford World Map* (London: British Library, 1996), p. 389.

6 Texts and translations of the Hereford map are taken from Scott D. Westrem, *The Hereford Map: A Transcription and Translation of the Legends with Commentary* (Turnhout: Brepols, 2001).

7 Hartmut Kugler, *Die Ebstorfer Weltkarte: Kommentierte Neuausgabe in zwei Bänden,* 2. Vols. (Berlin: Akademie Verlag, 2007), vol. 1, p. 42. The translation follows Marcia Kupfer, 'Reflections in the Ebstorf Map: Cartography, Theology and *dilectio speculationis*', in *Mapping Medieval Geographies: Geographical Encounters in the Latin West and Beyond, 300–1600,* ed. Keith D. Lilley (Cambridge: Cambridge University Press, 2013), pp. 100–26, p. 102.

8 Dom David Knowles, 'The Mappa Mundi of Gervase of Canterbury', *The Downside Review* 48:3 (1930), 237–47.

treated in categoric isolation. It is not certain that the integrity of the corpus treated in this book would have been recognised in the Middle Ages. Even in contemporary scholarship, these cartographic genres are not ordinarily studied together. While maps of the tripartite type, whose forms are epitomised by the well-known Ebstorf and Hereford Maps, have been assigned their place in literary and art historical enquiry, the more schematic maps, which sometimes require a more technically astute mode of interpretation, are usually seen as the province of historians of science. An inclusive view of these medieval images, however, enables us to apprehend the wide range of conventions for visualising the world available to medieval Icelanders, and reveals layers of sophistication in these maps that are not obvious when cartographic genres are inspected separately.

Unfixing the Map

Approaches to the Icelandic maps have invariably centred on their 'geographies', in the easy assumption that a map's value as a source lies in its potential for reconstructing past geographic understandings. The focus has been on what they can tell us about the Viking voyages of exploration and settlement across the North Atlantic and more distant seas, as recorded in Old Icelandic literature contemporary with the map. To be sure, this literature expresses a lively interest in the Icelanders' place in the world, and vividly recalls the geographic range of their activities then and in earlier centuries.[9] But assumptions about the kinds of literatures that are relevant to maps make a powerful argument for their meaning that can distort our perspective on their role in medieval culture. We interpret the maps as windows onto the histories of 'geography' and 'cosmology', modern analytical categories that did not exist as distinct or separate disciplines in the European Middle Ages.[10] To see the maps as their medieval viewers would have seen them requires that we loosen

9 Judith Jesch, 'Geography and Travel', in *A Companion to Old Norse Icelandic Literature and Culture*, ed. Rory McTurk (London: Blackwell, 2008), pp. 119–35. On the 'mental maps' of the Vínland voyages see Gísli Sigurðsson, *The Medieval Icelandic Saga and Oral Tradition: A Discourse on Method* (Cambridge, MA: Harvard University Press, 2004), pp. 253–301. On the culmination of the Icelanders' inherited and observed geographical knowledge see Sverrir Jakobsson, *Við og veröldin: Heimsmynd Íslendinga 1100–1400* (Reykjavík: Háskólaútgáfan, 2005).

10 Andy Merrills, 'Geography and Memory in Isidore's *Etymologies*', in *Mapping Medieval Geographies: Geographical Encounters in the Latin West and Beyond, 300–1600*, ed. Keith D. Lilley (Cambridge: Cambridge University Press, 2013), pp. 45–65. See also Edward Grant, *The Foundations of Modern Science in the Middle Ages: Their Religious, Institutional, and Intellectual Contexts* (Cambridge: Cambridge University Press, 1996), p. 7.

their association with geography in the narrowest sense, and realise that the relationship between them is subtler than is generally supposed.

Two nineteenth-century titles exemplify the course that interest in the Icelandic maps has taken. Carl Christian Rafn's *Antiquitates Americanae sive Scriptores Septentrionales rerum ante-Columbianarum* (1837) is a monumental summary of the evidence for Scandinavian presence on the North American Continent before Columbus. This volume cites the Icelandic hemispherical world map, which it reproduces in facsimile, as evidence for medieval Icelanders' familiarity with mainstream European theories about the existence of lands beyond the ocean, a learned context in which the landfalls related in the so-called Vínland sagas, *Eiríks saga Rauða* and *Grænlendinga saga*, may have been understood in later centuries.[11] Rafn's follow-up *Antiquités Russes d'après les monuments historiques des Islands et des anciens Scandinaves* (1852) is a companionate summary of Scandinavian sources for the histories of the Baltic Sea regions and Russia, called the *austr vegr* ('the eastern way') in Old Norse literature.[12] This volume features facsimile reproductions of all the Icelandic maps, but is especially interested in the Icelandic tripartite world map, which bears three place-names in its depiction of the Baltic Sea region. The map names Rus', the Scandinavian- and Slavic-inhabited region on the eastern Baltic littorals, and its capital Kiev, locating them next to the provinces of Sweden – Götaland and Svealand – on the Scandinavian Peninsula. A legend south-east of these, 'eronei' ('vagrants'), may indicate the nomadic peoples of the Eurasian Steppe, the Mongols who lay siege to Kiev in 1240. This map has become a staple in Russian scholarship on Scandinavian sources for the origins and western European perceptions of Kievan Rus'.[13] An interest in the

11 Carl Christian Rafn, *Antiquitates Americanae sive Scriptores Septentrionales rerum ante-Columbianarum* (Copenhagen: Schultz, 1837). This evidence was summarised and translated for a wider Anglophone readership in Joshua Toulmin Smith's *The Discovery of America by the Northmen in the Tenth Century* (London: Charles Tilt, 1839). Although this volume reproduces no maps *per se*, it does feature a reconstructed rectilinear 'chart of the world, according to Icelandic MSS. of the thirteenth century', p. 339. Toulmin Smith does not identify the Icelandic manuscripts on which he based his map, but its Old Norse legends are extracted primarily from the so-called Icelandic *Geographical Treatise*, with an isolated borrowing, 'synrri byggð' ('southern inhabitable land'), from the Icelandic hemispherical world map.

12 Carl Christian Rafn, A*ntiquités Russes d'après les monuments historiques des Islandais et des anciens Scandinaves*, vol. 2 (Copenhagen: De l'imprimerie des frères Berling, 1852).

13 For earlier work, see Omeljan Pritsak, *The Origin of Rus', vol. 1: Old Scandinavian Sources other than the Sagas* (Cambridge MA: Harvard University Press, 1982) and Elena A. Mel'nikova, *Drevneskandinavskie Geograficheskie Sochineniia: Teksty, Perevod, Kommentarii* (Moscow: Nauka, 1986). For the most recent authoritative account, see Leonid S. Chekin, *Northern Eurasia in Medieval*

history of European exploration and discovery, and the sometimes-related desire to re-write or shore up historiographic 'nationalisms', have converged on these maps to ensure their interpretation as graphic expressions of the Icelanders' geographic awareness.

These approaches focus our attention on those parts of the map that might contain new Scandinavian intelligence about the world's places, particularly in Scandinavia and on the Baltic littorals. They accept the map as a mirror of the world that reflects on its surface the visions and experiences of those who made it. If, however, the Icelandic maps have sometimes proven disappointing as sources it is because, as Halldór Hermansson implied, they are mostly silent on the Norse explorations of northern waters in the Viking Age and earlier centuries. The most complex and large-scale representation of the world to come down to us from medieval Iceland, the larger Viðey Map, contains 104 geographic legends showing the world's regions and countries, cities, rivers, mountains, and people, but does not show Orkney, Shetland, the Faroe Islands, or Greenland, places that were certainly known to the Icelandic map-maker. This map, as we shall see in Chapters 3 and 4, is more interested in historical regions in the Levant and Central Eurasia than the islands of the North Atlantic, and was not drawn to provide a complete and unobstructed view of the map-maker's experience of their own local geography.

The most comprehensive attempt to understand these mentalities and the place of maps within them is Rudolf Simek's important *Altnordische Kosmographie* (1990), a source-book and commentary on Old Norse texts and images on 'cosmographical' subjects.[14] Simek situates the Icelandic maps alongside related items of cosmographic study and speculation, such as Old Norse geographic treatises, itineraries, and descriptions of far-travel in saga literature. This creates a new context for the Icelandic maps that draws attention away from their historical depictions of territory and refocuses it on their place in a history of geographic mentalities. This approach enables us to contextualise the maps, but it also settles the terms by which we try to understand them. It lets us see how the maps relate to contemporary intellectual currents in Iceland and the rest of Europe, but their relationships to other literatures and histories remain unseen.

The implications for having accepted 'geography' as the most appropriate intellectual context in which to view these maps become clear when we turn to the manuscript books in which they were drawn. Simek's

---

Cartography: Inventory, Text, Translation, and Commentary (Turnhout: Brepols, 2006).

14 Rudolf Simek, *Altnordische Kosmographie: Studien und Quellen zu Weltbild und Weltbeschreibung in Norwegen und Island vom 12. Bis zum 14. Jahrhundert* (Berlin: Walter de Gruyter, 1990).

Altnordische Kosmographie situates the Icelandic maps alongside thematically related texts and images but shines little light on the items that accompany them in their thirteenth- and fourteenth-century manuscripts. These unseen texts and images are an inalienable condition of these maps' historical reception, but have found no mention in commentaries on them. The items that accompany maps in their manuscript books are not restricted to items of 'cosmological' interest , but cover a diversity of themes. They include a text about the three Biblical kings – Melchior, Balthazar and Caspar – treatises on the revolutionary movements of the planets, and a list of Icelandic priests' names attributed to the preeminent Icelandic historian Ari Þorgilsson the wise (1067–1148). These companion items conditioned audiences' historical encounters with these maps, but have had no place in modern anthologies and source-books that register the maps alongside geographically-interested literatures. 'Cartography' and 'geography', that is, are not interchangeable terms.[15]

Modern source-books of cosmographical works make a powerful argument for the meaning of maps that can distort our thinking about their role in medieval culture. The difference between the contents of modern source-books and the manuscripts from which the maps were taken points towards a discontinuity between modern and medieval assumptions about their natures. Until now, the maps have been seen in almost total isolation, so that information about their origins and purposes have been sought only within the narrow horizons of their own outlines. It has been assumed that meaning is a quality intrinsic to the map image, and that the map was drawn to be consulted and understood entirely on its own terms. But this was rarely so. Inattention to the contents of the books in which we find maps can cause us to overlook what their makers tried to achieve in drawing them.

In earlier scholarship the Icelandic maps have been seen as relevant to a relatively narrow range of historical questions, their analysis limited to a handful of place-names on and around the Scandinavian Peninsula. This book is an attempt to free the map from these expectations in ways that enable us to out-think the narrow and persistent 'geographic' framework that has set the terms for its interpretation. It starts, then, with the maps themselves. It does not impose 'geography' as theme *a priori*, but looks to the maps and the items that frame them in their manuscript books for what they can tell us about the interests and preoccupations of their makers. We shall see that they were used to explain medieval theories about the tides (Chapter 1), theorise about the revolutionary motions of the planets (Chapter 2), conceptualise the origins of Icelandic literature

15 On re-theorizing cartography's relationship to geography, see Denis Cosgrove, 'Cultural Cartography: Maps and Mapping in Cultural Geography / Les Cartes et la Cartographie en Géographie Culturelle', *Annales de Géographie* 660/661 (2008), 159–78.

and culture (Chapter 4), and comment on Icelandic social institutions in the Commonwealth period (Chapter 5). The Icelandic maps intersect more textual worlds than has previously been supposed.

The cultural contexts of maps are in the foreground of this book. As the map historian J. B. Harley noted, 'maps redescribe the world – like any other document – in terms of relations and power and cultural practices, preferences, and priorities'.[16] A map is not a neutral or transparent record of its maker's sense of geographical reality, drawn in the objective and detached pursuit of scholarship. This is equally true of the technically astute hemispherical and zonal maps, which represent theories about the form and structure of the universe; 'representation is never neutral, and science is still', as Harley averred, 'a humanly constructed reality'.[17] The wider cultural discourses in which maps were drawn and through which they moved have gained wider attention in recent decades. *Mappae mundi* in medieval England have been interpreted as witnesses to incipient English 'nationalisms', as interventions into contemporary political cultures that work by visualising the territorial aspirations and sovereign fantasies of the elites who commissioned them.[18] The Icelandic maps also intersect major historical processes and cultural movements, but have not, before now, been analysed as products of the cultural and social transformations of the thirteenth and fourteenth centuries. This book reveals that the Icelandic *mappae mundi* are pioneering works of Icelandic historical writing that show how Icelandic thinkers were able to manipulate cartographic space to address contemporary anxieties about the place of Iceland in Scandinavia, and attendant questions of Icelandic history and identity.

Mappae mundi and Medieval Literature

Evidence for cartographic knowledge in medieval Iceland can be sought outside the corpus of known survivals. In the High Middle Ages, maps were an uncommon form of expression, far fewer in number than geographic descriptions in treatises on natural philosophy or in literature. The degree to which this stems from actual rarity or subsequent loss is unclear. At least one Icelandic map has perished but is known to us anecdotally through the letter correspondence between Árni Magnússon (1663–1730), the preeminent collector of medieval Icelandic manuscripts

16 J. B. Harley, *The New Nature of Maps: Essays in the History of Cartography*, ed. Paul Laxton (London: John Hopkins University Press, 2001), p. 35.

17 Harley, *New Nature*, p. 46.

18 Daniel Birkholz, *The King's Two Maps: Cartography and Culture in Thirteenth-Century England* (London, 2004); Kathy Lavezzo, *Angels on the Edge of the World: Geography, Literature, and English Community, 1000–1534* (Ithaca: Routledge, 2006).

in the eighteenth century, and the antiquarian Þormóður Torfason (Thormodus Torfæus) (1636–1719). In two letters, dated 3rd and 22nd April 1690, Árni refers to a 'gamla landcorte a pergament' ('old land map on parchment') and a 'töblu' ('document', or 'map', cf. Latin *tabla, tabula*), that appears to have been drawn into the compilation of annalistic and astronomical items called the Codex Resenianus 6.[19] What we know of this map comes down to us in passing remarks on its mapped place-names, which apparently included 'Þrasnes', which was situated, according to Árni, on the 'promontorium Celticum i Spanien strax vid þad nes, sem skipsfolk kallar Cabo d'Ortegal' ('on the *Promontorium Celticum* in Spain, immediately next to that headland that sailors call the Cape of Ortegal'); and 'hafið dauða' ('the Dead Sea').[20] This manuscript was destroyed, together with a copy drawn by the priest Hjalti Þorsteinsson (1665–1754), in a fire in Copenhagen in 1728. The map, as Simek observes, must have been more detailed than any other that survives. The appearance of the name *Þrasnes* on the map, which Árni recognised as a coastal place-name known to 'skipsfolk' ('sailors'), might suggest that this map incorporated information from sea charts, or that Icelandic sailors had begun to make their own oral contributions to cartographic knowledge in later centuries. None of its known place-names appear on any other map from medieval Iceland. What we do know about the map does not conclusively indicate that it was a *mappa mundi* and not an early map in the Ptolemaic tradition. We know that it was copied into a book, like most other maps from the Middle Ages, but are otherwise poorly informed about its contents, date and provenance.

Since so many medieval maps have been lost or perished, church book-lists and inventories have proven valuable sources for their study.[21] Though there are no inventories of maps alone until the sixteenth century, they do occasionally appear as separate items in more general inventories. A map image may have been described as a *mappa mundi* ('world cloth') or *pannus depictus* ('painted cloth'), or, more ambiguously still, in terms

19 This manuscript appears in a seventeenth century catalogue, where its contents are described as containing Icelandic annals, genealogical materials, 'et varia plura'. The map itself is referred to in its catalogue as 'mappam geographicam totius orbis tunc temporis cogniti'. See Stéfan Karlsson, 'Resenshandrit', *Opuscula* 4 (1970), 269–78, 271.

20 Kristian Kålund, ed., *Arne Magnusson. Brevveksling med Torfæus (Þormóður Torfason)* (Copenhagen: Christiania, 1916), pp. 33, 37. See Simek, *Altnordische Kosmographie*, pp. 60–61.

21 For an outline of the methodological problems relating to the use of book-lists and inventories, their imprecision, and the high dispersal rate of libraries, see Michael Lapidge, *The Anglo-Saxon Library* (Oxford: Oxford University Press, 2008), p. 55. Examples of inventory entries that relate to maps have been assembled in Leo Bagrow, 'Old Inventories of Maps', *Imago Mundi* 4 (1948), 18–20; and Birkholz, *Two Maps*, p. xvii.

of its documentary support, as a *tabula* ('document', 'board', 'panel') or *rotulus* ('little roll', 'wheel').[22] The *máldagar*, inventories of books and other possessions held by Icelandic churches and religious houses, are seldom explicit, but contain no obvious references to map artefacts.

Evidence of Icelanders' familiarity with maps and their conventions abounds in manuscript illumination, in which we find preserved numerous depictions of the simple T-O map in the form of the sovereign orb. There are six such examples in the Icelandic Teiknibók (Reykjavík, Stofnun Árna Magnússonar, AM 673a III 4to), a model book for manuscript illuminators that combines the works of four artists working from the middle of the fourteenth to the beginning of the sixteenth centuries.[23] This book contains five depictions of kings who hold the *orbis terrarum* as emblem of their sovereign status and dominion over the world.[24] Twice, on folios 12r (*c.* 1450–75) and 18r (*c.* 1500), the orb is in the hands of the Norwegian royal St Ólafr (Ólafr II of Norway, 995–1030), locally canonised in 1031 and confirmed by Pope Alexander III in 1164. In Ólafr's possession, the orb may combine associations of earthly rule with a statement of Christianity's globalising mission, suggestive of his role in Scandinavia's conversion and this region's new place in European Christendom. The most complex composition in Teiknibók to feature a T-O map shows Christ – 'Lord of Lords, and King of Kings' (Revelation 17:14 and 19:16) – enthroned amid representations of the Four Evangelists, and with the simple T-O map, representing all miniaturised creation, at his feet (see Chapter 5). These images are simple and iconic and, since they contain no outlines or geographical nomenclature, commonly overlooked as maps. Their appearance in manuscript illumination, however, demonstrates that artists were familiar with graphic conventions for depicting the world in visual form, and understood their associations with power and rule.

The most valuable – and difficult to interpret – evidence for cartographic knowledge outside the maps themselves is supplied by the written texts that show their influence. C. S. Lewis examined the cosmological substrate in medieval literature in *The Discarded Image*, in which he

22 Bagrow, 'Old Inventories', p. 18; David Woodward, 'Medieval *Mappaemundi*', in *The History of Cartography*, vol. 1: *Cartography in Prehistoric, Ancient, and Medieval Europe and the Mediterranean*, ed. J. B. Harley and David Woodward (London: Chicago University Press, 1987), pp. 286–370, p. 292. Michael Lapidge makes no mention of maps in his reconstruction of the *Anglo-Saxon Library*, though Loredana Teresi has since established that maps are not mentioned in early English or Anglo-Norman inventories. Teresi Loredana, 'Anglo-Saxon and Early Norman England Mappaemundi', in *The Foundations of Learning: The Transfer of Encyclopaedic Knowledge in the Early Middle Ages*, ed. R. Bremmer and C. Dekker (Paris: Peeters, 2007), pp. 341–377, p. 341.

23 Guðbjörg Kristjánsdóttir, *Íslenska Teiknibókin* (Reykjavík: Crymogea, 2013).

24 Simek, *Altnordische Kosmographie*, pp. 121–122.

understands cosmology as the ultimate synthesis of medieval theology, science, and history 'in which most particular works were embedded, to which they constantly referred, from which they drew a great deal of their strength'.[25] While little concerned with the roles played by cartographic images in the transmission of this model, Lewis demonstrated that literary texts could be opened up with reference to the implicit cosmological models that underlie them. More recent scholarship – in the wake of the spatial turn in literary study – has re-theorised cartography's relation to literature. Tom Conley argues that literatures can be considered cartographic in so far as both maps and written texts are shaped by 'tensions of space and of figuration'.[26] Patrick Gautier Dalché likewise sees 'the tacit influence of the cartographic medium' on written descriptions of world geography, in so far as they evoke, and in some instances rely upon, contemporary map images.[27] An interest in 'geography', however, is not by itself enough to make a text 'cartographic', and tells us little, on its own, about the presence of maps in a culture. More certain evidence of literary authors' familiarity with map images can be detected in examples of cartographic *ekphrasis* – literary descriptions of map artefacts – and verbal reminiscences of map conventions in geographical descriptions.

Descriptions of map artefacts in literary texts are few, but valuable as sources for the presence of maps in antiquity and the Middle Ages. The earliest known allusion to a map in the ancient world is the ekphrastic description of the design that adorns the shield that Achilles uses in his battle with Hector in Homer's *Iliad* (18.478–608), a concentric design with a depiction of the Earth, sea, sky and heavenly bodies at its middle.[28] The earliest allusion to a map in medieval Europe takes literary form in Jonas of Bobbio's *Life of Columbanus*, written in Northern Italy *c.* 643. In the life, the Irish missionary Columbanus intends to travel to Slavic lands to preach Christianity, but quickly reconsiders when an angel shows him how far away these territories lay on a world map. The mention of this map, which may have been akin to those simple T-O maps that circulated with the writings of Isidore of Seville (560–636), antedates

25 C. S. Lewis, *The Discarded Image: An Introduction to Medieval and Renaissance Literature* (Cambridge: Cambridge University Press, 1964), p. 12.

26 Tom Conley, *The Self-Made Map: Cartographic Writing in Early Modern France* (Minneapolis: University of Minnesota Press, 1996), p. 3.

27 Patrick Gautier Dalché, 'Maps in Words: The Descriptive Logic of Medieval Geography from the Eighth to the Twelfth Century', in *The Hereford World Map: Medieval World Maps and their Context* (London: British Library, 2006), pp. 223–42, p. 225.

28 Germaine Anjac, 'The Foundations of Theoretical Cartography in Archaic and Classical Greece', in *The History of Cartography*, vol. 1: *Cartography in Prehistoric, Ancient, and Medieval Europe and the Mediterranean*, ed. J. B. Harley and David Woodward (London: Chicago University Press, 1987), pp. 130–47, p. 131.

the earliest extant example of such a map by around 150 years.[29] The literary evidence for the use of maps at sea likewise antedates our earliest extant nautical charts by around twenty years. This appears in the French chronicler Guillaume de Nangis's (d. 1300) life of the French royal Saint Louis IX (1214–70), in an episode that relates Louis' troubles aboard a ship bound for Tunis in around 1270. When a storm forces the ship to change course, its captain shows the king a 'mappa mundi', which must in the circumstances have been a portolan chart.[30]

We encounter an Icelandic example of cartographic ekphrasis in *Alexanders saga*, the Old Norse prose translation of Walter of Châtillon's *Alexandreis*, a twelfth-century Latin epic poem about the imperial exploits of Alexander the Great. In both Walter's *Alexandreis* (7.420–77) and the Icelandic *Alexanders saga*, the tomb of the Persian King Darius is adorned with a world map.[31]

> Vppi yvir stolpunum var hvalf sva gagnsett sem gler. Þvilict vaxet sem himinn til at sia. áþvi hvalve var scrifaðr heimrenn allr greindr isina þriðiunga. oc sva hver lond liggia ihveriom þriðiunge. eða hverirr ágetir staðer erv íhverio lande. oc þar með nattura. beðe landanna oc þeira þioða er londin byggia. oc sva eyiar þér er i hafino liggia. Þar var oc markar hversu vthafet gerðer vm oll londin. eða hversv miðiarðar siar er allar ár falla í.[32]

> (Up above the pillars was a vault as clear as glass, and just as wide as the sky. On this vault was inscribed all the world, divided into its thirds, so that each land lies in its third, and each noble place in its land, and with the natures of both the lands and the peoples that lived there, and also the islands that lie in the ocean. There was also marked how the ocean that girds all lands, and the Mediterranean Sea into which all rivers flow.)

The image inscribed onto this vault recalls the tripartite or T-O map, the world's lands divided into 'thirds' ('þriðjungur') – the three continents –

29 Patrick Gautier Dalché, 'De la glose à la contemplation: place et fonction de la carte dans les manuscrits du haut Moyen Age', *Testo E Immagine Nell'alto Medioevo* (Spoleto: Presso la sede del Centro, 1994), pp. 693–764, pp. 697–98; and more recently Thomas O'Loughlin, 'Map Awareness in the Mid-Seventh Century: Jonas' *Vita Columbani*', *Imago Mundi* 62:1 (2010), 83–85.

30 Campbell, 'Portolan Charts', p. 439.

31 David Townsend, trans. *Alexandreis: A Twelfth Century Epic* (Philadelphia: University of Pennsylvania Press, 1996), p. 156, fn. 1. For the description of the tomb see pp. 156–57; and Maura K. Lafferty, 'Mapping Human Limitations: The Tomb Ecphrases Walter of Châtillon's Alexandreis', *The Journal of Medieval Latin* 4 (1994), 64–81.

32 *Alexanders saga: Islandsk Oversættelse ved Brandr Jónsson, ed. Finnur Jónsson* (Copenhagen: Kommissionen for det Arnamagnænske Legat, 1925), p. 112. My translation. On the Icelandic tomb ekphrasis see Simek, *Altnordische Kosmographie*, p. 62.

according to a prominent hydrographic framework: an ocean girds all lands ('úthafit gerðir um öll löndin'), and the world's rivers empty into the Mediterranean Sea (named in Old Norse with the calque 'miðjarðar sjár'). The geographic terms of this description are conventional in written descriptions as well as cartographic images. However, the object of this description is not the world *per se*, but an image inscribed ('skrifaðr') with geographic and ethnographic details about 'the natures of both the lands and the peoples that lived there'. The described object is a complex amalgam of word and image that evokes the monumental wall-mounted maps that may have been known to the Icelandic translator and his audiences. One such map was the thirteenth-century Ebstorf Map, displayed at a convent in Northern Germany, which was painted on thirty goatskins and measured 3.6 by 3.6 meters. Before its destruction in the allied raids on Hanover in 1943, this was the largest medieval map to survive to modern times, containing more than 2,300 texts and images. The map at England's Hereford cathedral (*c.* 1300), drawn onto a single piece of vellum measuring 1.58 by 1.33 meters, bears 1,091 written inscriptions that relate information in fields as diverse as geography, ethnography, zoology and history. The so-called 'donor inscription' in the map's lower corner urges its communities of owners, viewers, readers and listeners to pray for the soul of its patron, the clergyman Richard de Bello; the map, like the one that adorns Darius's tomb, commemorates its patron. To be sure, the Icelandic translator of the *Alexandreis* derives their description of the tomb from a literary source, not a map, but audiences may well have recognised that the described image, containing written information about the world's places and the peoples who inhabited them, had parallels in the monumental world maps that sometimes adorned European churches and religious houses.

Verbal reminiscences of map images, descriptions that are not ekphrases but only imply their authors' familiarity with maps, are harder to identify with certainty. Geographically themed introductions and digressions, historiographical and narrative structures that Icelanders knew from Latin writings, are numerous in Old Norse literature.[33] A prominent example is the introductory chapter to *Heimskringla* (*c.* 1230), a collection of sagas about the Norwegian kings whose authorship is usually attributed to the Icelandic statesman and literary magnate Snorri Sturluson (1179–1241). *Heimskringla* begins with *Ynglinga saga*, which relates the legendary origins of the Swedish Yngling dynasty at Uppsala by the god Freyr. The world description on which the saga opens avers the centralist origins of the Scandinavian royal families, who, as we shall see in Chapter 4, traced their origins to the heroes who migrated westwards following the

33 On the geographical digression as a historiographical structure in antiquity, and its afterlife as a literary device, see Andrew H. Merrills, *History and Geography in Late Antiquity* (Cambridge: Cambridge University Press, 2005).

legendary Trojan war. The author of *Ynglinga saga* claims that the Norse gods, the alleged progenitors of the Scandinavian royal lines, were not truly divinities but exceptionally cunning people from Central Eurasia, and contextualises their primeval migrations into Northern Europe with a geographical description.

> Kringla heimsins, sú er mannfólkit byggir, er mjǫk vágskorin. Ganga hǫf stór ór útsjánum inn í jǫrðina. Er þat kunnigt, at haf gengr frá Nǫrvasundum ok alt út til Jórsalalands. Af hafinu gengr langr hafsbotn til landnorðrs, er heitir Svartahaf. Sá skilr heimsþriðjungana. Heitir fyrir austan Ásíá, en fyrir vestan kalla sumir Európá, en sumir Eneá. En norðan at Svartahafi gengr Svíþjóð in mikla eða in kalda. Svíþjóð ina miklu kalla sumir menn eigi minni en Serkland it mikla, sumir jafna henni við Bláland it mikla. Inn nørðri hlutr Svíþjóðar liggr óbygðr af frosti ok kulða, svá sem inn syðri hlutr Blálands er auðr af sólar bruna. Í Svíþjóð eru stórheruð mǫrg, þar eru ok margskonar þjóðir ok margar tungur. Þar eru risar ok þar eru dvergar, þar eru ok blámenn, ok þar eru margskonar undarligar þjóðir, þar eru ok dýr ok drekar furðuliga stórir. Ór norðri frá fjǫllum þeim, er fyrir utan eru bygð alla, fellr á um Svíþjóð, sú er at réttu heitir Tanais. Hon var forðum kǫlluð Tanakvísl eða Vanakvísl. Hon kømr til sjávar inn í Svartahaf. Í Vanakvíslum var þá kallat Vanaland eða Vanaheimr. Sú á skilr heimsþriðjungana. Heitir fyrir austan Ásíá, en fyrir vestan Európá.[34]

> (The circle of the world, which humanity inhabits, is much indented with bays. Large seas extend out of the ocean into the earth. It is known that the sea extends from the Straits of Gibraltar and all the way out to the Levant. From the sea extends a long gulf to the northeast, which is called the Black Sea; this divides the world's three regions: the east is called Asia, and the west is called by some Europe, and others *Enea*. To the north of the Black Sea extends Sweden the Great or the Cold [North-Western Eurasia]. Some people say that Sweden the Great is no smaller than Serkland the Great [North Africa], some equate it with Bláland the Great [Saharan Africa]. The northern part of Sweden is not settled because of the frost and cold, just as the southern part of Bláland is empty because of the scorching sun. In Sweden there are many great realms, and many kinds of people and many languages. There are giants and there are dwarves. There are black people and there are many kinds of wonderful people, there are wonderfully large animals and dragons. Out of the north from those mountains that are outside all inhabited regions a river flows into Sweden [the Great, i.e. north-western Eurasia], which is correctly called the Tanais. It is previously called the Tana-river branch or Vana-river branch. It comes to the sea in the Black Sea. Around the Vana-tributary was then called Vana-land or

34 Snorri Sturluson, *Heimskringla,* Íslensk Fornrit 26:1, ed. Bjarni Aðalbjarnarson (Reykjavík: Hið Íslenska Fornrit Félag, 1941), p. 10.

Vana-home. This river separates the three parts of the world: called to the east Asia; and to the west Europe.)

The world in which *Ynglinga saga* develops its euhemeristic argument is mapped onto the circular T-O framework, the world circle divided into its continental thirds by its connecting waters. *Kringla heimsins* ('circle of the world') appears to be a calque of the Latin *orbis terrarum* ('circle of lands').[35] A cognate expression, *hringr jarðar* ('circle of the Earth'), appears in the Norwegian *Konnungs skuggsjá*, a vernacular educational handbook compiled *c.* 1250 for the Norwegian prince Magnús Hákonarson, but subsequently popular in Icelandic manuscript copies. The lesson concerns the basic fact of the Earth's sphericity: 'bǫllóttur er jarðar hringur, og ber eigi ǫllum stǫðum jafnnær sólu' ('the circle of the earth', we are told, 'is ball-shaped, and not all places are equally close to the sun').[36] In *Ynglinga saga*, the hydrographic framework familiar from the T-O map can be seen in Snorri's description of the circumambient ocean, referred to metonymically as the bays that indent the world circle. Simek observes that the description of a world circle frayed at its edges by inlets and bays ('mjǫkvágskorin') is a feature unattested in the Latin geographies, but often depicted on world maps, and is perhaps evidence that Snorri had a map image in mind.[37] The description of the Mediterranean Sea, which extends between the Straits of Gibraltar ('Nǫrvasund') and the Levant ('Jórsalaland'), however, does have parallels in Latin geographical writings, William of Conches informing us that 'inter Calpem et Atlantem, usque juxta Hierusalem, Mediterraneum vocatur' ('between the Calpe Mountains and the Atlas Mountains [i.e. the Strait of Gibraltar], nearly as far as Jerusalem, is called the Mediterranean Sea').[38] In other aspects, the introduction to *Ynglinga saga* evidences its author's familiarity with theories about the Earth's climatic zones, which underpin, as we shall see in Chapters 1 and 2, maps of the hemispherical and zonal types. Snorri describes climatic extremes in the southern part of Bláland, Saharan Africa, and in the northern part of Sweden that form the limits of human inhabitability on the spherical Earth.

The introduction to *Ynglinga saga* provides a space for the writing of Scandinavian history into a geographical framework. Its author's argument that Scandinavians came to Northern Europe from Troy can be detected in the name *Enea*, a name unattested outside the geographical introductions to *Ynglinga saga* and *Snorra Edda*. This name seems to

35 Rudolf Simek, 'Snorri als Kosmograph', in *Snorri Sturluson: Beiträge zu Werk und Rezeption*, ed. Hans Fix (Berlin: Walter de Gruyter, 1998), pp. 255–66, p. 259.

36 *Konungs Skuggsjá: speculum regale, ed.* Magnús Már Lárusson (Reykjavík: H. F. Leiftur, 1955), p. 21.

37 Simek, 'Snorri als Kosmograph', p. 264.

38 William of Conches, *De philosophia mundi*, III.xiv. *PL* 172, col. 80.

have been invented to associate Europe with an eponymous founder in the Trojan hero *Aeneas*. [39] The River Tanais, also, which divides Europe and Asia, is called by several names. Tanakvísl and Vanakvísl – and the related regional names Vanaland and Vanaheimr – seem also to have been created in order to confect a link between Central Eurasia and the subgroup of Norse divinities called the *Vanir*, whom chapter five of *Ynglinga saga* informs us originated in the area surrounding the Tanais river.[40] The name Tanakvísl, which the author of *Ynglinga saga* presents as an archaic version of the name Tanais ('var forðum kǫlluð Tanakvísl'), appears on a world map drawn in medieval Iceland. In Chapter 4, we shall see how the legendary geography that accommodated the Trojan diaspora into Northern Europe and Scandinavia described in Icelandic saga literature found cartographic expression.

Another example of verbal reminiscence in which we can detect an author's familiarity with map cultures is in the prologue to *Snorra Edda*, an Icelandic *ars poetica* written *c.* 1220 and also conventionally attributed to Snorri Sturluson. The prologue comprises a euhemeristic account of the primeval migrations of the Norse divinities, the Æsir, out of Asia and into Northern Europe. Snorri links the Æsir to their ancestral home in Asia on etymological grounds, claiming a connection between the *Æsir* (singular *Ás*) and *Asiamenn* ('Asians').[41] The geographical introduction contextualises the primeval migrations of the Æsir out of Asia, and avers the centralist origins of Icelandic vernacular poetics, which are implied once more to have originated in Troy.

> Verǫldin var greind í þrjár hálfur. Frá suðri í vestr ok inn at Miðjarðarsjá, sá hlutr var kallaðr Affrica. Hinn syðri hlutr þeirar deildar er heitr ok brunninn af sólu. Annarr hlutr frá vestri ok til norðrs ok inn til hafsins, er sá kallaðr Evropa eða Enea. Hinn nyðri hlutr er þar kaldr svá at eigi vex gras ok eigi má byggja. Frá norðri ok um austrhálfur allt til suðrs, þat er kallat Asia. Í þeim hlut veraldar er ǫll fegrð ok prýði ok eign jarðar ávaxtar, gull ok gimsteinar. Þar er ok mið verǫldin, ok svá sem þar er jǫrðin fegri ok betri at ǫllum kostum en í ǫðrum stǫðum, svá var ok mannfólkit þar mest tignat af ǫllum giptum, spekinni ok aflinu, fegrðinni ok alls kostar kunnustu.
>
> (The world was divided into three regions. From south to west and in at the Mediterranean Sea was called Africa. The southern part of that division is hot and burnt by the sun. The second part from west to north and in at the sea is called Europe or Enea. The more northerly part is so cold that grass does not grow and none may settle there. From the

39 Simek, 'Snorri als Kosmograph', p. 264.

40 Snorri Sturluson, *Heimskringla*, p. 10. The Vanir are a subgroup of the Æsir, the Old Norse divinities, whose number includes Njörðr, Freyja and Freyr.

41 Snorri Sturluson, *Snorra Edda: Prologue and Gylfaginning*, ed. Anthony Faulkes (London: Viking Society for Northern Research, 2005), p. 6.

> north, around the eastern half, and all to the south is called Asia. In that part of the world all is beautiful and magnificent, and rich in the fruits of the earth, gold and gemstones. The middle of the world is also there, and just as the earth there is more beautiful and better in all ways than other places, so are the people there most noble and most possessed of all gifts, in wisdom, body, and beauty and kinds of knowledge.)[42]

This geographical description condenses the elements laid out in greater detail in *Ynglinga saga*. The three continents of the known world are disposed in a circle, as on the tripartite and T-O maps, their divisions marked by the four cardinal points and the Mediterranean Sea. Onto this has been mapped the traditional distinction of the climatic zones, depicted on zonal and hemispherical world maps, so that the southern reaches of Africa are consumed by the intense heat of the sun, and the northern reaches of Europe are stripped by the cold. Asia is emphasised as the most bounteous of the world's regions and the place where the world has its middle.

In these descriptions, we can see that an awareness of maps and the traditions from which they emanate is at large in medieval Icelandic literature and culture. Icelandic seafarers who sailed southern waters may have encountered sea charts from the late thirteenth century, while their ecclesiastic counterparts who studied at urban schools abroad will have encountered maps in the books they read, and seen them displayed on walls in schools and churches. Simple maps, in the form of the tripartite sovereign orb, are numerous in Icelandic manuscript illumination, and Icelandic literature conveys a clear sense of the conventions that underpin them. In these literatures, maps did not simply exist to fix and transmit information about the Earth's surface, but were understood as potent symbolic registers of sovereignty, empire and possession. This assembled evidence shows that Icelanders knew about maps, and understood their singular value as tools with which to show and think alike. In the chapters that follow, we turn to the evidence of the maps themselves.

The chapters assembled in this book examine the maps in turn. Explanations of what the maps show are necessary because little has been said about them explicitly or in detail. The Icelandic hemispherical and zonal maps present complex visual arguments about the nature and structure of the physical universe and its clocked processes. The paired tripartite and T-O maps also possess complex symbolic structures that are initially difficult to interpret. The maps must be understood before they can be analysed. Each chapter then explains the map's origins, and its subsequent history up to its translation into Old Norse. Finally, each chapter examines the map's relationships with its companion texts and

42 Snorri Sturluson, *Snorra Edda*, p. 4. My translation.

images, so that they are understood in terms consistent with period assumptions about their natures. These chapter analyses situate the maps in their previously unseen manuscript contexts, and reveal how they relate to natural philosophical, theological, and literary discourses.

Formal and generic similarities between maps in the Icelandic corpus lend this book a diptych structure. Chapters 1 and 2, on the hemispherical and zonal maps, are built around how these technical maps relate to contemporary theological and natural philosophical discourses, while Chapters 3 to 5, on the maps from Viðey, centre on the ways in which cartographic space is manipulated by Icelandic map-makers to think about Icelandic culture and society in the thirteenth century. While this book's two halves can be read as standalone, it is vital that these map genres are studied in concert so that the full range of conventions for depicting the world in visual form available to medieval Icelanders can be seen.

Chapter 1 presents the Icelandic hemispherical world maps, preserved in two early fourteenth-century fragmentary manuscripts (Copenhagen, Arnamagnæan Collection, AM 736 I 4to, f. 1v, *c.* 1300 and AM 732b 4to, f. 3r, *c.* 1300–25). This map was conceived as a work of literary criticism. In his *Commentary on the Dream of Scipio*, the Roman author Macrobius drew the map to demonstrate that cosmic descriptions in the works of two of Rome's most celebrated literary icons, Cicero and Virgil, did not contradict one another. Its simple outline underwent many medieval revisions, through such intermediaries as Lambert of Saint Omer (d. *c.* 1125). The Icelandic witnesses contain few place-names, but do bear inscriptions that identify important geographic concepts such as the habitable middle latitudes in the northern and southern hemispheres (the 'antipodes'), and the Old Norse terms for the equator, tropics, and polar circles. The primary purpose of its Icelandic iteration was to explain the seasonal variation in the tides. These are preserved in illustrated miscellanies that combine items – written in Latin and Old Norse – on time, the calendar, and the cosmos. This chapter traces the history of the map, from its origins in Macrobius's *Commentary on the Dream of Scipio* to the translation of its inscribed contents into Old Norse. It then restores the map to its Icelandic manuscript contexts to analyse the conditions of its Icelandic reception, and the local effects of drawing a hemispherical world map in Iceland.

Chapter 2 presents the Icelandic zonal map, preserved in a manuscript written in the fourteenth century (Reykjavík, Stofnun Árna Magnússonar, GkS 1812 I 4to, f. 11v, 1315–*c.* 1400). This map was also in pan-European circulation, and seems to have come to Iceland with the natural philosophical treatises of the French scholastic philosopher William of Conches (*c.* 1090–*c.* 1154). This simple map expresses medieval theories about the extent of the Earth's habitability, and marks in words the temperate middle latitudes in both the northern and southern hemispheres. This map is an understated witness to the intense medieval debates about the theological implications of antipodal habitation: if humanity descends from Adam, from where could peoples south of the impassable equator have originated? If the equator is impassable, how can Christians fulfil Christ's injunction to the Apostles to spread Christianity to all nations? In its Icelandic manuscript, the map is situated amid a suite of planetary diagrams that explain the revolutionary motions of the naked-eye planets around the Earth. This chapter traces the history of this map, through comparison with its European precursors and parallels, and then examines its role in an Icelandic astronomical handbook to understand the intellectual work it did in its Icelandic iteration.

Chapter 3 introduces the Icelandic tripartite and T-O maps, drawn in a manuscript fragment bound into the Viðey book and dated to the second quarter of the thirteenth century (Reykjavík, Stofnun Árna Magnússonar, GkS 1812 III 4to, ff. 5v–6r and f. 6v, *c.* 1225–50). These two maps form a pair, drawn together on the recto and verso of the same manuscript folio. The larger of the two maps, occupying a whole manuscript opening, shows the geographical dispersion of the world's places and people, its surface inscribed with the names of 136 regions and cities, islands, mountains, rivers, peoples, and cosmological matters. The smaller map is an iconic shorthand, a simple T-O that bears only the names of the three continents, located amid a diagram of universal harmony. Until now, these maps have been analysed in perfect isolation from each other and their immediate manuscript context. This chapter explores the relationship between these two maps, probably drawn at the Augustinian monastery at Viðey in south-western Iceland, and is a touchstone for the two chapters that follow. Together, the two Viðey Maps are the most complex cartographic monument to survive from medieval Iceland. Their extensive geographical nomenclature – from Iceland in the west to China and India in the east – is presented here and analysed.

Chapter 4 concerns the depiction of Europe on the two maps from Viðey, with a particular focus on their depictions of Iceland and Scandinavia. This is the only map in the Icelandic corpus to show Iceland, which it places in an ambiguous relation to its occasional double, Thule. The map is an important work of cartographic self-portraiture, which shows how Icelanders perceived their global position at the time it was drawn. This chapter complicates the map's view of this region through looking at the convention of calling Iceland by the name *Thule*, a motif that surfaces in medieval European literatures as well, famously, in the prologue to the Icelandic *Landnámabók*. I review the occurrences of this convention in Icelandic and other European literatures in order to characterise its contemporary historiographical usage, and ask whether or not the Viðey Maps are a visual statement of these names' equivalence.

Chapter 5 alters the lens to focus on a prominent structural device that recurs in both Viðey Maps, and the one other item preserved alongside them in their manuscript fragment. These complex maps situate the world amid a conceptual frame that correlates the four cardinal directions – north, south, east and west – with other 'fours' that appear in nature, the four seasons, the four principal winds, and the four ages of man. The one other item preserved alongside them in their manuscript fragment is a list of forty highborn Icelandic priests allegedly compiled in 1143, and attributed to the preeminent Icelandic historian Ari Þorgilsson the wise. This document is a statement of the geopolitical organisation of the Icelandic Commonwealth, organised into administrative Quarters under the jurisdiction of the Icelandic chieftains (*goðar*), and a single elected official, the lawspeaker (*lögsögumaðr*). I argue that the two maps and their companion register were purposefully combined to allow the Icelandic map-maker to compare the fourfold nature of the physical universe with that of the Icelandic Commonwealth. This chapter examines the Viðey Maps as witnesses to the most unstable moment in Icelandic political history, the period of civil war that culminated in Iceland's loss of independence to Norway.

The purposes to which these maps were put are, like the maps themselves, diverse. It is the aim of this book to rehabilitate the Icelandic *mappae mundi* as sources for the literary and cultural histories of medieval Iceland. These maps emanate from a period of intense literary and historiographic activity in thirteenth- and fourteenth-century Iceland. They created a space for Icelanders to explore their culture and history, and thus provide a window onto Icelandic self-perception at the time they were drawn. They do not, however, do their work in isolation. The Icelandic *mappae mundi* were combined with other texts and images to form composite statements about Icelandic history and society, and relate more broadly to the Icelanders' literary output at this time. This book demonstrates that the Icelandic *mappae mundi* intersect more textual worlds than we have previously known.

CHAPTER 1

The Icelandic Hemispherical World Maps

The Icelandic hemispherical world map is a representation of global space, and the clocked processes that measure time on Earth. It shows the terrestrial sphere divided into climatically distinct regions along lines of equal latitude, or *parallels*: the two polar circles, the two tropics, and the equator. In the northern hemisphere, the temperate region between the Earth's frozen poles and the scorched equator is anatomized to show the relative positions of the three continents: Asia, Africa, and Europe. The Earth's southern hemisphere – an unknown but theorized mirror image of the northern hemisphere – bears an Old Norse legend that denotes the temperate region south of the equatorial ocean. The sun and moon are shown in two configurations in their orbits around the Earth: in conjunction (on the left of the diagram, where sun and moon are in the same region of the sky) and in opposition (on the right of the diagram, where the moon stands alone opposite the sun). These two depictions of the sun are connected by a narrow band inscribed with the signs of the zodiac, the series of constellations through which it moves in its annual orbit around the Earth. The map is inscribed with twenty-one legends, written in a combination of Latin and Old Norse. In both manuscript versions, the map accompanies two short Old Norse texts: a note on the error in the Julian calendar, and a note on the ebb and flow of the tides and the influences of the sun and moon upon them.

This map survives in two fragmentary Icelandic manuscripts of the early-fourteenth century, now held in Copenhagen's Arnamagnæan Institute. The older version, drawn *c.* 1300, preserved in a single bifolium, accompanies a geographically-focussed collection of texts and images (AM 736 I 4to, f. 1v). The younger version, drawn *c.* 1300–25, belongs to a fragmentary illustrated encyclopaedia of nine folios (AM 732b 4to, f. 3v).

The Icelandic hemispherical world map has a long history in literary and map historical scholarship. It first came to attention in the second volume of Rafn's *Antiquités Russes* (1852), where it was presented, with a slender commentary, as evidence for Icelanders' familiarity with basic geographical principles such as the Earth's sphericity, the division of its surface into climatic zones, and the positions of the three continents.[1] Its description in Marcel Destombes' *Catalogue* is also perfunctory. Destombes attributes to the map just three Latin names, those of the three continents, and a legend in Icelandic that designates the habitable region in the southern hemisphere.[2] Kålund characterises it as 'typical', but is otherwise silent on the genre to which it belongs, while Destombes numbers it among those maps whose authors are unknown to us.[3] The map's genre and origins have not been properly scrutinised. Simek observes that the manuscripts in which we find the Icelandic hemispherical world maps are 'obviously remnants of copies of an illustrated encyclopaedic MS modelled closely on Latin illuminated encyclopaedias', but their precise character – their thematic interests and the editorial policies that shaped them – have not yet been seen.[4]

The map's legends describe the lunisolar motions that cause the seasonal variation in the ocean's tides, but no attention has been directed to the map's function as an explication of medieval tidal theory. The map's contexts are also unknown. The assumption that a map's meaning is a quality intrinsic to the cartographic image has lead scholars to overlook the intellectual programmes to which it originally contributed and for which it was created. The map illustrates, in both manuscripts in which it is found, a note on the error in the Julian calendar, attributed to the otherwise unknown computist Master Walter ('Meistari Galterus'), and an anonymous note on the seasonal variation in the ocean tides.[5] In 736 I these two texts are preserved alongside the map, on folio 1v; in 732b they are preserved on the facing verso, on folio 2v. These companion

1 Kålund reproduces the map in his edition of the Old Norse *computus* treatise Rímbegla II, whose principal witness is the text preserved in Reykjavík, Stofnun Árna Magnússonar, AM 624 4to (*c.* 1490–*c.* 1510). Kålund footnotes a facsimile of the map and a transcription of its legends in his edition of an Old Norse text about the tides, which accompanies the hemispherical world map in both of its manuscripts. The tidal note accompanies a hemispherical world map in both the manuscripts 736 I and 732b, but not in 624. *Alfræði Íslenzk: Islandsk encyklopædisk litteratur*. 3 vols, ed. Kristian Kålund (Copenhagen: Møllers bogtrykkeri, 1908–18) (hereafter *AÍ* I–III). *AÍ* II, pp. 118–19.

2 Marcel Destombes, ed., *Mappemondes A.D. 1200–1500. Catalogue préparé par la commission de Cartes Anciennes de l'Union Géographique Internationale* (Amsterdam: N. Israel, 1964), p. 175.

3 *AÍ* II, p. 118, fn. 3. There are, according to Destombes, forty such anonymous maps drawn before 1200, and a further ninety from the period up to 1500. Destombes, *Mappemondes*, pp. 7, 164.

4 Rudolf Simek, 'Cosmology', in *Medieval Scandinavia: An Encyclopaedia*, ed. Philip Pulsiano (New York: Garland, 1993), pp. 110–11, p. 110; Simek, *Altnordische Kosmographie*, p. 70.

5 The note on the error in the Julian calendar is transcribed in *AÍ* II, pp. 237–239; the tidal note in *AÍ* II, pp. 117–118, §§ 67–68.

texts have occasioned no mention in previous commentaries on the map, even though they instruct the reader explicitly to consult the accompanying *figura* ('shape', 'image'). In what follows, we shall see that the map has a history that can be reconstructed through comparison with earlier maps and diagrams. Then, we shall examine the map in its Icelandic environment to understand the uses to which its lesson on the nature of the physical world was put by a medieval reader.

Origins

The hemispherical world map originated as a work of literary criticism, a diagrammatic attempt to harmonise two potentially contrary poetic descriptions of the cosmos in Cicero's *De re publica* (54–51 BCE) and Virgil's *Georgics* (29 BCE).

The *Dream of Scipio* (*Somnium Scipionis*) is the last part of Cicero's *De re publica*, a dialogue on Roman politics that concludes with the cosmic vision of the Roman military tribune Scipio Aemilianus.[6] In a dream, the young soldier Scipio is taken up into the heavens, to a high place among the stars, to look upon the cosmos, and take in the cosmic smallness of the world known to the Roman Empire. Scipio's cosmic vision created the narrative space for a digression on the nature and structure of the physical universe, within which the temperate middle latitudes of the northern hemisphere, and all those lands known to Rome and its enemies, were but a small part. In the fifth century CE, the Roman provincial Macrobius Ambrosius Theodosius wrote a *Commentary on the Dream of Scipio*, which explained Cicero's literary allusions to the shape and nature of the cosmos through a series of diagrams. Macrobius provides instructions for the drawing of four diagrams. The first shows the Earth in relation to the seven planets (1.21.3–5); the second shows how rain falls on the spherical Earth (1.22.11–12); the third depicts its zonal divisions (2.5.13–17); and the fourth shows the correlation between lines of celestial and terrestrial latitude (2.7.3–6).[7] Because these diagrams are integrated into the text through detailed instructions on how to draw them, they have been noted as remarkable for their consistency.[8] In addition to these four canonical diagrams, Macrobius alludes to a world map that shows the Earth's climatically optimal zones, and onto which several cities and waters

6 Denis Cosgrove, *Apollo's Eye: A Cartographic Genealogy of the Earth in the Western Imagination* (Baltimore: John Hopkins University Press, 2001), pp. 68–69.

7 Alfred Hiatt, *Terra Incognita: Mapping the Antipodes before 1600* (Chicago: University of Chicago Press, 2008), p. 48.

8 Destombes, *Mappemondes*, p. 85. On the relationship between text and diagram in the *Commentary* see Hiatt, *Terra Incognita*, p. 49.

might be plotted.[9] Versions of this zonal map are preserved in thirty-five extant manuscripts of the *Commentary* written before 1100, and 150 before the end of the fifteenth century.[10]

The *Commentary* was enormously popular throughout the High Middle Ages, and its manuscript transmission indicates that it was particularly prized for its cosmographical contents. It survives in two distinct versions: one complete, and the other an abridgement that enabled its cosmological contents to circulate independently.[11] Macrobius was certainly renowned as an authority in Iceland, where translated summaries of his work bear the author attributions 'sva telr Makrobius' ('so reckons Macrobius') and 'Macrobius dicens' ('Macrobius says').[12] An Old Norse description of the sun's course through the zodiac refers to a work by Macrobius called the 'tractatus philosophie de spera'.[13] Macrobius did not write a work by the title *De Sphera* (*On the sphere*), though many astronomical works by that name existed, and the reference may imply that his work was known through an abridgement of its cosmological contents.

In setting out Scipio's cosmic vision, Cicero had adapted the theories of Crates of Mallos (second century BCE), a Stoic commentator on Homer who theorised that the world was divided into four habitable regions, separated from one another by the ocean, which ran from pole to pole as well as across the equator.[14] The entirety of the *orbis terrarum*, the world known to the Roman Empire, was just one of these four habitable spaces, located in the Earth's northern hemisphere.

The Icelandic hemispherical world map has its origins in Macrobius's fourth diagram, the so-called celestial-terrestrial zone diagram (2.7.3–6, figure 1). The purpose of this diagram was to demonstrate that the zonal division of the sky, which had been poetically described by Virgil in the *Georgics* (I.233), did not contradict Cicero's allocation, in the *Dream of*

9 On this map, see Alfred Hiatt, 'The Map of Macrobius before 1100', *Imago Mundi* 59:2 (2007), 149–176. See also Konrad Miller, *Mappaemundi: die ältesten Weltkarten*, vol. 3: *Die kleineren Weltkarten* (Stuttgart: Roth, 1895), pp. 122–26.

10 Hiatt, *Terra Incognita*, pp. 69, 153.

11 On the manuscript transmission of the *Commentary* and the independent circulation of its cosmological components see Hiatt, 'Macrobius', p. 154.

12 *AÍ* II, pp. 239–42; 239–41.

13 *AÍ* II, pp. 96–97.

14 Crates of Mallus's quadripartite globe is known to us through a description in the *Geographica* (2.5.10) of the Greek historian and geographer Strabo (64/63 BCE – *c.* 24 CE). On the so-called Cratesian division of the globe see Germaine Aujac, 'Greek Cartography in the Early Roman World', in *The History of Cartography* vol. 1: *Cartography in Prehistoric, Ancient, and Medieval Europe and the Mediterranean*, ed. J. B. Harley and David Woodward (London: University of Chicago Press, 1987), pp. 161–76, esp. pp. 162–64. For a more recent discussion of the implications of the Cratesian division for medieval representations of the antipodes see Hiatt, *Terra Incognita*, p. 17.

Scipio, of these zonal divisions to the Earth (cited by Macrobius, 2.5.2). Macrobius explains that the lines of equal latitude, or parallels, that delineate the zones on the celestial sphere (the concavity of the sky, against which we can see the stars and planets) are projected onto the terrestrial sphere (the Earth's surface) in the same way that a large object is reproduced in a small mirror, on a smaller scale but in its correct proportions (2.7.3). The climatic conditions experienced within these regions on the Earth's surface are caused by the physical nature of the upper air, which is conducted to the portion of the Earth directly below (2.7.2).

The text of the *Commentary* contains detailed directions for the construction of this diagram, and Macrobius's definitions of its parts illuminate the basic form of the Icelandic map.

> Sed hic quoque adserendi quod dicitur minuemus laborem oculis subiciendo picturam. Esto enim caeli sphaera ABCD, et intra se claudat sphaeram terrae, cui adscripta sunt SXTV, et ducatur in caeli sphaera circulus septentrionalis ab I usque in O, tropicus aestivus a G in P et aequinoctialis a B in A et tropicus hiemalis ab F in Q et australis ab E in R; sed et zodiacus ducatur ab F in P; rursus in sphaera terrae ducantur idem limites cingulorum quos supra descripsimus in N in M in L in K.

> (Once again, we shall lessen the difficulty of proving our point by using a diagram. Let the circle ABCD represent the celestial sphere, and include within it the circle SXTV representing the earth. Draw upon the celestial sphere the line IO to represent the Arctic Circle, GP to represent the Summer Tropic, BA to represent the Equator, FQ to represent the Winter Tropic, and ER to represent the Antarctic Circle. Then draw the zodiac line from F to P. Next draw upon the earth the same demarcations for the zones mentioned above, lines terminating at N, M, L, and K.)[15]

The diagram that Macrobius drew to explain the parity of these literary descriptions underwent many medieval revisions, through such intermediaries as the Flemish scholar Lambert, canon of St Omer (*c.* 1050–1125?), and the French scholastic philosopher William of Conches (*c.* 1090–*c.*1154). Although Macrobius did not tell his reader to add a geographical nomenclature to his diagram, the maps derived from his empty template sometimes bear legends that name the parallels of latitude and the climatic zones. One such example is preserved in the large illustrated encyclopaedia called the Liber Floridus ('Book of Flowers'), completed in 1121 by Lambert of Saint-Omer. The autograph manuscript of the

15 Ambrosius Theodosius Macrobius, *Commentarii in Somnium Scipionis*, ed. J. Willis (Leipzig: Teubner, 1963), 2.7.3–4, p. 117–18. Translations from Macrobius's *Commentary* are drawn from Macrobius, Ambrosius Theodosius, *Commentary on the Dream of Scipio*, trans. William Harris Stahl (New York: Columbia University Press, 1952), p. 208.

Liber Floridus (Ghent, University of Ghent Library, MS 92) comprises 287 numbered folios that contain a universal history, assorted items on natural philosophy, and fifteen maps and diagrams.[16] The hemispherical world map on folio 227v (figure 2) is clearly based on Macrobius's template, but has been expanded with inscriptions that identify cosmological features identified in Macrobius's text. It shows the parallels by their Latin names, with the names of the celestial and terrestrial zones they divide written between them.[17]

The parallels of latitude are astronomically defined in Macrobius's *Commentary* (1.15.12–13) – as well as widely circulating epitomes of ancient learning such as Isidore's *Etymologies* (3.64), and Martianus Capella's *On the Marriage of Philology and Mercury* (8.818–822) – as markers by which to locate phenomena on the celestial and terrestrial spheres. The Arctic Circle (*circulus septentrionalis*, 'northern circle') is the parallel north of which there is at least one twenty-four hour period of continuous daylight, and one twenty-four hour period of continuous night, per year. The Summer Tropic (*tropicus* or *circulus aestivus*, 'summer tropic') marks the northernmost point at which the sun can be directly overhead: the sun is on this parallel on the summer solstice. The Equator (*circulus aequinoctialis*, 'equinoctial circle') girds the Earth at its middle, and marks the latitude at which sun passes on the two equinoxes, while the projected Winter Tropic (*tropicus* or *circulus hiemalis*, 'winter tropic') and Antarctic Circle (*circulus australis*, 'southern circle') mirror the Summer Tropic and Arctic Circle in the southern hemisphere.

The Icelandic hemispherical world map is a reworked version of Macrobius's fourth diagram that appears to have been copied from a map similar to the one preserved in the Liber Floridus.[18] The Icelandic map depicts the traditional distinction of the climatic zones, with inscriptions written in a combination of Old Norse and Latin. In the northern hemisphere, the temperate region has been marked to show the positions of the three continents – Africa, Asia, and Europe – while the southern hemisphere contains the Old Norse inscription *synnri bygð* ('southern inhabitable area') that denotes the projected temperate region south of the Equator. The Old Norse names used to designate the parallels on the Icelandic hemispherical world map can be compared with their Latin

16 The autograph Liber Floridus has been edited by Albert Derolez, *Lamberti S. Audomari Canonici Liber Floridus: Codex Autographus Bibliothecae Universitatis Gandavensis* (Ghent: Story-Scientia, 1968). Citations from the Liber Floridus are taken from this volume; page numbers refer to the text from the plates, which are transcribed at the back of this volume. An index to its contents is contained in Albert Derolez, *The Autograph Manuscript of the Liber Floridus: A Key to the Encyclopedia of Lambert of Saint-Omer* (Brepols: Turnhout, 1998). On Lambert's maps, see Miller, *Mappaemundi*, vol. 3, pp. 43–53.

17 Derolez, *Liber Floridus*, p. 98.

18 Miller, *Mappamundi*, vol. 3, p. 125; Simek, 'Cosmology', p. 110.

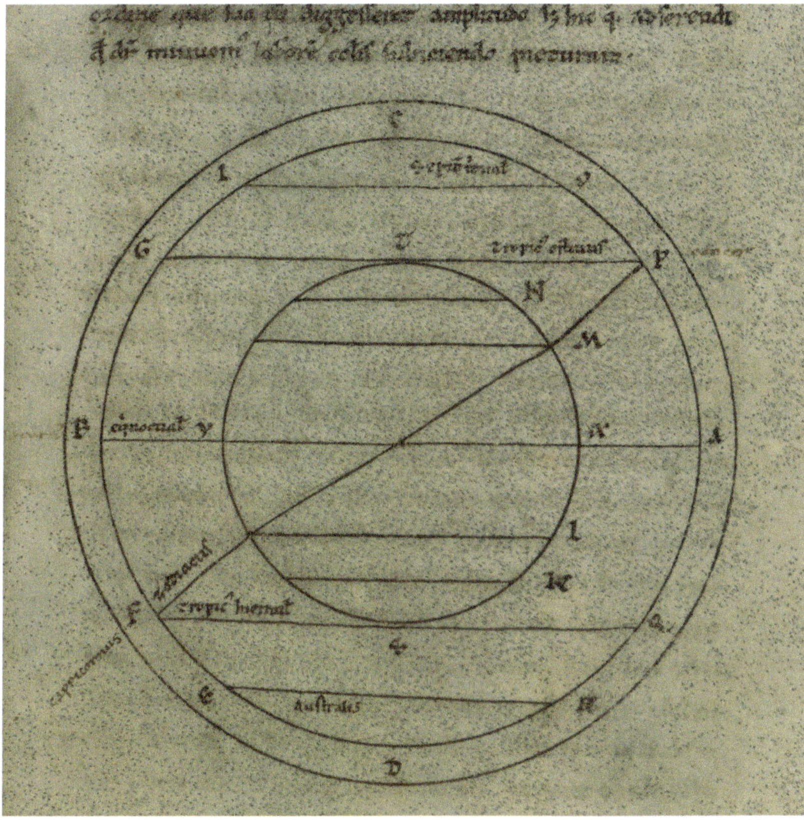

FIGURE 1: Macrobius's celestial-terrestrial zone diagram, the fourth diagram in his *Commentary on the Dream of Scipio*. This copy illustrates a text of the *Commentary* produced in north-eastern France. The parallels and zodiac have been named, as well as the sun's position in the two tropical signs, Capricorn and Cancer (though Macrobius gave no instructions to do so). This version of the diagram has north at the top, though the Icelandic hemispherical world maps have been turned to place south at the top. Baltimore, Walters Art Gallery, Walters Ms. W.22, f. 41r, *c.* 1175–1200.

equivalents on the map in the Liber Floridus. The Old Norse *vetr hringr* ('winter circle') and *sumar hringr* ('summer circle') are loan translations from the Latin *tropicus* (or *circulus*) *hiemalis* and *tropicus aestivus*, and the Old Norse *jafndægris hringr* ('Equator') calques the Latin *circulus aequinoctialis*. The names of the polar circles are more original productions. Old Norse possessed its own name for the 'sun that shines at night' (*nátt sól*, 'night sun'), and the Icelandic translator has combined it with the noun *hringr* to form *nátt sólar hringr hinn nerðri* and *nátt sólar hringr hinn syðri* (the 'northern' and 'southern night sun's circle') in analogy with the other parallels. That the Old Norse inscriptions are loan translations from Latin

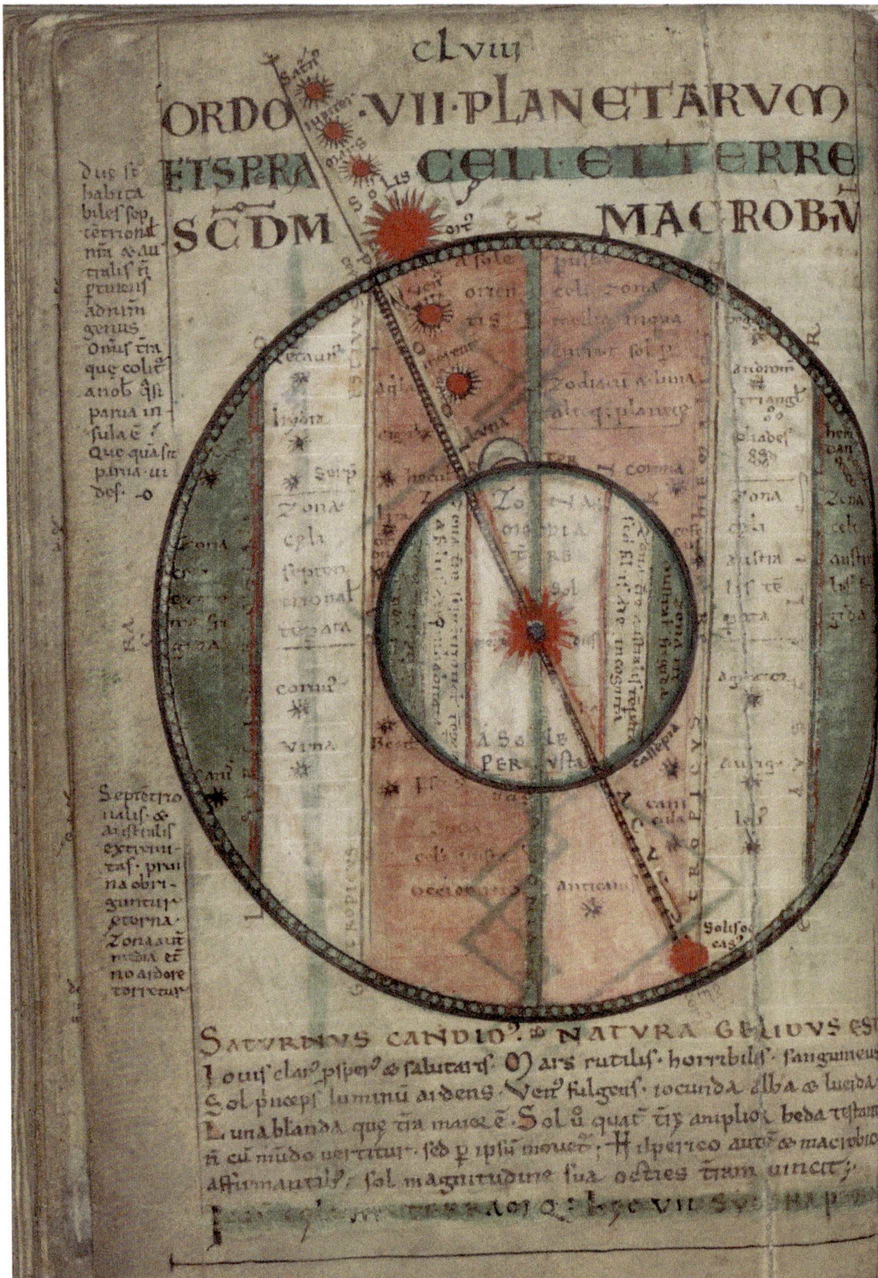

FIGURE 2: A hemispherical world map from the autograph Liber Floridus, modelled on Macrobius's fourth diagram. The *titulus* 'ordo vii planetarum et spera celi et terre secundum Macrobium' ('order of the seven planets and the celestial and terrestrial spheres according to Macrobius') indicates that the map's purpose was to show the structure of the physical universe. Ghent, University of Ghent Library, MS 92, f. 227v, 1121 (Ghent University Library, BHSL.HS.0092).

originals demonstrates that the Icelandic hemispherical world map is not an original composition, but rather that its textual inscriptions had been incorporated into the Macrobian outline before the map had arrived in Iceland; the Icelandic map-maker translating these technical Latin phrases into a more easily understandable vernacular.

Another feature of the Icelandic hemispherical map illuminated by Macrobius's description is the zodiac. Inclined across the map is a narrow band onto which six signs of the zodiac – Gemini, Taurus, Aries, Pisces, Aquarius and Capricorn – are named. The zodiac is the sequence of twelve constellations through which the sun moves on its apparent yearly orbit of the Earth, and occupies the portion of the celestial sphere 8–9° north and south of the ecliptic. The ecliptic is the course taken by the sun through the middle of the zodiac belt, and is so-called because an eclipse occurs when the sun and moon travel along this line at the same time. The map shows the sun in Cancer on the left of the map, where it rises on the summer solstice; and in Capricorn on the right of the map, where it rises on the winter solstice. The tropical signs (Cancer and Capricorn) are those in which the sun appears to stand still in its course around the Earth and reverse direction, and contain the sun's course within their bounds. As Macrobius says, 'the solstices lie on either side of the sun's path, preventing farther progress and causing it to retrace its course across the belt beyond whose limits it never leaves' ('in utraque obviante solstitio ulterius solis inhibetur accessio, et fit ei regressus ad zonae viam cuius terminos numquam relinquit') (1.12.1). The six signs named within the band represent the course of the sun over the half of the year between the summer and winter solstices, centred on the vernal equinox, when the sun rises in Pisces.[19] The ecliptic inclines between the two tropics because the Earth's orbit is not perpendicular to its celestial pole. In modern terms, we understand that the Earth tilts on its axis by 23.4°, but the antique explanation was that the plane of the ecliptic was set at an angle to the celestial equator, so that the zodiac inclines between the tropics ('de tropico in tropicum zodiacus obliquatus est') (2.7.17).

The zodiac is the only circle that Macrobius recognises as having breadth as well as length (1.15.8). Although the zodiac is represented on the fourth diagram as a single line, Macrobius explains that it is better imagined as region bound by two lines, which accommodate the constellations within (1.15.10). In the autograph Liber Floridus, three maps show the ecliptic, and while the 227v and 228r maps depict it as a single line that connects depictions of the sun, moon, and other planets, the map on folio 24v (figure 3) features an ecliptic with breadth enough to accommodate the signs.[20] Macrobius tells us that with the addition of the

19 John North, *The Fontana History of Astronomy and Cosmology* (London: Fontana Press, 1994), p. 259.

20 Derolez, *Liber Floridus*, p. [10].

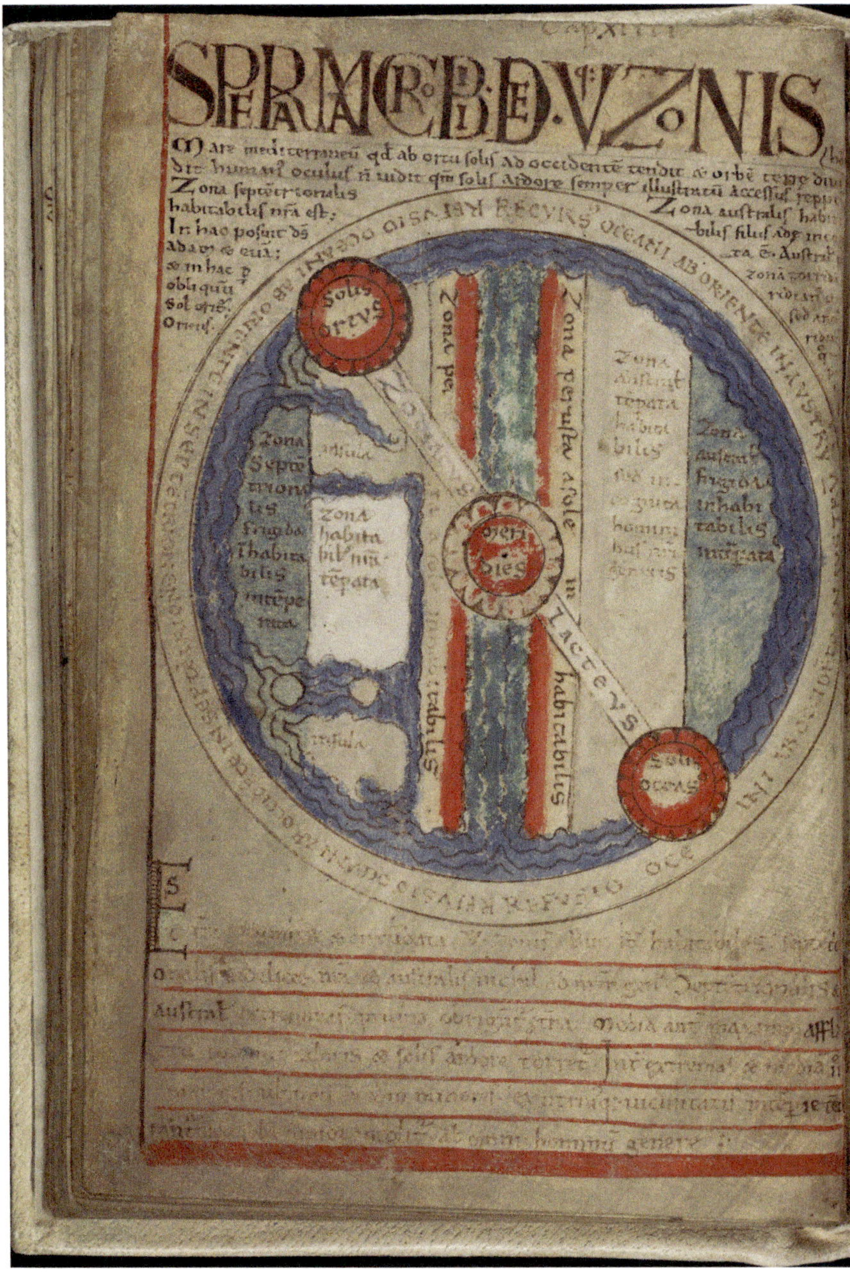

Figure 3: A zonal map from the Liber Floridus. This map features a broad belt inscribed 'zodiacus… lacteus' ('zodiac … Milky Way') (Ghent, Ghent University Library, MS 92, f. 24v, 1121 (Ghent University Library, BHSL. HS.0092).

ecliptic, the number of lines used to represent the zodiac can increase to three: 'et tertia ducta per medium ecliptica vocatur' ('in addition, a third one drawn through the middle is called the ecliptic') (1.15.10). This is another point on which Macrobius is eager to assure us that Cicero and other Roman poets did not err in their poetical descriptions: 'quamvis igitur trium linearum ductus zodiacum et claudat et dividat, unum tamen circum auctor voluit antiquitas' ('although three lines are required to bound and divide the zodiac, the men of antiquity, the authors of our vocabulary, wished it to be spoken of as one circle') (1.15.10). A similar discussion over the nature of the ecliptic might be evidenced by its depiction on the older version of the Icelandic hemispherical map in AM 736 I 4to. Here, a third line – drawn in the same red ink as the sun and the illuminated face of the moon – intersects the names of the zodiac written within this band. Perhaps this line represents the ecliptic, the route taken by the sun, which is also coloured red, through the zodiac. This third line is absent from the version in AM 732b 4to, but might be understood as a visual exegesis on the map's form that shows the map-maker's familiarity with Macrobius's extended description of the zodiac.

The parallels and ecliptic on the Icelandic hemispherical world map are imprecisely drawn. The ecliptic should extend between the two tropics, the *sumar hringr* and *vetr hringr*. On Macrobius's fourth diagram, the ecliptic inclines between points F and P, located on the winter and summer tropics respectively. Macrobius is explicit in his statement that the sun never deviates in its motions from the region between the two celestial tropics (2.5.11–15; 2.7.2–6). However, on the Icelandic map the ecliptic line falls considerably short of the points at which the *vetr* and *sumar hringar* touch the celestial circle. Furthermore, while the names of the parallels have been written onto the version of the map in AM 732b 4to, the lines have not been drawn. These minor oversights might evidence the map-maker's unfamiliarity with the finer cosmographical doctrines that, had the map been drawn more accurately, it might have illustrated.

The Antipodes

The Icelandic hemispherical world map exceeds its Macrobian template most in its depiction of the world's habitable regions in the northern and southern hemispheres. The map shows the known world, the *orbis terrarum*, divided into its continental thirds, while its temperate counterpart in the southern hemisphere bears an inscription that identifies it as the southern habitable area. These two landmasses are separated from one another by the equatorial ocean, the *megin haf*. The version of the map in AM 732b 4to contains a number of minor errors and omissions, lacking the line that separates Africa and Europe, as well as that which marks the limit of the temperate region in the southern hemisphere.

The conjecture that the Earth contained habitable regions unknown to the Roman Empire arose in late antiquity out of mathematical proofs of the Earth's sphericity and size, which confirmed that the known world did not amount to its entire surface.[21] Macrobius explains the division of the Earth into these four inhabitable regions by the circumambient equatorial and meridional oceans, whose intersection divides the Earth's surface into four regions (2.9.5–6).[22] The hemispherical world map shows two of these habitable regions – sometimes known as the *ecumene* and the *antoecumene* – with the Earth's remaining two habitable regions on the vertical hemisphere hidden from view.[23] In the tidal note that accompanies the map, which we shall examine shortly, the northern habitable region is called 'vorri byggð' ('our inhabited region'). The southern temperate region is labelled on the map as 'synnri bygð ('southern inhabited land'), separated from the northern temperate region by the *megin haf*. The antipodes are seldom mentioned in Icelandic literature, but in *Alexanders saga*, they are referred to as 'annarr heimrinn' ('the other' or 'second world').[24]

The antipodes are commonly granted cartographic expression in maps of this type, where the putative temperate region south of the equator stands as a mirror to the world known to medieval Europeans. In the Liber Floridus, four such maps identify this region as 'zona terre Australis temperata sed filiis Ade incognita' ('the southern temperate zone unknown to the sons of Adam'), to which another adds the Latin adjective *habitabilis* ('habitable').[25] Simek determines that the Old Norse inscription 'synnri byggð' is a loan translation of the Latin *zona habitabilis*, which sometimes labels the putative southern continent on Macrobian zonal maps.[26] Simek posits that the *synnri byggð* is 'one big inhabited island', but such an interpretation depends on the Icelandic translator's

21 Hiatt, *Terra Incognita*, p. 14.

22 This passage from Macrobius is paraphrased in William of Conches' *De philosophia mundi*, which exists in an Old Norse paraphrase and translation alongside the zone map presented in GkS 1812 I 4to. The Cratesian division will also be examined in Chapter 2 of this book.

23 Martianus Capella, *De nuptiis Philologiae et Mercurii*, ed. James Willis (Leipzig: Teubner, 1983), 6. 604–05.

24 David Ashurst, 'Journey to the Antipodes. Cosmological and Mythological Themes in *Alexanders saga*', in *Proceedings of the Eleventh International Saga Conference*, ed. Margaret Clunies Ross and Geraldine Barnes (Sydney: University of Sydney, 2000), pp. 1–13, p. 5.

25 Those on folios 24v, 92v–93r, 225r, and 227v and 228r.

26 Simek, *Altnordische Kosmographie*, p. 199. Hiatt's study of the toponyms of Macrobian maps before the twelfth century has shown that the antipodes are more commonly referred to in terms of their inhabitants, the antipodeans. Inscriptions that refer to the climatic qualities of these regions, such as habitable or temperate, are rarer. See Hiatt, 'Macrobius', p. 156.

familiarity with the debates that surrounded antipodal habitation in the Middle Ages, an issue that divided medieval authorities.[27] In his *Commentary* (2.5.23–36), Macrobius allows that the Earth's distant climatically optimal regions may be inhabited, but Bede, in *On the reckoning of time* (*De temporum ratione*) discredits these notions as *fabulae*. In *The City of God* (*De civitate dei*) (*c.* 413–26), Augustine rejected the idea of antipodal habitation on theological grounds: if the Earth's distant habitable regions are peopled but unreachable, then Christ's injunction to his Apostles to spread Christianity to all nations would be impossible.[28] Augustine's rejection of antipodal habitation was known in medieval Iceland through *Stjórn*, a collection of translated material from the Old Testament and other authoritative works, which denies antipodal habitation on the basis 'at engin iarðneskr madr ma þagat komaz or uarri byggiligri uerolldu sakir solar hita ok margrar annarrar umattuligrar ufaeru' ('that no earthly person can go there from our habitable world because of the heat of the sun and many other impassable obstacles'). The compiler of *Stjórn* further argues that 'allt mannkyn er fra Adam komit ok allt hans afkuemi hefir bygt þessar þrennar fyrr nefndar heimsins haalfur medr þeim ollum laundum ok eyjum utöldum sem i nordrhaalfunni liggia' ('all humankind is descended from Adam and all his descendants have settled these three named regions of the world [the three continents] and with them all the lands and countless islands that lie in the northern region').[29] Despite the controversial implications of antipodal habitation, the Old Norse *byggð* seems to imply, as Simek sees it, that these regions are not just habitable but settled, and is thus an uncertain witness to these debates.[30]

The Ocean's Tides

A considerable proportion of the map's legends pertain to the tides, but its function as an explication of tidal processes has garnered little attention. The Old Norse *missong* ('spring tide') appears twice on the map, in association with its two representations of the moon. The full moon, which marks the middle of the lunation at around fourteen nights, is

27 Simek, 'Cosmology', p. 110.

28 For a comprehensive account of late antique and early medieval theories about antipodal habitation see Hiatt, *Terra Incognita*, p. 38.

29 *Stjorn*: *Gammelnorsk Bibelhistorie fra verdens skabelse til det Babyloniske fangenskab*, *ed.* C. R. Unger (Christiania: Feilberg & Landmarks Forlag, 1862), pp. 99–100.

30 This is related to the verb *byggja* ('to inhabit' or 'settle') and the noun *bygging* ('inhabitation' or 'colonisation'). By contrast, its antonym *úbygðir* includes deserts, mountains, and wooded areas. This definition is supported by examples assembled in the ONP. See also Richard Cleasby and Guðbrandur Vigfússon, *An Icelandic-English Dictionary* (Oxford: Clarendon Press, 1957), p. 89.

shown on the right of the map: 'tungl xiiij natta / missong mikil' ('moon [at] fourteen nights, high spring tide'). The new moon, which marks the end of one lunation and the start of another at around thirty nights, is shown on the left of the map: 'tungl xxx missong' ('moon [at] thirty [nights] / spring tide').[31] These legends have no direct equivalents on the maps presented in the Liber Floridus. While a number of Lambert's maps show the *refusiones* ('tides') caused by the collision of the equatorial and meridional oceans,[32] there is no implication of the moon's involvement in their ebb and flow. The Icelandic map is accompanied by an Old Norse note on the tides in both 736 I (on the same folio, 1r) and 732b (on the facing verso, folio 2v). This short text describes the augmentation and diminution of the tides as the result of the motions of the sun and moon, and culminates in the directive 'þessa hluti máttu gjör prófa í þessi figuru' ('these matters can be proven in this diagram'). The reference to this *figura* ('diagram') formalises the connection between the map and the tidal note, and makes explicit the map's function as an illustration of these processes.[33]

Both the map and tidal note describe how the motions of the sun and the moon cause the augmentation and diminution of the tides.[34] Monthly variations are caused by the relative positions of the sun and moon in the sky, principally, those times at which the sun and moon stand opposite one another, with the Earth between them, and those times at which they stand in the same part of the sky, on the same side of the Earth. These are the celestial arrangements that produce the full and new moons, respectively, and are clearly depicted on the Icelandic map. When the sun and moon are thus aligned their influences on the ocean are compounded, and produce high tides that are higher, and low tides that are lower, than average, a phenomenon called the spring tide (Old Norse *missong*).

The tidal note further elaborates on the annual variation in the tides produced during the solstices and equinoxes, at which times the spring tides are greatest.

31 Macrobius was interested in these points in the lunar month because they are the points at which a solar or lunar eclipse can occur, when the sun and moon travel along the ecliptic at the same time (1.15.10–11).

32 The repeated inscription *refusio* features on the maps on ff. 24v, 225r, and 228r. See Derolez, *Liber Floridus*, pp. [10], [94], [99].

33 As we have seen, any word meaning 'picture' could refer to a map. The term *figura* was used to designate a map in Classical Latin, and by the medieval period had come to signify a smaller illustration or diagram in the service of an accompanying scientific text. See Woodward, 'Medieval *Mappaemundi*', p. 287; Harvey, *Mappa Mundi*, p. 389.

34 My transcription is based on the text in AM 736 I 4to. It can be compared with Kålund's in *AÍ* II, pp. 117–118, §§ 66–68.

Þa *er* tungl isomu ætt af iorðu at lita 7 sol. En solin gengr iþ*vi* marki *er* vær kaullum krabba mark 7 norðaz *er* isolar hring minkaz meir en [vant] *er* hiti solarinnar i megin hafinu af [fiar]lægd solarin*n*ar. þ*vi* at zodiacus gengr en*n* af vestri iafnt iaustr sva sem megin hafit girðer irorðina. Heldr af ættingi vestrs 7 utsuðrs iætting æustrs 7 l*an*dnorðss. verða þa missaung at nyi meiri en aðr. 7 þa gengr ny hest sem sol af varri byggveligre halfu at sia. En þa er tunglit stende gegnt solu i marci steingeitar. 7 solen er ikrabba marci sem fyr gatu*m* vær gengr tunglit lægzt fullt. þ*vi*at þa gengr tu*n*glit iþvi marki er sun*n*az er izodiacus 7 fi[arlægaz] er voRi bygð. 7 af nalægd tungls vaxa flæðar. þa er solin gengr iruzt liki eða iskalamarki 7 jafndæg*ri* er um alla v*er*old Enntunglit stendr gegnt sol e*r*u missaung af vellu solar hita. þviat solin er þa iþeim lut zodiaco er iafnt stendr yfir megin hafino. þa gengr tungl hæst vaxanda. en lægz þu^e^randa. 7 þa er ny verðr i þessum maurkum. skytur tunglit meire sinn e*r* vauku ahafit en vant er. þ*vi*at þa stendr tungl gengr yfir hafino. þessa lute ma[tu] gior pba iþesse figuru.[35]

(When the moon is seen from the Earth in the same house[36] as the sun, and the sun goes into the sign we call Cancer and is northernmost in its cycle, then the heat of the sun over the ocean lessens more than usual due to the distance of the sun,[37] because the zodiac extends as far to the west and east as the ocean that encircles the world, extending rather from a west-south-westerly direction to an east-north-easterly direction.[38] The spring tides then rise higher during the new moon than before. The moon is then seen to rise as high as the sun, in the region we inhabit. When the moon stands in the sign of Capricorn opposite the sun, and the sun is in Cancer as we conjectured previously,[39] the moon becomes lowest in the sky. This is because the moon then goes into that sign which is southernmost in the zodiac and farthest away from the lands we inhabit, and the tides increase due to the nearness of the moon.[40] When the sun declines into Capricorn or Libra and there

35 *AÍ* II, pp. 117–18.

36 The 360° of the night sky is divided into twelve astrological houses (Old Norse *ættingar*) of 30°, each assigned a zodiacal sign.

37 The sun rises and sets to the north ('solin… nordaz er i solar hring') in the half of the year centred on the summer solstice, during which there are longer days and shorter nights in the northern hemisphere. The summer solstice occurs when the sun is directly overhead the Summer Tropic, which is identified on the map.

38 This confusing statement seems to describe the inclination of the zodiac, which inclines relative to the celestial equator, under which lies the equatorial ocean.

39 The note previously mentioned the position of the sun on the summer solstice when, in the northern hemisphere, the sun rises to its highest point in the sky, causing its influence on the equatorial ocean to lessen.

40 These are the high or midsummer springs that occur around the summer solstice (when the sun in Cancer) when the moon is full.

is a solstice throughout the world.[41] When the moon stands opposite the sun, spring tides result from the boiling heat of the sun,[42] because the sun is then in the part of the zodiac which stands over the ocean. Then the moon becomes at its most waxed and least waned. And when it becomes new in these signs, the moon exerts its influence more than usual, because then the moon's course stands above the ocean. This matter can be proved in this diagram.)

There are three verbal echoes between the tidal note and the map's written inscriptions. The phrases 'megin hafit' ('the ocean'), 'solu i marci steingeitar' ('sun in Capricorn'), and 'jafndæg*ri* er um alla v*er*old' ('solstice throughout the world') bind the tidal note to the hemispherical world map; either the map derives its textually inscribed contents from the text, or the two items were paired because of their shared 'hlutir' ('matters'). Of particular interest is the connection between the legends showing the sun in the two tropical signs (Cancer and Capricorn), which represent the points in the sun's orbit around the Earth when it appears to stand still and reverse direction. The map's 'sól í steingeitarmarki' ('sun in Caprincorn') and 'sol í krabbi marki' ('sun in Cancer') echo the tidal note's 'sólu í marki steingeitar' and 'solin gengr iþvi marki er vær kaullum krabba mark' ('the sun goes into the sign we call Cancer').[43] The map bears legends indicating multiple zodiacal signs, but the naming of this one in the vernacular, when all others are written in Latin, strengthens the verbal correspondence between map and text, emphasising that they were designed to be inspected together. The position of the sun when it

41 The sun is directly above the equator on the autumn equinox when the sun rises in Libra. Libra is named the scales, hence *skálamarki*, because the sun is balanced mid-way in its course in this sign. See Isidore of Seville, *Etymologiarum sive originum libri xx*, ed. W. M. Lindsay, 2 vols. (Oxford: Clarendon Press, 1911), III.lxxi.29. All references to the *Etymologiae* are to this edition.

42 The verb *vella* ('to well over, to boil') frequently appears in Old Norse writings on the tides. This might originate in a translation from Latin, the noun *aestus* has the dual meaning of 'heat' and 'tide.' Another possible origin is Hermann of Carinthia's twelfth-century translation of Abū Maʿshar's *Introductorium in astronomiam* (III.5), which used the Latin verb *efferventes* ('to boil up') to describe the ocean's flow. See below for more details on the similarities between the tidal theory in Icelandic encyclopaedias and the *Introductorium*. See Charles Burnett, 'Does the Sea Breathe, Boil, or Bloat? A Textual Problem in Abū Maʿshar's Explanation of Tides,' in *Mélanges offerts a Hossam Elkhadem par ses amis et ses élèves*, ed. Frank Daelemans et al. (Brussels: Archives et bibliothèques de Belgique, 2007), pp. 73–79, esp. p. 76.

43 The sun in Cancer is marked only on the younger version of the map in AM 732 b 4to. The map-maker may have expanded upon or completed the map by drawing from the tidal note preserved alongside it, which explains the sun's course between the two tropical signs.

is in Capricorn (on the winter solstice) is identified to show the seasonal variations in the tides caused by the solstices and equinoxes.

The tidal note describes the midsummer springs that occur when the sun is in Cancer, and the equinoctial springs when the sun is in Libra. This is not the theory expounded by Macrobius. Most discussions of the tides in the works of ancient writers are cursory, since the tides in the Mediterranean Sea exhibit considerably less range than they do in the Atlantic and other oceans. Aristotle, for instance, attributes the ocean's ebb and flow to the size of the sound or basin into which the water flows.[44] Macrobius says little about the tides, but adduces their origin to the collision of the equatorial and meridional oceans, as they appear on the Cratesian globe (2.9.14). William of Conches, as we shall see in Chapter 2, elaborates upon Macrobius's theory, and illustrates these colliding waters with a zonal map. Maps in the Liber Floridus similarly contain the repeated inscription *refusio* ('tide') at the intersections between the equatorial and meridional oceans.

The main proponents for the moon's influence on the tides were Bede and the ninth-century Persian astronomer Abū Maʿshar (Albumasar), whose works became available in the Latin West in the twelfth century. In *On the reckoning of time* (*De temporum ratione*) (Ch. 29), Bede revised and enlarged the earlier tidal theories of Pliny and seventh-century Irish cosmographers, and advances a theory, which he supports with observational evidence, that the moon was the principal influence on the ocean's tides. Bede's particular interest was in their periodicity, and in determining the precise figure for the daily retardation of the moon.[45] However, from the twelfth century onwards tidal theories were based primarily on Abū Maʿshar's (787–886 CE) *Introduction to astronomy* (*Introductarium in astronomiam*), a text originally written in Arabic in the ninth century, and translated into Latin by John of Seville in 1133 and Hermann of Carinthia in 1140.[46] This treatise exerted considerable influence on later thinkers,

44 Aristotle, *Meteorologica*, Loeb Classical Library, trans. H. D. P. Lee (Cambridge, MA: Harvard University Press, 1952), II.354a. See Marina Tolmacheva, 'Geography, Chorography', in *Medieval Science Technology and Medicine: An Encyclopedia*, ed. Thomas F. Glick, Steven Livesey, and Faith Wallis (London: Routledge, 2005), pp. 186–91, esp. p. 188.

45 North, *Astronomy and Cosmology*, p. 228. For a discussion of Bede's contribution to tidal science see the commentary in Bede, *Bede, The Reckoning of Time*, trans. Faith Wallis (Liverpool: Liverpool University Press, 1999), pp. 310–14; and Faith Wallis, 'Bede', in *Medieval Science Technology and Medicine: An Encyclopedia*, ed. Thomas F. Glick, Steven Livesey, and Faith Wallis (London: Routledge, 2005), pp. 81–83, esp. p. 81.

46 References to the Latin translation of Abū Maʿshar's *Introductorium* are taken from Hermann of Carinthia, trans. *Introductorium in astronomiam Albumasaris Abalachi octo continens libros partials* (Venice: Sessa, 1506).

including the prolific Bishop of Lincoln, Robert Grosseteste (1175–1253).[47] In the *Introductorium* (III.6), Abū Maʿshar enumerated the causes of the tides and their seasonal variations in considerable detail and placed particular emphasis on the influences of the sun and moon upon them.[48] The Icelandic tidal note summarises how both the relative positions of the sun and moon and the moon's celestial latitude influence the ocean. The alignments of the sun and moon, in conjunction and opposition, are clearly depicted on the Icelandic map. Of these two alignments, Abū Maʿshar asserts that opposition exerts the stronger influence (III.6). This view is maintained on the Icelandic map, which contrasts the *missong* ('spring tide') produced when the sun and moon are in conjunction, with the *missong mikil* ('high spring tide') produced when they are in opposition.[49] The Icelandic tidal note further describes the moon's latitude, and its inferred proximity to the *megin haf* ('the equatorial ocean'). On the strength of the tides in the northern hemisphere, Abū Maʿshar maintains that 'quandiu luna in latitudine sua ascendit accessus vis augetur quandiu descendit recessus' ('when the moon ascends in its latitude its strength increases and when it descends it ebbs') (III.6). The implication, as understood by Edgar S. Laird, is that when the moon declines southwards it approaches the equatorial ocean, which increases its influence over the tides. The ostensible purpose of the hemispherical world map is to illustrate these processes, and relate them, as we shall see, to other clocked processes.

The Older Version of the Icelandic Hemispherical World Map (Copenhagen, Arnamagnæan Collection, AM 736 I 4to, f. 1v)

The Icelandic hemispherical world map was conceived as a work of literary criticism, as an explanation of Cicero's *Dream of Scipio* in Macrobius's *Commentary*. This template was elaborated upon by medieval anthologists of late antique learning such as Lambert of Saint-Omer, who recognised the value of diagrams in teaching the cosmological doctrines medieval thinkers had inherited from antiquity. Comparisons between

47 Edgar S. Laird, 'Robert Grosseteste, Albumasar, and Medieval Tidal Theory', *Isis* 81:4 (1990), 684–94, 684. For further information about Albumasar on the tides see Pierre Duhem, *Le système du monde. Histoire des Doctrines Cosmologiques de Platon à Copernic*, 10 vols. (Paris: Hermann et Fils, 1913–59), vol. 2, pp. 369–86.

48 Tolmacheva, 'Geography, Chorography', p. 187.

49 Grosseteste omits Abū Maʿshar's statement that opposition causes greater tidal range than conjunction from his paraphrase of it in *De fluxu*. The Icelandic encyclopaedias in which the tidal note appears, therefore, must have received their information about the tides from some other channel than *De fluxu*. See Laird, 'Medieval Tidal Theory', 690.

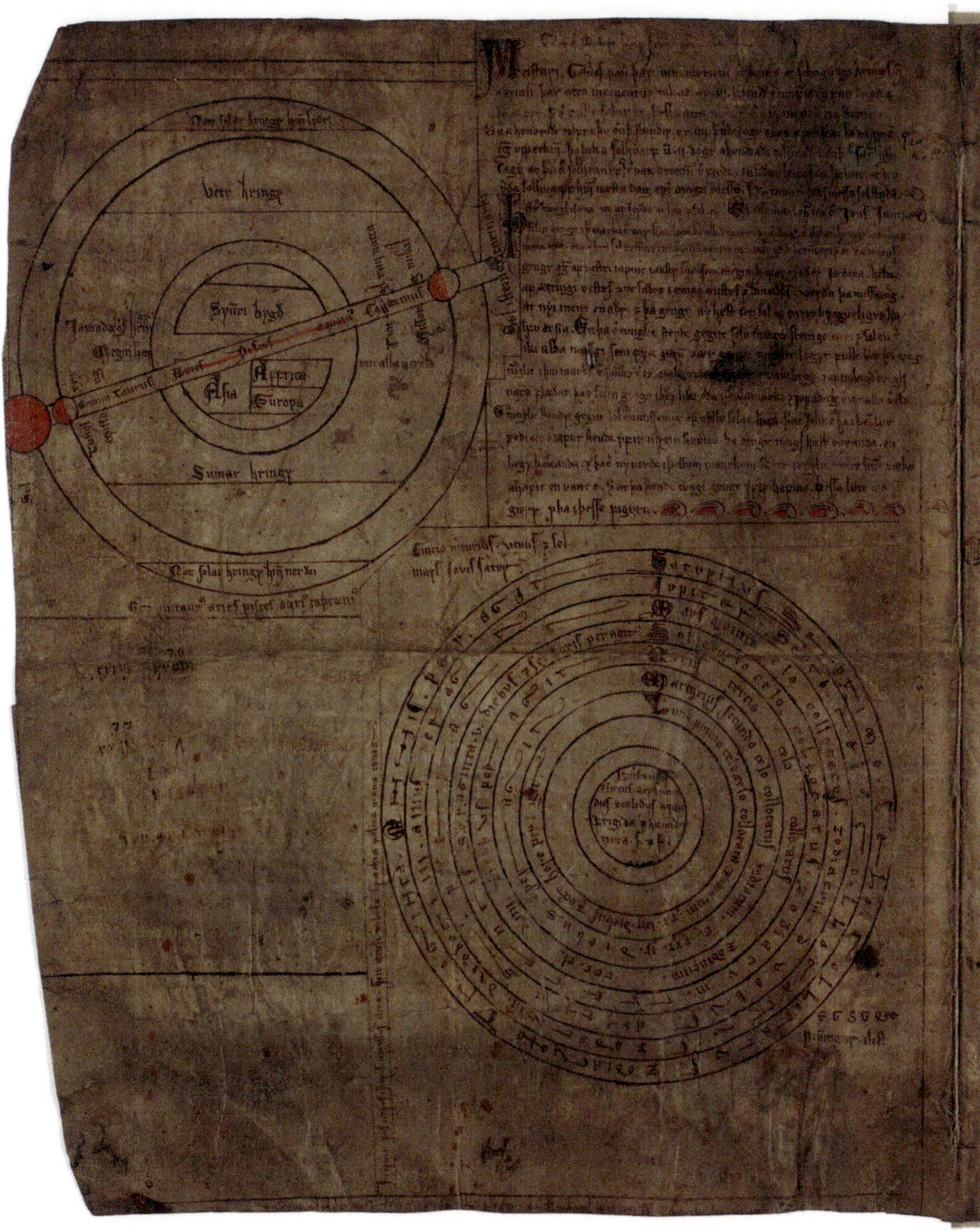

Figure 4: The older version of the Icelandic hemispherical world map is preserved alongside two notes, on the error on the Julian calendar and the seasonal variation in the tides (*right*), and a circular diagram showing the planets' orbits around the Earth (*lower right*). This folio faces a map of Jerusalem (f. 2r). Copenhagen, Arnamagnæan Collection, AM 736 I 4to, f. 1v, *c.* 1300. Photograph Suzanne Reitz. Reproduced with permission from the Arnamagnæan Institute.

the Icelandic map and its diagrammatic antecedents shows us that its function has changed from a visual statement of the relationship between celestial and terrestrial latitude, to an illustration of clocked processes on Earth: the lunar and solar calendars, and the tides. To this end, the Macrobian template has been elaborated upon with a suite of written inscriptions, originally in Latin and then translated into Old Norse.

The map does not produce meaning in isolation, but rather was drawn in deliberate sequence with the texts and images that accompany it in its manuscripts. The version of the hemispherical world map preserved in AM 736 I 4to – a single bifolium with texts written in two hands – is probably the older of the two, the contents of folios 1r–2r, which includes the map, having been written *c.* 1300, and the description of the Church of the Holy Sepulchre on folio 2v having been added several decades later in *c.* 1350. The other Icelandic manuscripts in which we find maps (AM 732b 4to and GkS 1812 III and I 4to) are both large encyclopaedic compilations, and this bifolium was probably intended for inclusion in, or has become separated from, a similar anthology of natural philosophical works. The items in AM 736 I 4to contemporary with the hemispherical world map comprise a description of the three continents, sometimes called the *Icelandic Geographical Treatise* (f. 1r); an Old Norse text – now for the most part illegible – on the three Biblical Magi (f. 1r); an Old Norse note on the error in the Julian calendar (f. 1v); the Old Norse tidal note (f. 1v); a *rota* diagram that shows the orbits of the seven planets (f. 1v); and a circular map of Jerusalem (f. 2r). The later addition, dated *c.* 1350, is a description of the Church of the Holy Sepulchre and the environs of Jerusalem (f. 2v). It is productive to examine these items, serially, for what they can tell us about the map's intellectual environment and its inclusion in this anthology.

The note on the error in the Julian calendar accompanies the hemispherical world map in both manuscripts that preserve it, and calls to attention the disparity between the lunisolar and ecclesiastical calendars that formed the basis of *computus*, the medieval science of calendar construction. The observation is attributed to a Master Walter – a computist apparently unknown outside Iceland.[50]

50 Kålund conjectures that Galterus is a computist unknown outside Iceland, and notes that this observation is more commonly attributed to the computist Magister Chonrad, *c.* 1200. *AÍ* II, p. 238 fn. 3. The name Galterus is also occasionally cited as an authority, alongside Isidore and Lucretius, in the bookish *riddarasögur*, Icelandic romance sagas that make liberal use of encyclopaedic lore. Barnes presumes this Galterus is Walter de Châtillon. Geraldine Barnes, *The Bookish Riddarasögur: Writing Romance in Late Mediaeval Iceland* (Odense: University Press of Southern Denmark, 2014), p. 17. See also Annette Lassen, 'Indigenous and Latin Literature', in *The Routledge Companion to the Medieval Icelandic Sagas*, ed. Ármann Jakobsson and Sverrir Jakobsson (London and New York: Routledge, 2017), pp. 74–87, p. 82.

Galtervus Meistari segir solar gang.

Meistari. Gal*ter*us fan*n* þær. viij. moment*ur* e*r* skortir at solin gangi hring sin*n* a arinu. þær atta momentur takaz af. vi. stund*um* er um f*ra*m e*r*u fim daGa 7 .lx. 7 ccc. þ*at* er fullt solar ar. þessa atta momentur e*r*u a fim aru*m* ein stund. En*n* a hundraði tolfræþu e*r*u þat stundir. xx. vij. þ*at* v*er*ðr dagr 7 nott 7 firi þvi at sadagr e*r* *eigi* upptekin*n*. þa þoka a solhvarf u*m*. ij. dogr ahundraði tolfræður. en þ*at* e*r* sanliga sagt at þa v*oro* solhvarf iola nott e*r* vaR drotin*n* var fædr. en siðan hafa sva þokat. at nu v*er*ða solhvarf hin*n* næsta dag ept*ir* magn*us* messo. I þan*n* tima v*ar* jons messo sol staða asumar. en*n* nu er Idus Ivnij.

Master Walter says about the course of the sun

Master Walter discovered those eight moments[51] that are lacking as the sun moves in its circle over the course of the year. Those eight moments are equivalent to the six hours [*sic*] that exceed the 365 days of the full solar year. Over five years, these eight moments add up to one hour, and over 120 years that becomes twenty-seven hours,[52] a day and a night, because that day is not used in calculation. Therefore, the solstice moves one day in every 120 years. It is said truthfully that there was a solstice on the night the Lord was born, but it has since moved so that now the solstice occurs the day after the feast of St Magnús [13th December]. At that time, the feast of St John the Baptist was on the summer solstice, but now it falls on the Ides of June [13th June].

The aim of *computus* was to synchronise local and universal time, ensuring that movable feast days, most importantly Easter, were celebrated simultaneously throughout Europe. Icelandic *computus* materials – calendars, Easter tables, and instructive texts on how to use them – are found in large compendia modelled on European illustrated encyclopaedias, alongside other texts on subjects connected to the reckoning of time, such as mathematics, astronomy, and history. These educative volumes enabled the computist to enter Iceland into a clocked community whose observance of the liturgical calendar was consistent across European Christendom.

51 A moment (Latin *momentum*, Old Norse *momenta*) is a medieval unit of time equivalent to 1.5 minutes. Stephen C. McCluskey, *Astronomies and Cultures in Early Medieval Europe* (Cambridge: Cambridge University Press, 1997), p. 150. Bede derives his understanding of the moment from Isidore's *Etymologiae* (5.29.1). In *De temporum ratione* (Ch. 3) he states that: 'they name *momenta* after the swift motion [*motu*] of the stars, when it was observed that something moved and succeeded itself in a very brief space of time' (*The Reckoning of Time*, p. 15). There are forty moments in an hour.

52 The text in AM 736 I 4to states that there are twenty-seven ('xx. vij') hours in a day, which in Kålund's edition has been silently corrected to twenty-four (the version in AM 732b has instead 24, written in Arabic numerals). *AÍ* II, p. 238.

There are two substantial errors in the note. The text states that Galterus discovered the eight moments that are lost over the course of the sun's annual revolution around the Earth. This is incorrect, as the Julian year was in fact too long, not too short.[53] The Julian year comprised 365 days in which the changes of the seasons, the solstices and equinoxes, fell on fixed dates.[54] However, the solar year is a few minutes shorter than the 365.25 days for which the fourth-year embolism, or leap year, was intended to compensate. As a result of the disparity between the lunisolar and calendar years, the Julian calendar gained, as the Old Norse text correctly asserts, around one day every 120 years. In the year 1200, the summer solstice was taking place ten days before it was being celebrated by the Church. Certainly, the short astronomical notes on folio 1v are not sufficiently technical to train a cleric in *computus*, and such confusions are likely indicative of the ability and education of the scribe, who was probably not a skilled computist. In the twelfth and thirteenth centuries, such anthologists sought 'to gather the knowledge of their time for non-specialist audiences'.[55] Rather than being a technical handbook for the instruction of *computus* operations, this manuscript is an introduction to the fundamental concern of *computus*, the disparity between the lunisolar and calendar years, for non-specialist readers. The note relates its observations on the disparity between the lunisolar and calendar years with a northern frame of reference, by using a native saint's day to illustrate the retardation of the solstice. Magnúss messa, the feast day of St Magnús Erlendsson, earl of Orkney (d. 1115), falls on 13th December, and Jóns messa, the feast day of John the Baptist, falls on 24th June. This reference to St Magnús relates universal observations on the error in the calendar with the veneration of local cults, bringing Icelandic local time into harmony with the universal calendar.

Below the notes on the error in the Julian calendar and on the tides is a circular diagram that shows the Earth, inscribed with the names of the four elements and their qualities, amid the nested planetary spheres, which guide the orbits of the moon, Mercury, Venus, the sun, Mars, Jupiter, and Saturn around the Earth.[56] This diagram locates the hemispherical world map, which concerns mostly the Earth, moon, and sun, in its wider cosmic environment. This diagram also supplies more information about the duration of the sun and moon's orbits, depicted on the hemispherical world map and described in the note on the error

53 *AÍ* II, p. 238.

54 McCluskey, *Astronomies and Cultures*, p. 24.

55 B. Ribémont, 'Encyclopaedias', in *Medieval Science Technology and Medicine: An Encyclopaedia*, ed. Thomas F. Glick, Steven John Livesey, and Faith Wallis (New York: Routledge, 2005), pp. 159–61, p. 159.

56 The contents of this diagram are edited in Simek, *Altnordische Kosmographie*, p. 418.

in the calendar, telling us that 'sol quarto celo collocatus zodiacum trecentis sexaginta .v. diebus & sex horis peragit' ('the sun stationed in the fourth heaven completes the zodiac in 365 days and six hours') and 'luna primo celo collocatus zodiacum xxvij diebus & octo hore peragit' ('the moon stationed in the first heaven completes the zodiac in 27 days and eight hours'). The hemispherical world map is a succinct statement of the relationship between the world and time. It illustrates the clocked processes that measure time on Earth – the courses of the sun and moon, and the seasonal ebb of the tides – to contextualise those *computus* operations whose aim was to bring Iceland into a European community centred on its observance of the ecclesiastical calendar.

Continents and Kings

The map illustrates the clocked processes that must be reconciled in the construction of a calendar and the generation of the tides, and thus combines the study of the globe with the study of time. This combination of themes can also be detected in an Old Norse description of the three continents, written on the preceding folio (f. 1r), sometimes called the *Icelandic Geographical Treatise*. [57] This text, which begins 'svá er kallat sem þrideíld se íorð at nofnum' ('the three parts of the world are named thus'), describes the three continents of the *orbis terrarum* that are depicted schematically on the map that follows it. The *Treatise* contains more than forty place-names, the preponderance drawn from the gazetteers assembled in authoritative and widely-read works such as Isiodre's *Etymologies*. The treatise relates the locations of ancient cities such as Nineveh, Babylon, Antioch, and Alexandria, usually with reference to their place in the early history of Christianity. 'Asia en mínni' ('Asia Minor'), for example, 'heitir land ihinni miklo asia þar kendi ion postoli tru oc þar er grof hans iborg þeiRi er effesus heitir' ('is the name given to that land in Asia where John the Apostle taught the Faith, and there lies his grave, in a town there called Ephesus'). Antioch is similarly celebrated as the place where 'setti petrus postoli biskups stol oc þar song hann fyst

57 The *Icelandic Geographical Treatise* is preserved in eleven manuscripts written between the early fourteenth and eighteenth centuries, but only three are medieval. For editions see *AÍ* I, pp. 8–12; and Simek, *Altnordische Kosmographie*, pp. 428–29. An English translation of the *Treatise* first appeared in Toulmin Smith, *Discovery*, pp. 335–37. This entire folio was reproduced in a drawn facsimile in Rafn, *Antiqitates Americanae*, p. 502. Tenney observed that most of the information contained in the treatise about Southern Europe originated in Isidore's *Etymologies*, however the works are of a vastly different scale and character, necessitating that the impulses that shaped the appropriation of those elements that were adopted is understood. See Frank Tenney, 'Classical Scholarship in Medieval Iceland,' *The American Journal of Philology* 30:2 (1909), 139–52, esp. 145–46.

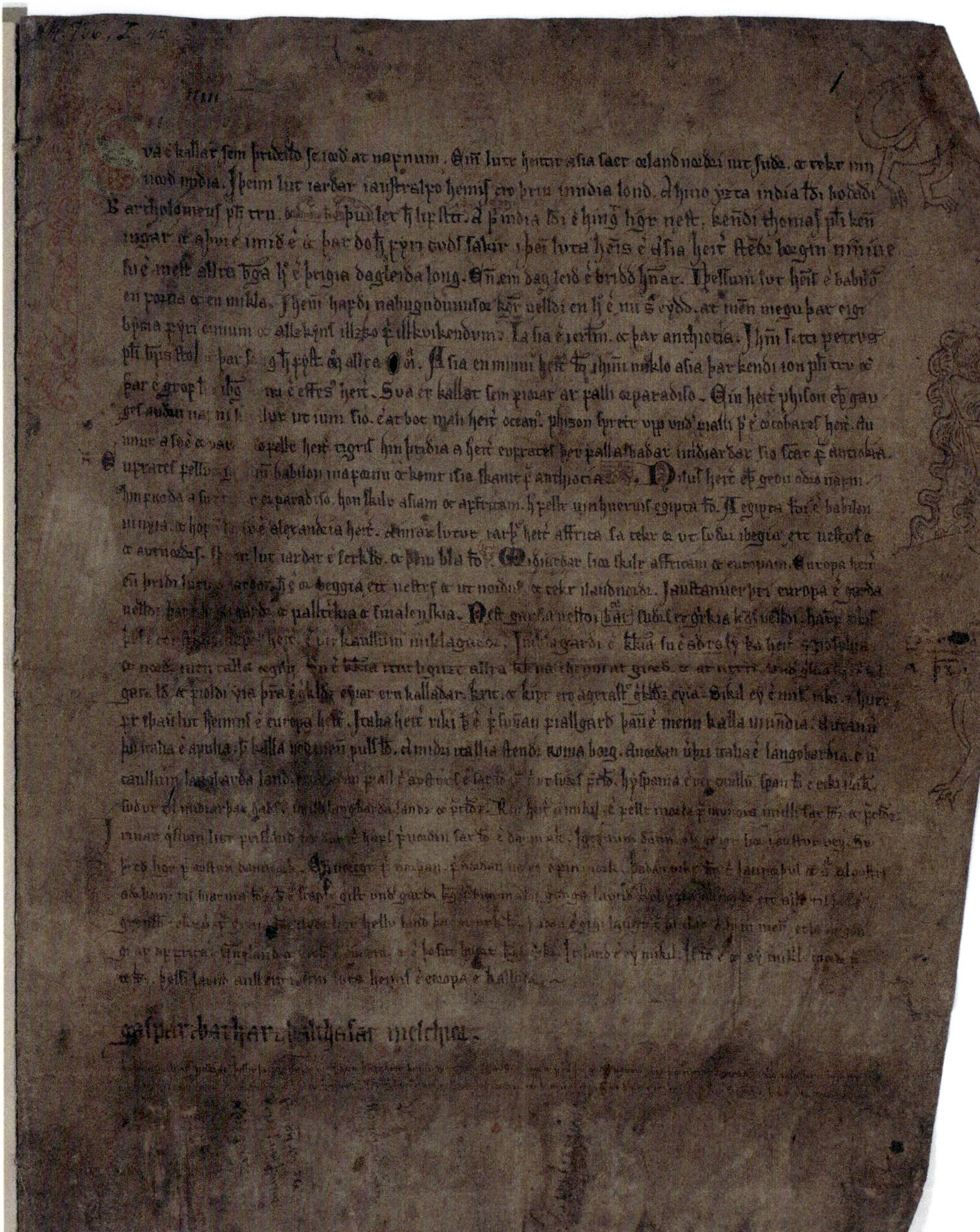

FIGURE 5: An Icelandic geographical treatise above a now faded text headed with the names of the three Biblical Magi. This folio precedes the older version of the Icelandic hemispherical world map. Copenhagen, Arnamagnæan Collection, AM 736 I 4to, f. 1r, *c.* 1300. Photograph Suzanne Reitz. Reproduced with permission from the Arnamagnæan Institute.

messo allra manna' ('Peter the Apostle established an episcopal seat and sang the first mass'); and Babylon as the place where 'hafði nabugudunusor konungr uelldí en hon er nu sva eydd' ('King Nebuchadnezzar ruled but it is now destroyed').[58]

The treatise shows keener topographical awareness in its description of north-western Europe: England and Scotland are described as 'ei eín' ('one island') and Iceland as 'ey mikil i nordr fra irlandi' ('a large island to the north of Ireland'). The world that this treatise presents is calibrated to the Icelanders' own perspectives on and knowledge about distant lands. The author frequently combines Latinate or traditional place-names with their Old Norse vernacular equivalents. The treatise guides a reader through 'ríkis þes er constantinopel heitir, er ver kaullum míklagarðr' ('that kingdom called Constantinople, which we call Miklagarðr'); 'agiosofia oc norðr menn calla agisif' ('the Hagio Sofia, which the Norse call Agisif'); 'apulía þat kalla norðmenn pulsland' ('Apulia, which the Norse call Pulsland'); 'langobardia, er uer caullum langbardaland' ('Lombardy, which we call Langbardaland'); and 'hyspania er ver caullum spanland' ('Spain, which we call Spanland'). Through putting these place-names in apposition, the treatise's Icelandic author demonstrates Icelanders' participation in pan-European geographical discourses; these translation moments are not markers of the Icelanders' distinctiveness or exceptionalism, but rather of their familiarity with the world's places near and far. The repetition of the formula 'sem ver kǫllum' ('which we call') and 'sem norðmenn kalla' ('which the Norse call') does similar work to the citation of local saints' days in the calendrical note on the folio overleaf, both harmonising the local Icelandic with the universal.

The treatise's topographical nature can be apprehended at a glance because it is rubricated in parts with pen decorations that throw certain place-names into relief against the surrounding text. Through rubrication, pen ornamentation, or bold initials, the place-names 'Indialand' (India), 'Anthiocía' (Antioch), 'Nilus' (the River Nile), 'Blálönd' (Saharan Africa), and 'Miðjarðarsior' (the Mediterranean Sea) are emphasised. India, we are told, was where the apostle Bartholomew preached, while the patriarchate of Antioch was an important crusader state and pilgrimage destination. The Nile and the Mediterranean were two of the vital waters that separated the lands of the *orbis terrarum*. The treatise and the hemispherical world map complement one another in their visualisation of the known world, and its positioning in its wider cosmic environment in the globe's northern hemisphere, on an Earth circled by the sun and moon. In its manuscript, the rubricator has invested the treatise with an additional layer of shape and figuration; the treatise may be its own map in so far as its contents can be seen, as well as read.

58 The place the Apostle Peter preached, Caesarea, is marked on the Thorney Map, 'Cesaria, hic Petrus predicavit'. See Chekin, *Northern Eurasia*, pp. 65–66.

Beneath the *Geographical Treatise* on folio 1r are two lines of text titled 'gaspar balthalar melchior' ('Caspar, Balthazar, Melchior'), the apocryphal names given to the three Biblical kings who attend Christ's Nativity in Matthew 2. 1–12.[59] The three kings are distinguished astronomers and travellers from the east who come to Judea having observed the appearance of a new star, or a *nova*. The reason for the journey of the Magi and their number is explained in the Old Norse *Elucidarius*.[60]

> Disciplus: Hvi kallaði gvð til sin þria avstr vegs kononnga meðr fornum? Magister: Þvi at hann villdi til sin leiða alla þriðiunga heims sins Asiam ok Affricam ok Evropam ok goðverk.
>
> (Student: Why did God call to himself the three eastern kings, with their gifts? Teacher: Because he wanted to lead to himself all the three parts of the world, Asia, Africa, and Europe, and good deeds.)[61]

The three kings, then, stand as representatives of the three continents described in the treatise above. In their capacity as archetypal travellers and representatives of the three parts of the world, the three kings are a common motif on medieval maps and globes. They appear prominently on the Catalan World Atlas (*c.* 1375), commissioned by Charles V of France and generally attributed to the Catalan book illuminator Abraham Cresques (figure 6), as well as the sea-chart produced by the Catalan cartographer Gabriel de Vallseca in 1439. The Biblical Magi also appear on the anonymous chart in Florence, Biblioteca Nazionale Centrale di Firenze, portolan 16, *c.* 1440; and the chart produced by Juan de la Cosa in *c.* 1500 (Madrid, Museo Naval, no. 270); the Behaim Globe, produced at Nürnberg in 1492; and the Schöner Globe, produced at Frankfurt and Weimar in 1515.[62] The inclusion of this text on the three Biblical kings in AM 736 I 4to may have been motivated by their role as representatives from the world's three continents, and by their status as the archetypal travellers to Judea and the lands venerated by medieval Christianity.

The Map of Jerusalem

The older hemispherical world map is preserved on a manuscript opening that includes a planetary diagram, tidal and calendrical notes, and opposite it, on f. 2r, a map of Jerusalem (figure 7). The circular

59 The text is scarcely legible. Isolated words can be read, e.g. 'voro þeir þrir konungar' ('they were three kings'), but not enough to produce a reading.

60 The *Elucidarius*, attributed to Honorius Augustodunensis, was written in Latin in the early twelfth century and translated into Old Norse in the same century. It survives in several independent Old Norse translations.

61 Evelyn Scherabon Firchow, ed., *The Old Norse Elucidarius* (Drawer: Camden House, 1992), pp. 34–35.

62 I am grateful to Gerda Brunnlechner for bringing these examples to my attention, by personal communication, in August 2014.

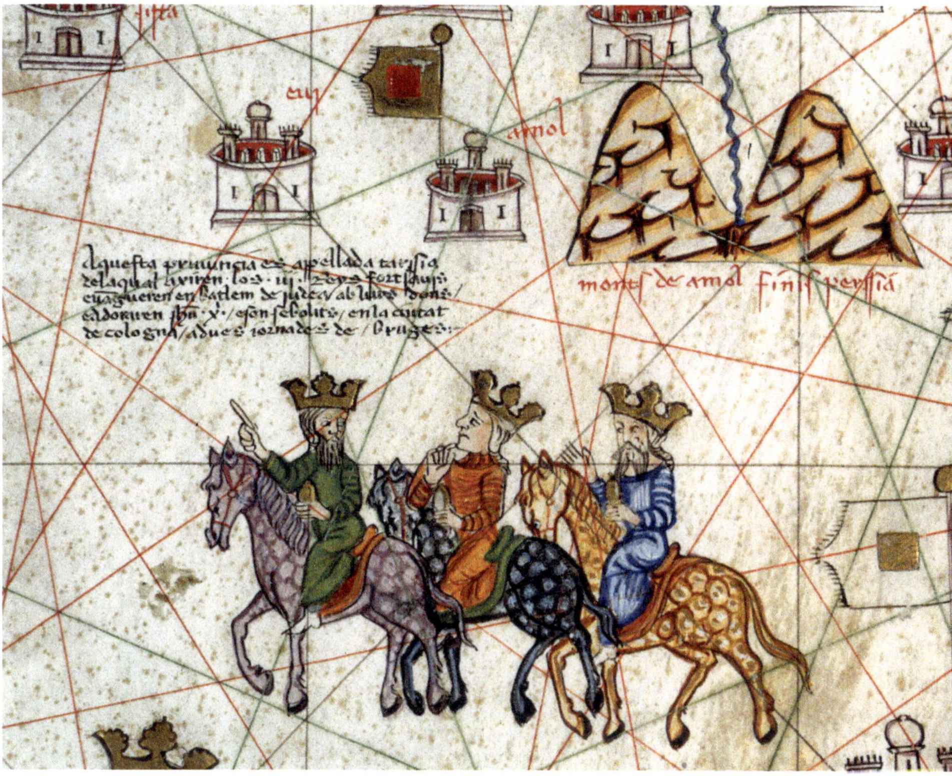

FIGURE 6: The Biblical Magi following the star on the Catalan World Atlas, attributed to Abraham Cresques. Paris, Bibliothequè Nationale de France, *c*. 1375.

forms of the map, the planetary diagram, and the map of Jerusalem compel the viewer to compare and relate three images of the world, the cosmos, and its spiritual centre. There are thirteen maps of Jerusalem in medieval European manuscripts, three of which are Icelandic. The Icelandic examples are preserved in the two manuscripts that contain the hemispherical world map, here and in AM 732b 4to (f. 8v), and in Hauksbók (f.19r).[63] The three Icelandic examples are very similar and plainly derive from a single exemplar. Other maps of Jerusalem generally maintain the city's distinctive circular form and cruciform street plan, but vary stylistically and in terms of their written inscriptions.[64] Briefly, three

63 Finnur Jónsson, ed., *Hauksbók, efter de Arnamagnæanske Håndskrifter no. 371, 544 og 675 4to* (Copenhagen: Thieles Bogtrykkeri, 1892–96), p. 186.

64 The Icelandic map of Jerusalem has been compared on stylistic grounds (the representation of battlements, the cruciform street plan) with the representation of Jerusalem near the centre of the English Hereford Map. See Birkholz, *Two Maps*, p. 99.

of these maps are found in crusade chronicles that bear the title *Gesta* or *Historia Hierolymitana*, two are found in versions of the Liber Floridus, and five appear in psalters and theological compilations.[65] Since two of the Icelandic examples are found in illustrated encyclopaedias, it is most likely, as Elizabeth Ashman Rowe suggests, that the map came to Iceland in an illustrated compilation akin to the Liber Floridus.[66]

The Icelandic map of Jerusalem contains Latin place-names for important sites in and around the city. Prominent buildings include the Dome of the Rock ('Templum Domini') and the Temple of Solomon ('templum salomonis'). The map contains a single vernacular place-name, which, like those in the geographical treatise on folio 1r, is placed in apposition to its traditional form; the vernacularized 'Jorsalaborg' inscribed in a prominent position above 'Jerusalem' at the map's centre.[67] The map of Jerusalem continues the suite of circular diagrams that began with the hemispherical world map and the planetary diagram on folio 1v. The circular form that repeats across this manuscript opening enables the viewer to picture, in series: the globe centred on the equator, the physical universe centred on the spherical Earth, and the world centred on Jerusalem. The series locates Jerusalem at the centre of both the world and time, locating the city geographically at the middle of the world's lands, and historically, through Christ's Incarnation, at the epicentre and focal point of human history.

As the hemispherical world map complements the description of the three continents on folio 1r, the map of Jerusalem is paired with an Old Norse description of the Church of the Holy Sepulchre and the environs of Jerusalem on folio 2v. This text describes a number of locations named on the preceding map of Jerusalem, including 'Loco Calvarie' (Calvary), 'Templum Domini' (Dome of the Rock) and 'Valles Iosafhat' (Valley of Josaphat, which is vernacularized in the written description to 'iosafaðs dalr'), and extends out beyond Jerusalem's city walls to the Mount of Olives and the River Jordan. The description of Jerusalem was not originally included in this manuscript, written in a hand dated *c.* 1300, but was added later in a hand dated *c.* 1350. The description has been identified as a fragment of the so-called *Leiðarvísir*, an itinerary that is attributed in a colophon in its main witness (Copenhagen, Arnamagnæan Collection,

65 Simek, *Altnordische Kosmographie*, p. 70. See also Rudolf Simek, '*Hierusalem civitas famosissima*: Die erhalten Fassungen des hochmittelalterlichen *Situs Jerusalem* (mit Abbildungen zur gesamten handschriflichen Überlieferung', *Codices Manuscripti* 12 (1992), 121–51; and Miller, *Mappaemundi*, vol. 3, pp. 61–67.

66 Elizabeth Ashman Rowe, 'Literary, Codicological, and Political Perspectives on *Hauksbók*', *Gripla* 19 (2008), 51–76, 69.

67 The vernacularized form does not appear on the version of the map in 732b.

AM 194 8vo, 1387) to an Icelandic Abbot Nikulás.[68] The complete itinerary describes a journey from Iceland to Rome and Jerusalem supposed to have taken place in the middle of the twelfth century, but evidently having made use of later medieval written sources.[69] The preponderance of the excerpt written in AM 736 I 4to relates to the interior of the Church of the Holy Sepulchre ('pulkro kirkia').[70] Built in the 1140s and consecrated in 1149, the Church brought a number of places traditionally associated with the Passion, including Calvary, Golgotha, and the place of the anointing, under a single roof.[71] The description's focus is on important shrines and reliquaries inside the church, and places associated with the Incarnation. It describes, for example, 'kapella þar fannz kross drottinns & ero þar markaðir krossarnir a golfinu a marmara steini sem krossarnir lagu' ('the chapel where the Lord's cross was found, and there are crosses incised into the marble floor where the cross lay'), as well as the place 'er hann uar pindr bundinn & bardr aðr hann ueri kross festr ('where he was tortured, bound, and beaten, before he was fastened to the cross'). In the description of the three continents (f. 1r), the topographical description provided an Icelandic author with a space

68 The *Leiðarvísir* were edited and translated into Latin in Erich Christian Werlauff, *Symbolas ad Geographiam Medii Ævi* (Copenhagen: Shultz, 1821), pp. 56–59; into German in Simek, *Altnordische Kosmographie*, pp. 484–96; and into English in Joyce Hill, 'From Rome to Jerusalem: An Icelandic Itinerary of the Mid-Twelfth Century', *The Harvard Theological Review* 76:2 (1983), 175–203. On the text's dating see Tommaso Marani, '*Leiðarvísir*. Its Genre and Sources, with Particular Reference to the Description of Rome' (PhD diss., University of Durham, 2012); and Arngrímur Vídalín, 'Óláfr Ormsson's *Leiðarvísir* and its Context: The Fourteenth-Century Text of a Supposed Twelfth-Century Itinerary', *JEGP* 117:2 (2018), 212–34. Identifications of this fragment with the *Leiðarvísir* have been made in Simek, *Altnordische Kosmographie*, pp. 479–90; and Benjamin Z. Kedar and Chr. Westergård-Nielsen, 'Icelanders in the Crusader Kingdom of Jerusalem: A Twelfth-Century Account', *Mediaeval Scandinavia* 11 (1978–79), 193–211. For an overview of the evidence for Scandinavian participation in crusade and pilgrimage see P. Riant, *Expéditions et pèlerinages des Scandinaves en Terre Sainte au temps des Croisades* (Paris: Lainé et Harvard, 1865). The *Leiðarvísir* are edited in *AÍ* I, p. 54.

69 Marani has shown that its compiler made extensive use of written sources, and that aspects of the itinerary would be anachronistic at this time. *Leiðarvísir*', pp. 233–35; Kedar and Westergård-Nielsen, 'Crusader Kingdom', p. 194–95; and Arngrímur Vídalín, 'Óláfr Ormsson's *Leiðarvísir*'.

70 The word 'pulcro' / 'pulkro' is a near phonetic rendering of the word sepulcro which would likely have been used by locals and adapted by pilgrims to the Holy Lands. See Kedar and Westergård Nielsen, 'Crusader Kingdom', p. 200.

71 Tomasso Marani, 'Contextualising Leiðarvísir: Sant' Agnese, The Catacombs, The Pantheon and The Tiber', *Saga-Book* 33 (2009), 44–66, 47; Denys Pringle, 'Architecture in the Latin East 1098–1571', in *The Oxford Illustrated History of the Crusades*, ed. Jonathan Riley-Smith (Oxford: Oxford University Press, 1995), pp. 160–83, 164.

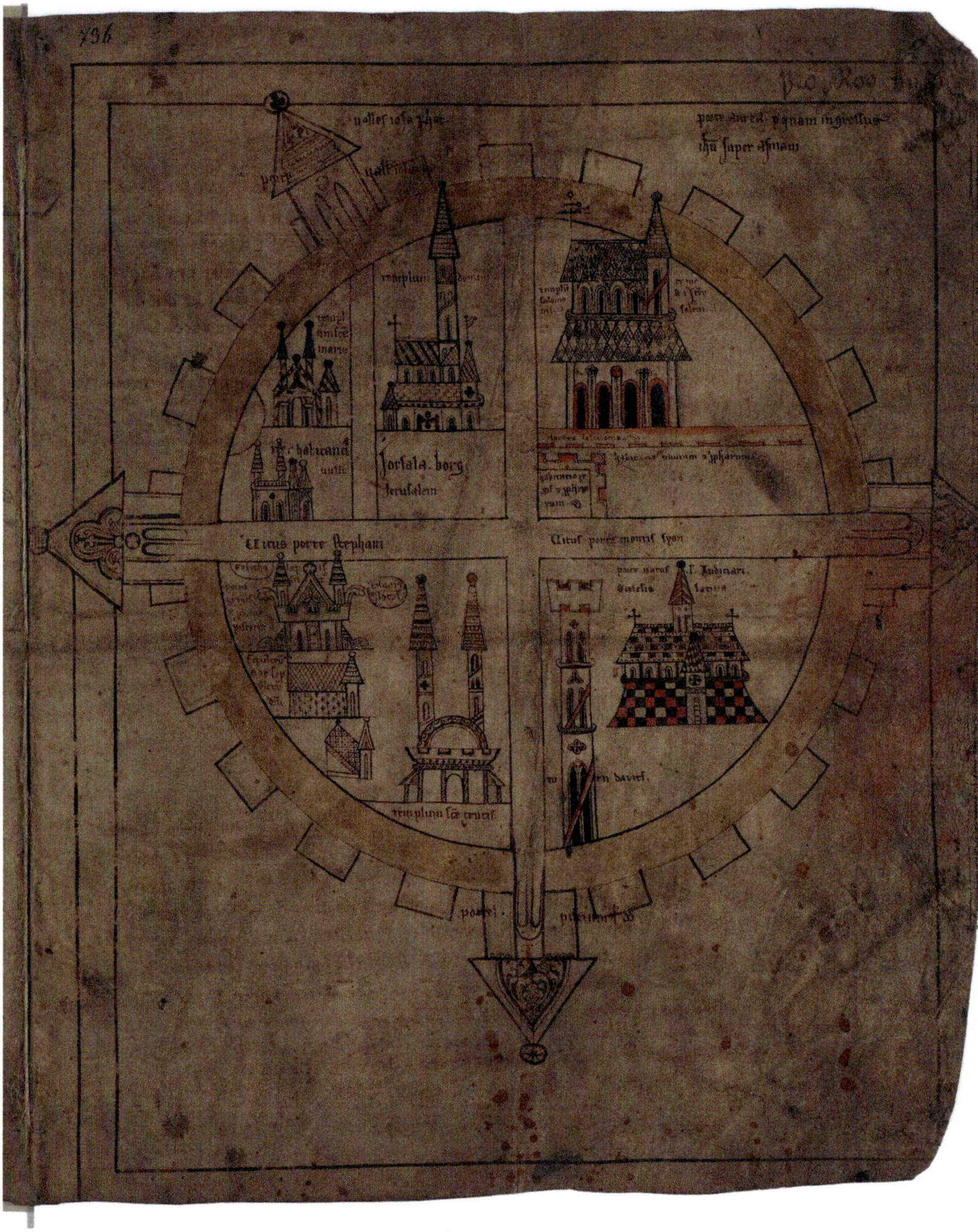

FIGURE 7: A map of Jerusalem on folio 2r faces onto the hemispherical world map. Its inscriptions mark places important to Christ's Incarnation. This has been complemented on the verso overleaf with a description of the Church of the Holy Sepulchre, added to this bifolium in a later hand dated *c.* 1350. Copenhagen, Arnamagnæan Collection, AM 736 I 4to, f. 2r, *c.* 1300. Photograph Suzanne Reitz. Reproduced with permission from the Arnamagnæan Institute.

in which to relate events from the lives of the Apostles, John in Ephesus and Peter in Antioch, and other historical events in the history of humanity's salvation. In the description of the Church of the Holy Sepulchre (f. 2v), the church's physical architecture likewise becomes a framework into which narrative episodes from the Incarnation can be related. The maps between these two treatises – one a description of the world's parts, one a description of its geographical and historical centre – focus the viewers' attention on the world and time. The lone bifolium with the shelf mark AM 736 4to presents a broadly diptych arrangement of texts and images; its first half (ff. 1r–1v) concerning the wide world, its second half (ff. 2r–2v), its centre.

The Younger Version of the Icelandic Hemispherical World Map (Copenhagen, Arnamagnæan Collection, AM 732b 4to, f. 3r, *c*. 1300–25)

The younger witness to the Icelandic hemispherical world map is preserved in a thematically similar compilation. AM 732b 4to comprises nine folios, with contents written in two hands. Folios 1r to 8v have been dated *c*. 1300–25, while 9r and 9v seem to have been written *c*. 1400.[72] The hemispherical world map is preserved on folio 3r. Like the stray bifolium AM 736 I 4to, this manuscript is also fragmentary. At least one quire is missing from the beginning of the manuscript: the text on folio 1r begins more than one page into an Old Norse account of the movement of the planets, an account that begins, in a fuller witness, '[s]vo seiger Beda prestur' ('so says Bede the priest').[73] The manuscript contains texts, written in Old Norse and Latin, on *computus*, cosmography, and astronomy.[74]

72 AM 732b 4to is notable for preserving the only two known examples of Latin poetry composed in skaldic meters. The most up-to-date general description of this manuscript accompanies the edition of these verses. See Jonathan Grove, ed., 'Anonymous, stanzas addressed to fellow ecclesiastics 1 & 2 (Anon Eccl 1-2)', in *Poetry on Christian Subjects, Part 2: The Fourteenth Century*, ed. Margaret Clunies Ross (Turnhout: Brepols, 2007), pp. 471–75, p. 471.

73 AM 624 4to. *AÍ* II, p. 85.

74 Although its contents have not been edited in their entirety, 732b appears in the variant apparatus of Kålund's edition of AM 624 4to (aforementioned because it contains the Old Norse note on the error in the Julian calendar and tidal note). *AÍ* II, pp. 81–178. Its contents are described in *AÍ* III, pp. ix–x. Its ciphered material is edited alongside that from the Codex Uppsaliensis in Finnur Jónsson, ed., 'Lønskrift og lejligedsoptegnelser fra et par islandske håndskrifter', *Småstykker* 1–16 (Copenhagen: S.L.Møllers, 1886), pp. 185–94. The texts that accompany the map in 732b are not included in Simek's *Altnordische Kosmographie*.

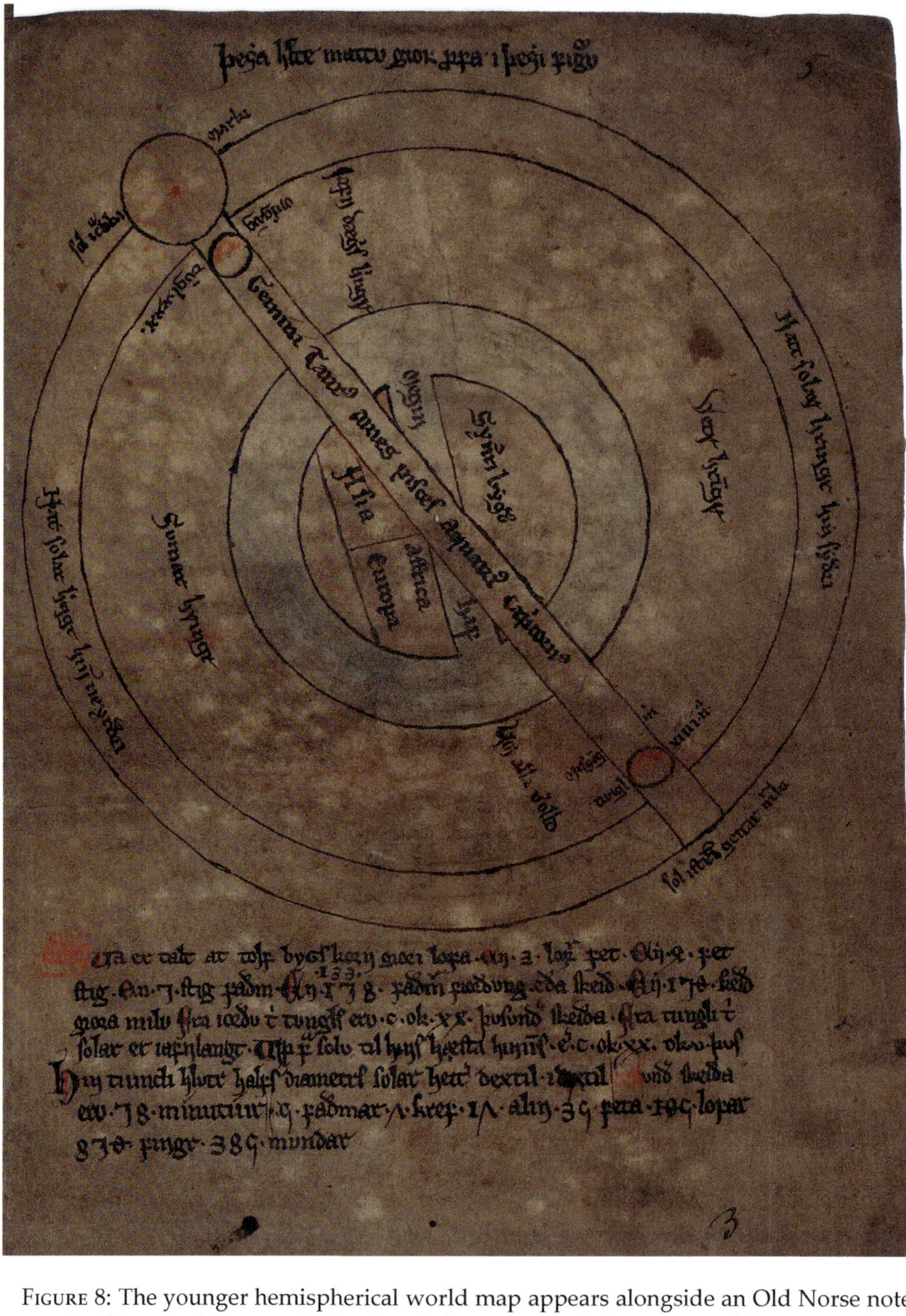

FIGURE 8: The younger hemispherical world map appears alongside an Old Norse note on the distances between the Earth and the moon, and the moon and the sun, and an Old Norse note on the diameter of the sun, with Latinate metrological vocabulary. Copenhagen, Arnamagnæan Collection, AM 732b 4to, fol. 3r, *c.* 1300–25. Photograph Suzanne Reitz. Reproduced with permission from the Arnamagnæan Institute.

As we have already seen, there are some minor differences between the two versions of the hemispherical world map. The map in AM 732b 4to map lacks a line separating Africa and Europe, as well as that which marks the southern limit of the temperate region in the southern hemisphere. On this younger map the ocean has also been coloured blue – an indication, perhaps, that the copyist understood enough of the map's iconography to elaborate upon a colourless exemplar.[75] The copying of these maps ought to be seen as a creative enterprise; differences between versions of the same map need not be framed in terms of accuracy or correctness, but can also be seen as purposeful decisions to enlarge, reduce, or redefine the map's contents. This can be seen in another significant difference between the two versions of the map, their orientations. The older is oriented, both in terms of its placement on the page and in the writing of its inscriptions, with south at the top. However, while the younger map's legends are written to be read with south at the top, the map has been placed on the manuscript folio – by design or accidentally – with east at the top. It is possible that the compiler of AM 732b 4to reoriented the exemplar to place east at the top in analogy with other east-oriented maps familiar to them. The greatest difference between the two versions of this map, however, is their relationship to their companion texts. While both maps are accompanied by notes on the error in the Julian calendar and on the generation of the tides, the *mise-en-page* relationships between them emphasise different aspects of the map's design.

The younger map's companion texts mostly pertain to calendar construction and the seasonal variation of the tides. The Old Norse text that begins *in medias res* on folio 1r describes how conjunctions of the sun and moon produce the tides.[76] Its contents are broadly similar to those summarised in the tidal note, above, and describe the cycle of *missong* ('spring tides') and *meðaldagar* ('neap tides') caused by the motions of the sun and moon.[77] The spring and neap tides were postulated by Bede in *De temporum ratione* (Ch. 29), to whom the treatise is attributed in its opening lines (missing from this version of the text).[78] The text then focusses on the solar cycle and measurements of ecliptical longitude, with notes on how many degrees along the ecliptic the sun moves each day following the solstice.[79] This treatise on the tides does not draw the entirety of its

75 The use of colour made its way into map production early, but its use was rare. See Ulla Ehrensvärd, 'Colour in Cartography: A Historical Survey', in *Art and Cartography: Six Historical Essays*, ed. David Woodward (London: University of Chicago Press, 1987), pp. 123–46, p. 126.

76 *AÍ* II, pp. 87–88, §§ 14–15.

77 The neap tides are those that occur at five and eleven nights into a lunation, at the midpoints between the full and new moons. At these times the tidal forces are weakest and produce the least tidal range.

78 *AÍ* II, p. 89, § 18.

79 *AÍ* II, p. 91, § 19.

information from Bede, but cites the observations of other astronomers, including the ninth-century computist Helpericus and the Icelandic computist Bjarni Bergþórsson (d. 1173). It describes the old Roman calendar and the Julian innovations of the golden numbers, epacts, and embolisms.[80] The treatise defines these terms and explains the need for the epact to be corrected by one day every nineteen years in order for the cycle to repeat, resulting in a *tunglhoppun* (the *saltus lune*, 'leap moon'). Although the treatise cites Helpericus's *De computo* as its authority, it is not a direct translation but a free and independent summary that cites the observations of other computists.[81] To complement Helpericus's definitions, the text cites the calculations of the length of the lunar month made by the Icelandic priest and computist Bjarni Bergþórsson, with the agnomen *enn tölvísi* ('the computist').[82] The text further describes lunar retardation and the intercalary leap moon in its explanation of the tides.[83] The text concludes with observations on the points of the lunar year when the moon is highest (at Christmas) and lowest (at Jóns messa), and whether there will be a spring or a neap tide on these dates.[84]

The reckoning of time is a theme developed throughout the items preserved in these folios. Folio 1v proceeds with a Latin explanation of the determination of the date of the Paschal full moon read from certain 'epistolis Grecorum' ('Greek letters'), in which it is described how God revealed the epact table below to the fourth-century Saint Pachomius.[85] Below, a nineteen-line alliterative poem in Latin provides a mnemonic means to recall the dates of the Paschal full moon.[86] The first half of each line gives the date of the Paschal full moon for each year in the nineteen-year cycle, while the second half gives the ferial regular (weekday displacement) of the Paschal full moon from the concurrent (the weekday of 24th March).[87] The poem is written out twice in two

80 The *aureus* or golden number is the number between one and nineteen that marks the position of the year in the nineteen-year Metonic or Paschal cycle. Every nineteen years, the lunar year and solar year achieve parity and begin on the same day. The epact refers to the age of the moon (from one to thirty days) on the 1st January in any given year. The embolism refers to the intercalary lunation inserted to keep the lunar year in sync with the solar calendar.

81 *AÍ* II, p. 92, fn.1. Helpericus's works make several appearances in Icelandic encyclopaedias (see Chapter 2). On Helpericus's contributions to medieval astronomy see Duhem, *Le Système du Monde*, vol. 3, pp. 71–76.

82 AÍ II, pp. 92–93, §§ 20–22.

83 *AÍ* II, p. 95, §§ 27, 28.

84 *AÍ* II, p. 96, § 32.

85 *AÍ* II, pp. 235–36.

86 The poem is well known, and probably dates to the late fifth century Wallis, *Bede, The Reckoning of Time*, p. xlvii, fn. 73.

87 When the embolism occurs in March (in the eighth and nineteenth years of the Paschal cycle) Paschal reckoning will be affected. See Wallis, *Bede, The Reckoning of Time*, p. xlvi.

columns: once with Roman numerals, and again with Arabic numerals for the ferial regular.[88] Beneath the table is an Old Norse note on how to count the year from the creation of the world,[89] and Old Norse notes on the disparity between the lunar and solar years, the concurrent, and the indiction cycle (a fifteen-year cycle developed by the Romans for taxation purposes).[90] The calendrical texts assembled on this folio culminate with the Old Norse note on the error in the Julian calendar attributed to Master Walter, which we saw preserved in association with the Icelandic hemispherical world map in AM 736 I 4to.

The immediate context of the younger map resembles that of the older. On folio 2r there is a wind diagram that shows the twelve winds and the directions with which they are associated. Below the diagram, there is a text about the winds excerpted from Isidore's *Etymologies* (13.11.3–18) that extends to folio 2v.[91] Folio 2v concludes, below the extract from the *Etymologies*, with the Old Norse tidal note, which is presented opposite the hemispherical world map on the facing recto. What is interesting about the relationship between the tidal note and the map in this manuscript is the position of its final line, 'these matters can be proved in this diagram' ('þessa hluti máttu giǫr prófa í þessi figuru'). In AM 736 I 4to, this interpretative directive – written at the bottom of the tidal note – formalised the connection between the hemispherical world map and a description of the flow and ebb of the ocean's tides. In AM 732b 4to, however, this line has instead been separated from the tidal note's main body and reassigned to the top of folio 3r, where it is centred and placed immediately above the younger version of the map. The relocation of this line strains the connection between the map and tidal note, which had been explicit in the older version. The repositioning of this directive can have implications for what we perceive the 'matters' ('hlutir') of the map to be. Below the map on folio 3r there is an Old Norse note on the distances between the Earth and the moon and the moon and the sun,[92] with an unedited Old Norse note on the diameter of the sun with Latinate

88 The ferial regular is repeated in Roman numerals in the third column of the poem. Common and embolismic years are marked *concurrents* and *embolismic*. *AÍ* II, pp. 236–37.

89 *AÍ* II, p. 125.

90 These latter two notes have been identified by Kålund, but have not been edited. *AÍ* II, p. 273.

91 This is the only wind diagram preserved in Icelandic encyclopaedias. I have described this wind diagram and the extract from Isidore's *Etymologies* it accompanies elsewhere. Dale Kedwards, 'Wind Diagrams in Medieval Iceland', *Quaestio Insularis* 16 (2015), 92–107.

92 *AÍ* II, p. 239. The origins of this figure, given as 120,000 stades between the Earth and moon and a further 120,000 thence to the sun, is not clear. Macrobius affirms that the sun is twice as far from the Earth as the moon (2.3.14). Pliny states in the *Historia Naturalis* that the distance between the Earth and moon

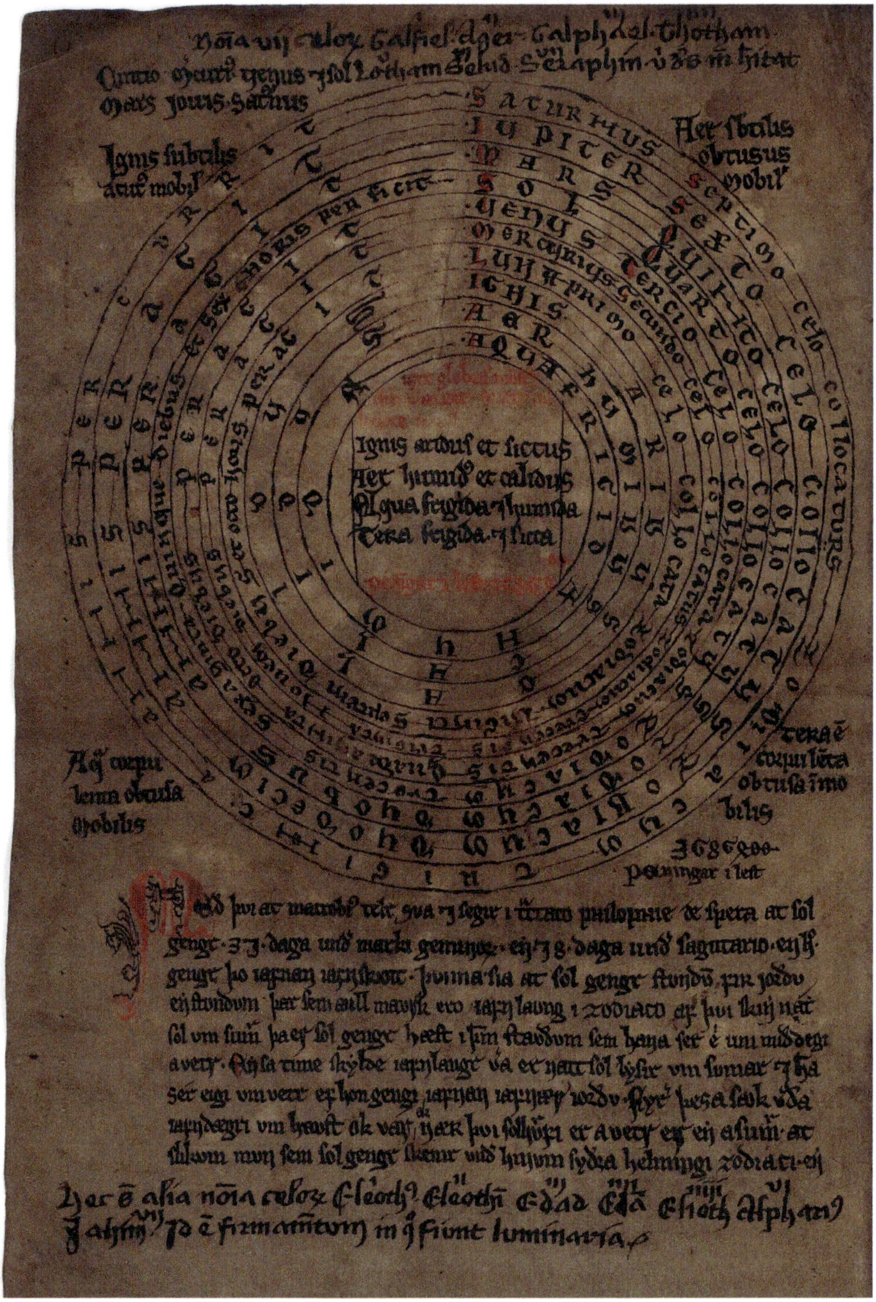

FIGURE 9: A diagram showing nested planetary spheres, with the Earth at their centre. This diagram has been drawn overleaf from the hemispherical world map, and shares the compass hole at its centre. Copenhagen, Arnamagnæan Collection, AM 732b 4to f. 3v, *c.* 1300–25. Photograph Suzanne Reitz. Reproduced with permission from the Arnamagnæan Institute.

metrological vocabulary.[93] The younger map thus appears in manuscript alongside notes on the sizes of and distances between the Earth, moon, and sun; planetary bodies prominently depicted in their relations to one another on the hemispherical world map. Depending on the reader's engagement with the map's wider context, therefore, the 'matters' proved by the map might be the relative dimensions of these bodies and the distances between them.

The diagram of the seven planetary spheres and their orbits that accompanies the older map also makes a reappearance with the younger map. This diagram (figure 9), which shows the order of the seven planetary spheres, the durations of their orbits around the Earth, and the four elements that compose all matter in the sublunary world, appears on folio 3v. The hemispherical world map and the planetary diagram are connected by the material circumstances of their production. The compass hole at the centre of the hemispherical world map has been reused to draw the diagram of the planetary spheres on the verso overleaf. These two diagrams have the same diameter (124mm), and were likely drawn at the same time using the same apparatus. The exigencies of manual techniques in the composition of such compilations must not be overlooked: while it is evident that texts and images were grouped on the basis of their conceptual or thematic resonances, they may also have been grouped on a more practical or material basis. The world map and planetary diagram relate to one another at least in part because their scribe preferred to reuse the hole punctured through the vellum on both sides of the manuscript folio. For the same reason, the unicursal labyrinth, called *völundarhús* ('house of Völundr', a calque of the Latin *Domus Dedali*), on folio 7r of this manuscript has been drawn overleaf from a diagram showing the lunar phases on folio 7v.[94] A similar technical parsimony will be seen in the gathering of medieval diagrams in the Viðey book (GkS 1812 4to), which we shall examine in Chapter 2.

The short text below the planetary diagram, concerning the tides associated with different lunar phases, also connects with the map thematically.[95] The text cites Macrobius directly, in order to disagree with his observations on the amount of time it takes for the sun to move

---

was 126,000 stades, a figure he attributes to Pythagoras (II.19). On the distance between the planets, see Duhem, *Le Système du Monde*, vol. 2, pp. 11–20.

93 The origins of this figure are likewise uncertain. However, it does fall in the range described by Stahl. On estimates of the sun's apparent size, see Stahl, trans. Macrobius, *Commentary*, p. 253.

94 The labyrinth appears in facsimile in *AÍ* III, p. 65. On this labyrinth and others in Icelandic manuscripts see Rudolf Simek, 'Völunderhús - Domus Daedali: Labyrinths in Old Norse Manuscripts', *Nowele* 21 (1993), 323–68. The lunar phase diagram is described in *AÍ* III, p. 66.

95 *AÍ* II, pp. 95–96, §§ 29–31.

through different constellations in the zodiac.[96] The text references a work by Macrobius described as 'tractatus philophie [*sic*] de spera', probably an abridgement of the *Commentary*'s astronomical sections, or a chapter in a florilegium such as the Liber Floridus, which contains chapter headings along these lines.[97] In AM 732b 4[to], the hemispherical world map is thematised in terms of the physical structure of the universe.

Conclusion

The Icelandic hemispherical world map has a history, the stages of which can be reconstructed through close comparison with its antecedent texts and maps. We can detect three phases in the map's development, two of them antedating its introduction to Iceland. Firstly, the map's outline was conceived in Macrobius's *Commentary on the Dream of Scipio*, its purpose to demonstrate that Macrobius's literary icons, Virgil and Cicero, did not contradict one another in how they wrote about the terrestrial and celestial spheres. Secondly, the Macrobian template underwent several revisions by medieval encyclopaedists, such as Lambert of Saint-Omer, who augmented Macrobius's empty template with Latin legends. Thirdly, the map was imported into Iceland and its Latin inscriptions were translated into Old Norse. The Icelandic map's written contents are loan translations of Latin originals that appear on Lambert's maps. Over the course of its history, the map has changed function. The diagram began as a statement of the equivalence between lines of celestial and terrestrial latitude; in its Icelandic form, the map functions primarily as a visual exposition of tidal theory.

The map does not mean, however, simply by itself, but through its thematic and formal correspondences with its companion texts and images. Both the older and younger maps accompany short notes on the error in the Julian calendar and the seasonal ebb of the ocean's tides. The older map belongs to a complement of circular diagrams that focus the viewer's attention on the world, the cosmos, and its centre; interceding between written descriptions of the three continents, on folio 1r, and of Jerusalem, on folio 2v. The hemispherical map allows us to see the three continents in their global context, and then, in the diagram of the planetary spheres on the same folio, in its wider cosmic context. The facing map of Jerusalem focusses the reader's attention on the world's centre. The

96 Kålund connected these observations with Macrobius's *Commentary* 1.6.51, which outlines the amount of time the sun spends in the constellation Gemini. The association is a lose one, the information might also have come from Martianus Capella, *De nuptiis Philologiae et Mercurii*, 8.848,865.

97 For example: 'ordo vii planetarum et spera celi et terre secundum Macrobium'. Derolez, *Liber Floridus*, p. [98].

younger map belongs to a compilation that more ostensibly thematises the structure of the physical universe and its clocked processes.

The map was conceived as a work of literary criticism, to explain descriptions of the cosmos in the works of Cicero and Virgil, and its Icelandic version is in turn a translated text whose authoritative description of geographical matters resonates across Old Norse literature. The equator and other parallels of latitude are described in the course of a lesson on the Earth's sphericity in the Norwegian educational handbook *Konungs skuggsjá* (*c.* 1260). Its author locates the parallels 'í híminum' ('on the heavens'), a formulation that probably derives from Macrobius's preoccupation with reconciling the wordings favoured by Cicero and Virgil.[98] Its author calls the Equator 'boginn rinng brænnannda vægar' ('the curved ring of the burning belt'), and the climatically hot zone through which it runs 'brennubeltið' ('the burning belt'). The two polar circles and Equator are together described as 'þrir vægir torfærileger' ('three belts difficult to cross'), in recognition of their capacity as the limits of terrestrial habitability. These global concepts also appear in literary works that do not otherwise seek to explain cosmological phenomena. In the opening lines of *Ynglinga saga*, the *kringla heimsins* is divided, as we have seen, into polar regions 'úbygðr af frosti ok kulda' ('unsettled because of the frost and cold') and equatorial regions 'auðr af sólar bruna' ('empty because of the sun's intensity').[99] The theory of the world's climatic zones and its intersection with ancient and medieval theories of race can be seen in *Gylfaginning*, where Snorri situates Muspell, the realm of fire described in the Old Norse cosmogenic poem *Völuspá*, and the fiery Surtr in the far south, harmonising Icelandic vernacular poetics with the medieval European world image that he understood from ancient authoritative texts.[100]

The lesson the map teaches is productive across literary genres, its understanding of global space invoked in sagas whose narratives traversed a wider world. The Icelandic *Nítíða saga*, an original romance composed in the fourteenth century, locates its action between the maiden-king Nítíða's Parisian court and those of her adversaries in Constantinople and India.[101] In the saga, Nítíða obtains three extraordinary stones from the

98 Dale Kedwards, 'The World Image of the *Konungs skuggsjá*', in *Speculum Septentrionale: Konungs skuggsjá and the European Encyclopaedic Tradition*, ed. Karl Gunnar Johansson and Elise Kleivane (Oslo: Novus, 2018), pp. 71–92.

99 Snorri Sturluson, *Heimskringla*, p. 10.

100 Richard Cole, 'Racial Thinking in Old Norse Literature: The Case of the *Blámaðr*', *SBVS* 39 (2015), 21–40, 29–30.

101 Sheryl McDonald, '*Nítíða saga*: A Normalised Icelandic Text and Translation', *Leeds Studies in English* 40 (2009), 119–45, 124. On this episode, see Dale Kedwards, 'Geography', in *A Critical Companion to Old Norse Literary Genre*, ed. Massimiliano Bampi, Carolyne Larrington, and Sif Rikharðsdóttir (Cambridge: D. S. Brewer, 2020), pp. 127–44.

island of Visio ('vision') located 'út undir heimsskautið' ('under the north pole'), which have the remarkable ability to conjure visions of the world's places. At the culmination of the saga's action, Nítíða uses these stones to reveal three interlocking visions of world geography, which reveal to her 'öll önnur lönd heimsins, og jafnvel um brúnabeltið, það sem ei er byggt' ('all the other lands of the world, and even around the burningbelt, which is uninhabited').[102] *Nítíða saga*'s geographic imaginary thus spans the latitudinal entirety of the inhabitable world, from an island located under the north pole, where Nítíða acquires her extraordinary stones, to the uninhabited areas at the Earth's equator. The world of this romance saga is the same world depicted on the hemispherical world map, its graphic elements – the world's three continents and climatic zones bounded by the polar region and the equator – reconstituted into narrative prose.[103] The hemispherical world map illustrates geographic principles commonly written into Icelandic literature, and the map's viewers may even have approached it with an awareness of its conventions drawn from saga literature. In the manuscripts in which we encounter them, these maps are not individual objects of attention, meant to be inspected in isolation. Rather, they interact with their companion texts and images, tasking their viewers with relating their perspectives on the world to those coded in other literatures.

102 *'Nítíða saga'*, 140–41.
103 Barnes, *Bookish Riddarasögur*, p. 37.

CHAPTER 2

The Icelandic Zonal Map

The Icelandic zonal map, preserved in the abundantly illustrated encyclopaedia in Reykjavík's Stofnun Árna Magnússonar with the shelf mark GkS 1812 4to, has much in common with the hemispherical world maps. This map, located within an incomplete diagram of the lunar phases, contains only three inscriptions, written in Old Norse, that designate the habitable areas in the northern and southern hemispheres and the ocean that divides them.

The map shows the division of the spherical Earth into climatic zones. As we saw in the previous chapter, the polar regions were theorised as uninhabitable due to the intense cold, while the equatorial region was thought uninhabitable due to the intense heat of the torrid zone. Between the frozen and torrid zones at extreme latitudes lay two habitable belts in the southern and northern hemispheres. In addition to its slender geographical nomenclature, the zonal map uses colour to distinguish between climatic regions: the polar regions are shaded in the same dark ink that was used to draw its outline, while the torrid equatorial regions either side of the ocean are coloured red. The use of colour is the only indication of the climatic characteristics of these regions, which bear no written inscriptions.

All three of the map's inscriptions will be familiar from the previous chapter, being very similar to those on the hemispherical world map. The legends 'sudr bygilig halfa' and 'nordr bygilig halfa' ('southern' and 'northern inhabitable region') are comparable with the inscription on the hemispherical world map *synnri bygð* ('southern inhabited region'). These legends calque the Latin *zona habitabilis*, which frequently identifies the habitable regions on other European zonal maps, such as those in the Liber Floridus.[1] As outlined in the previous chapter, notions of antipodal habitation were contentious in the Middle Ages. The adjective *byggiligr* ('habitable') used on the zonal map resembles the Latin *habitabilis*, but

1 Simek, *Altnordische Kosmographie*, p. 199.

conveys no certain pronouncement on whether or not these regions were actually peopled. The hemispherical map does not contain an equivalent term for the *norðr bygilig hálfa*, but instead divides this region to show the three continents Asia, Africa, and Europe. However, the Old Norse tidal note that accompanies the hemispherical world map in both manuscripts refers to the northern inhabitable region as *vorri bygð* ('our inhabitation', or 'the part of the world that we inhabit').

The ocean features prominently on the zonal map. The map shows the equatorial ocean that divides the Earth laterally into its northern and southern hemispheres, and the meridional ocean that divides the Earth vertically into its eastern and western hemispheres. These two arms of the ocean divide the world into four parts, each of which contains a region climatically suited to human settlement. The oceans on the zonal map and the version of the hemispherical world map in AM 732b 4to are coloured blue, which in both instances demonstrates the copyist's familiarity with the map's conventions.

This is the least studied of all the Icelandic *mappae mundi*.[2] Rafn provided a faulty description of its context, noting that it appears in the same manuscript as the arithmetical treatise *Algorismus*. However, while the *Algorismus* is bound into the present compilation (ff. 13v–16v) these folios are written in a different hand, and as we shall see below, might not have been bound into a compilation with this map until as late as the sixteenth century. Kålund describes the map and transcribes its inscriptions in his description of the manuscript GkS 1812 4to, but does not reproduce it in facsimile.[3] Simek follows Rafn in reproducing only the central zonal map, without the lunar phase diagram that surrounds it.[4] Previous facsimiles of the map have been partial: the lunar phase diagram has occasionally been described, but never properly recognised as an important part of the map's composition, so that its complete visual argument has never properly been seen or analysed. Nor have previous commentaries engaged with the other four cosmological diagrams drawn in sequence with the zonal map in its manuscript, tending to view the map, like others in this small corpus, in generic isolation. This chapter focuses on the map's relation to four cosmological *rotae* – diagrams showing the motions of the inner planets, Mercury and Venus (f. 10v); the outer planets, Mars, Jupiter, and Saturn (f. 11r); and the eclipses (f. 12v) – in a section of the encyclopaedia concerned with observations of the planets' revolutionary motions. These diagrams have not been studied or reproduced in facsimile until now, and the zonal map and its encasing

2 Rafn provided a description of the zonal map and its scheme of lunar phases, but only reproduced the central map in his drawn facsimile Rafn, *Antiquités Russes*, p. 390.

3 *AÍ* II, pp. ccxiv–ccxv.

4 Simek, *Altnordische Kosmographie*, p. 591.

lunar diagram on folio 11v, has never been seen in their company. The zonal map, however, maintains a complex suite of interractions with its three companion diagrams. This chapter describes their origins in medieval treatises on natural philosophy, before examining their roles in the present compilation. The Icelandic zonal map was not drawn to be looked at in isolation but in consultation with its companion texts and images, which together comprise the most detailed study of the globe that survives from medieval Iceland.

The Viðey Book

Three of the five world maps that survive from medieval Iceland are preserved in the manuscript with the shelf mark GkS 1812 4to, a voluminous illustrated encyclopaedia of thirty-six leaves. The items gathered in this manuscript pertain broadly to the science of *computus* and the tributary disciplines of mathematics and astronomy. This manuscript has on occasion been called the 'Viðey book', because of its association with the Augustinian monastery at Viðey in south-western Iceland.[5] The manuscript is not made up of regular gatherings. In its thirty-six leaves, Kålund detected contributions from four different hands dating from the late twelfth through to the fourteenth century.[6] These hands, and the folios for which they are responsible, are as follows.

GkS 1812 I (1315–*c.* 1400): ff. 1r–4v, 7r–12v*
GkS 1812 II (fourteenth century, Norwegian?): ff. 13r–23v
GkS 1812 III (*c.* 1225–50): ff. 5r–6v*, 35r–36v
GkS 1812 IV (*c.* 1200): ff. 24r–34v
*section contains world map(s)

It is unclear when these six fragments were bound together into the present compilation. They might have bound together, in seal skin, as late as the sixteenth century, when the ownership of the book becomes

5 The hands in 1812 are examined in Lilli Gjerløw, *Libri Liturgici Provinciae Nidrosiensis Medii Aevi* Vol. 11: *Ordo Nidrosiensis Ecclesiae (Ordubók)* (Oslo: Universitetsforlaget, 1968), p. 60; and Lilli Gjerløw, *Liturgica Islandica I: Text* (Copenhagen: C.A. Reitzels Boghandel A/S, 1980), p. 127.

6 It has been conventional to name 'parts I–IV' in the order youngest to oldest (IV being oldest). However, the hands have been conventionally termed 'hands 1–4' with 1 being oldest. These confusing designations have been abandoned for clearer descriptions. Kålund identifies the fifth quire as ff. 34–36, however the last quire constitutes ff. 35–36 only (a bifolium containing the first and second months in a calendar, and a time-keeping treatise at the rear). Since ten months are missing from the calendar, at least three bifolia are missing. *AÍ* II, p. ccx.

known.[7] The compilation is notable for its specially illustrated character, and the maps likely owe their survival to modern times to the compilator's interest in diagrams. Maps are preserved in parts I and III and, therefore, in the following studies of the maps' contexts, particular attention is directed towards these sections.

Three maps are preserved across two of the component parts of 1812. These maps are, in the order in which they are preserved in the present compilation:[8]

The larger Viðey Map (1812 III, ff. 5v–6r, *c.* 1225–50)
The smaller Viðey Map (1812 III, f. 6v, *c.* 1225–50)
The Icelandic zonal map (1812 I, f. 11v, 1315–*c.* 1400)

Besides these three maps, there are an additional three cosmological diagrams in 1812 I (on ff. 10v, 11r, and 12v) that show the Earth as part of their scheme. On these three diagrams – showing the revolutionary motions of the inner (f. 10v) and outer planets (f. 11r) and the solar and lunar eclipses (f. 12v) – the Earth is represented by a punctus at their centres, around the compass holes used to draw them, labelled *jörð* ('Earth').[9] Through its combination of maps and diagrams, this book is the most detailed study of the globe and its cosmic position that comes down to us from medieval Iceland.

Although some texts in this manuscript have a long history of reproduction in printed editions, the encyclopaedia as a whole has, until recently, received little critical attention. Its contents have been edited principally in the variant apparatus to Kålund's edition of the Icelandic astronomical and *computus* texts in the second volume of *Alfræði Íslenzk*. Other substantial editions include Ludvig Larsson's edition of the oldest section of the manuscript (1812 IV), and Piergiuseppe Scardigli and Fabrizio D. Raschellà's edition of the Latin-Icelandic glossaries in this

7 *DI* I, p. 183.

8 In Simek's *Altnordische Kosmographie*, the table of extant Scandinavian world maps provides the correct dates and foliation for the Icelandic maps, but there are errors in the headnotes that accompany their facsimiles and transcriptions in the same volume. The table is on p. 59. The smaller Viðey Map on folio 6v is wrongly attributed in the headnote that accompanies its transcription to folio 11r. The zonal map is correctly dated to the fourteenth century on the table but wrongly dated to the thirteenth century in the relevant headnote. These errors have confused later commentators (see Chapter 3). Simek corrected these errors in Rudolf Simek, 'Scandinavian World Maps', in *Trade, Travel, and Exploration in the Middle Ages: An Encyclopedia*, eds. John Block Friedman and Kristen Mossler Figg (London: Routledge, 2000), pp. 537–38.

9 These cosmological diagrams are not included in Simek's *Altnordische Kosmographie*.

same section.[10] A number of smaller items of interest to social and political historians of the Icelandic Commonwealth appear in the *Diplomatarium Islandicum*; these, and the maps in GkS 1812 III 4to, are considered in more detail in Chapter 3.

The section of this manuscript in which we encounter the Icelandic zonal map, 1812 I (ff. 1r–4v and ff. 7r–13r of the present compilation, eleven leaves) has been dated on palaeographic grounds to the fourteenth century. This can be narrowed with reference to the *Cisioianus* on folio 2r, a hexametrical mnemonic that aids memorisation of the most important immovable feastdays and holidays.[11] The version contained in 1812 I includes the feast day of the Icelandic bishop Guðmundr Arason (d. 1237) who was locally canonised in Iceland in 1315.[12] Since Guðmundr is named in this verse, the year 1315 can be established as a *terminus post quem* for the production of this part of the manuscript. In addition, 1812 I contains diverse items in Latin and Old Norse that pertain broadly to the determination of the calendar, and its tributary disciplines of astronomy and cosmology.[13] These include a Latin memorial verse on computation (ff. 1r–1v); a seven-line Latin memorial verse on the immovable feast days with an Old Norse prose explanation (f. 2r); Old Norse astronomical writings, with observations attributed to the Icelandic astronomer Stjörnu-Oddi ('Star-Oddi') Helgason (ff. 2r–2v); a series of nine roundels that contain the zodiacal signs with associated Latin descriptions of the constellations (ff. 3r–4r); a diagram that shows philosophy and its epistemological divisions (*Physica, Ethica, Dialectica, Rhetorica, Grammatica*) with associated Latin texts (f. 4v); an Old Norse account of the structure of the cosmos, attributed to Macrobius (f. 7r); a diagram showing the Macrobian planetary week (f. 7r); drawings of six of the constellations (*Centarus, Lupus, Cetus, Orion, Canis Maior, Canis Minor*) with associated Latin texts

10 Ludvig Larsson, *Äldsta delen af Cod. 1812, 4to Gml. Kgl. Samling på Kgl. Biblioteket i København* (Copenhagen: S. L. Møllers, 1883); Piergiuseppe Scardigli and Fabrizio D. Raschellà, 'A Latin-Icelandic Glossary and Some Remarks on Latin in Medieval Scandinavia', in *Idee, Gestalt, Geschichte: Festschrift Klaus von See: Studien zur europäischen Kulturtradition*, ed. Gerd Wolfgang Weber (Odense: Odense University Press, 1988), pp. 299–323.

11 This poem takes its name *Cisioianus* from the first two feasts whose dates it notes ('Feast of the Circumcision [January 1], January, Epiphany [January 6] …'). See Reginald L. Poole, *Medieval Reckonings of Time* (London: Macmillan, 1921), p. 19; and Bonnie Blackburn and Leofranc Holford-Strevens, *The Oxford Companion to the Year* (Oxford: Oxford University Press, 1999), p. 844.

12 *AÍ* II, pp. 225–28, esp. p. 226; Ashman Rowe, 'Perspectives', 66.

13 There is no reliable index to the contents of GkS 1812 4to, and many of its items, especially those written in Latin, are little studied or have yet to be edited. The following summaries are necessarily selective, but fuller explanations can be derived from *AÍ* II, where 1812 (Kålund's MS 'G') appears in the variant appartus. The map's immediate context will be more comprehensively detailed below.

(f. 7v); an Old Norse text excerpted from John Chrysostom's (whose name is calqued into Jón Gullmuðr, 'John golden-mouth') treatise on the Star of Bethlehem (f. 8r); a Latin note on the Nativity (f. 8r); a macaronic text about embolisms (f. 8r); an Old Norse text on the signs of the zodiac and other constellations (ff. 8v–10r); an Old Norse text about comets and the events they foretell derived from Bede (f. 10r); an explanation of Latinate metrological terms (f. 10r); an Old Norse text on the sidereal periods of the moon and Mercury and diagram showing the heliocentric orbits of Mercury and Venus (f. 10v); an Old Norse note attributed to Sacrobosco (f. 10v); an Old Norse note on the sidereal period of Saturn and a diagram showing the epicycles and deferent (f. 10v); Old Norse texts on the diameters of the Earth, moon, and sun (f. 10v); an Old Norse text about the leap moon attributed to the computist Helpericus (f. 11r); the zonal map and an Old Norse text about the division of the Earth into four parts by the ocean, attributed to Bede (f. 11v–12r); and an Old Norse text on the tides and a diagram showing the eclipses (f. 12v).

The section 1812 II (ff. 13v–23v of the present compilation, eleven leaves) has been dated on palaeographic grounds to the fourteenth century. This part of the compilation contains no maps, but is representative of the places we find them in medieval Iceland. Notes on folio 1v in the section 1812 I are written in the hand responsible for 1812 II, which indicates that 1812 II is younger, and moreover, that the zonal map would have been known to its compositor.[14] This part contains the Old Norse arithmetical treatise *Algorismus* (ff. 13v–17r); an Old Norse treatise on the use of a quadrant to determine latitude (ff. 17r–20r); an Old Norse text attributed to Macrobius (ff. 20r–21v); a numerical table that supplies Roman numerals and Latin terms for the numbers one to one hundred million, with some Arabic equivalents (f. 21v); a Latin mnemonic verse for *computus* operations with an Old Norse explanation (f. 22r); an Old Norse *computus* treatise (ff. 22r–22v); an Old Norse table of lunations with golden numbers (f. 23r); and an Old Norse text on the reckoning of days called *Bócarbót* (ff. 23r–23v). Norwegianisms in this section may indicate that the scribe responsible for it was Norwegian or had been trained in Norway.[15] However late the disparate sections of 1812 were bound together into the present compilation, additions to 1812 I folio 1v written in the same hand that produced 1812 II demonstrates that 1812 II has been in circulation with the Icelandic zonal map continuously or intermittently since the fourteenth century.[16]

1812 III (ff. 5r–6v and ff. 35r–36v of the present compilation, four leaves) is the second oldest part of the compilation and has been dated on palaeographic grounds to *c.* 1225–50. Two single leaves and one bifolium

14 *AÍ* II, p. ccx.

15 *AÍ* II, pp. ccxii–ccxxv.

16 *AÍ* II, p. ccx.

are written in this hand and have been bound into the compilation separately. Two single leaves (ff. 5r–6v) have been inserted together into the section 1812 I, between the diagram that shows *Philosophia* and its epistemological divisions (f. 4v) and a diagram of the planetary week (f. 7r). These two leaves contain three items: a list of forty highborn priests, attributed to Ari Þorgilsson, with a later (*c.* 1480s) list of Icelandic bishops (f. 5r); the larger Viðey Map (ff. 5v–6r); and the smaller Viðey Map (f. 6v). The bifolium in this hand (ff. 35r–36v of the present compilation) contains two items: two months from a calendar, January (f. 35r) and February (f. 35v), and a short treatise on time-reckoning (ff. 36r–36v). Since ten months are missing from the calendar, at least three bifolia are missing from this quire.[17] The maps in this fragment are dealt with in Chapters 3 to 5 of this book.

1812 IV (ff. 24r–34v of the present compilation, ten leaves) likewise contains no maps but, dating from the late twelfth century, may be one of the oldest extant Icelandic manuscripts. It contains two Latin-Icelandic glossaries (f. 24r and 34v), an Old Norse account of the six days of Creation and the Incarnation (ff. 24v–25r), and Old Norse treatise(s) on the reckoning of time and the calendar (ff. 25v–34v). This section contains a passage resembling Ari Þorgilsson's account of Þorsteinn Surtr's reform of the calendar in *Íslendingabók* (f. 25v).[18]

GkS 1812 4to has been tentatively associated with the Augustinian monastery at Viðey, established in 1226 by Snorri Sturluson and Þorvaldr Gizurarson, the brother of the bishop of Skálholt.[19] One reason for its association with Viðey is a note on the inside front cover that reads 'Bok Hakonar Ormssonar Anno…' ('Hákon Ormsson's book, year…' [no number follows]). It seems likely that this Hákon Ormsson was the *ráðsmaðr* (steward) and writer who lived at Skálholt (1614–56). This Hákon was the great-grandson of Alexius Pálsson, the last abbot of Viðey, who presided over the monastery's archives when it was dissolved in 1559. It may be then, that Hákon came to own this book from the library of his great-grandfather. The association with Viðey is strengthened by the inclusion of an earlier abbot of Viðey, Abbot Steinmóður (d. 1481), in a list of bishops added to folio 5r (1812 III) in the 1480s.[20] Since Steinmóður is the only abbot mentioned among these bishops, it has been suggested

17 It has been assumed that since these two fragments are written in the same hand they must have originally been part of the same manuscript composition. However, it is also possible that these folios derive from two different compositions written by the same scribe. Examinations matching the scribal agenda across these two fragments must be made tentatively.

18 The Old Icelandic computistical treatise *Rímbegla* I is edited in *AÍ* II, pp. 1–80, where 1812 (Kålund's MS 'G') appears in the variant appartus.

19 Larsson, *Äldsta delen*, p. iii.

20 *DI* VI, pp. 314–15.

that 1812 (or at least the fragment 1812 III) was associated with Viðey at this time. The hand that added this list is seen in other manuscripts associated with Viðey, including Reykjavík, Stofnun Árna Magnússonar, AM 680a 4to.[21] The addition of this list in the 1480s confirms that 1812 circulated in a monastic milieu into the late fifteenth century. Kristín Bjarnadóttir has further pointed out that tidal observations in the Old Icelandic computistical treatise *Rímbegla II* (composed *c.* 1275–*c.* 1300, preserved in 1812 II) associate the book with the south-western corner of Iceland, where Viðey is 'the only educational seat where these observations could have been performed'.[22]

The association with Viðey can be made more strongly for 1812 III than for the other components of GkS 1812 4to. Indeed, the oldest part of the manuscript (1812 IV, written *c.* 1200) antedates its foundation in 1226. After the book's association with Hákon Ormsson, its provenance becomes clearer. The Viðey book came into the possession of Bishop Brynjólfur Sveinsson (1605–75), whose hand is discernible in a note on the inside the front cover ('calendar membr. Islandicum') and on folio 1r ('calendarium Islandicum').[23] Bishop Brynjólfur sent the book, with other precious manuscripts, to the Danish King Frederick III in 1662.[24] It was returned to Iceland on 14th December 1984, where it is now held in Stofnun Árna Magnússonar.

An Icelander's Guide to Planetary Astronomy

The world maps that survive from medieval Iceland are all preserved in manuscripts that thematise calendar construction and the tributary disciplines of astronomy and cosmology. In the Middle Ages, natural philosophical and astronomical enquiry was sometimes stimulated by the need to understand the calendar.[25] In order to calculate the date of Easter in accordance with the terms decided at the Council of Nicaea in 325 CE, a science that could achieve parity between the lunar and solar calendars was required.[26] *Computus*-core encyclopaedias were handbooks that contained the basic *computus* materials (calendars, Easter tables, and instructive texts on how to use them) as well as supplementary cosmological treatises that explained the need for computistic enquiry by showing the structure of the cosmos and the planets' problematic motions.

21 Gjerløw, *Libri Liturgici*, pp. 59–60.

22 Kristín Bjarnadóttir, 'Mathematical Education in Iceland in Historical Context: Socio-Economic Demands and Influences' (PhD diss., Roskilde University, 2006), p. 42.

23 *DI* I, p. 183.

24 Larsson, *Äldsta delen*, p. iii.

25 Olaf Pedersen, 'Astronomy', in *Science in the Middle Ages*, ed. David C. Lindberg (London: University of Chicago Press, 1978), pp. 303–37, p. 304.

26 Pedersen, 'Astronomy', p. 307.

Computus was, as Faith Wallis puts it, 'a problem at once scientific, theological, and disciplinary', whose aim it was to harmonise observations of the sun and moon with the Church's observance of the calendar.[27] The cosmological treatises that accompanied *computus* documents were often elementary or preparatory, presenting their sources in a way that contextualised its operation but 'completely ignored both the methods by which the results had been obtained and the mathematical form in which they had been expressed'.[28]

The bound fragment in which we encounter the zonal map, 1812 I, is an assemblage of simple *computus* texts coupled with accessible descriptions of the stars and planets. The manuscript begins, on folios 1r–2v, with an assemblage of practically-focussed *computus* texts and verses that aided memorisation of important feast days.[29] In the folios that follow, the compositor assembles items on the tributary fields of stellar astronomy, planetary kinematics, and tidal science, subjects inducted into the encyclopaedia through their background association with *computus*. The book's items on the constellations and their mythologies have little practical use, but provided a basic perspective on the astronomical phenomena that troubled the ecclesiastical calendar and necessitated *computus*. The compilation's focus moves from simple *computus* verses, through a digression on the stars and constellations, to a focus on the planets' motions. It is in this section that the reader encounters the Icelandic zonal map.

There have been as yet few attempts to characterise Icelandic encyclopaedias, to analyse their intellectual programmes and the editorial policies that shaped them.[30] This chapter surveys the zonal map's immediate

27 Faith Wallis, 'Chronology and Systems of Dating', in *Medieval Latin: An Introduction and Bibliographical Guide,* ed. Frank Anthony Carl Mantello and A. G. Rigg (Washington DC: The Catholic University of America Press, 1996), pp. 383–87, p. 383.

28 Pedersen, 'Astronomy', p. 306. See also Grant, *Foundations,* pp. 12–13.

29 For example, on folio 1r and 1v are contained the *Cisioianus,* and a series of Latin computistical verses, with Old Norse prose explanations. This theme is maintained on folio 2r in a seven-line Latin mnemonic verse that indicates the immovable feast days. It provides information on how to use the seven ferial letters (A–G) to determine the day of the week on which feast days will fall when the dominical letter (the letter A–G that in that year that stands for Sunday) is known. This is accompanied by an Old Norse prose explanation. Also on folio 2r, there is an Old Norse treatise attributed to the Icelandic astronomer Stjörnu Oddi ('Star Oddi') on the length of the day and the places the sun rises and sets on the horizon over the course of the solar year. On folio 2v, there is statement about the durations of the planets' orbits, and on folio 7r there is a diagram that explains the origins of the planetary week.

30 A notable exception is Svanhildur Óskarsdóttir's work on the Universal History in AM 764 4to See Svanhildur Óskarsdóttir, 'Universal History in Fourteenth-Century Iceland: Studies in AM 764 4to.' (PhD diss.,University of

manuscript context in the encyclopaedia 1812 I to reconstruct the intellectual environment in which it worked. As we shall see, the map sustains a productive suite of relationships with its companion items, contributing to their description of the structure of the cosmos and the motions of the planets.

The Seven Planets

Folio 10v begins the exposition on the structure of the cosmos and the motions of the planets in which we encounter the Icelandic zonal map. This section follows on from a meditation on stellar astronomy and the mythologies of the constellations. The distinction between this and the previous section is explicitly stated in a short Old Norse text at the top of folio 10v, which describes the order of the planets and the durations of their orbits around the Earth.[31]

> Sio eru kollut lopt i bokum, þau er himin tungl hverfi um. Ok er tungl i neðsta lopti, ok er þat kallat minst himin tungla, ok synist þi mest, at þat er nest os. Þat gengr xxvii dagha hring sinn vndir zodiacum ok atta stundir. Merkurius er i audru lopti, hun gengr sinn hring ccc ok xxx dagha ok ix dagha.
>
> (There are seven heavens named in books, around which the planets turn. And the moon is in the lowest heaven, and is called the smallest of the planets, and seems greatest because it is closest to us. It completes its orbit under the zodiac in twenty-seven days and eight hours. Mercury is in the second heaven. It completes its orbit in 339 days.)[32]

This text appears twice in 1812 I. It appears alongside *computus* texts on folio 2v, where its details about the durations of the planetary orbits complement other texts on the motions of the sun and moon and their implications for the calendar. There are notable differences, however, in the presentation of the text across these two versions. The version

London, 2001); Svanhildur Óskarsdóttir, 'Writing Universal History in Ultima Thule: The Case of AM 764 4to', *Mediaeval Scandinavia* 14 (2004), 185–94; Svanhildur Óskarsdóttir, 'The World and its Ages: The Organisation of an 'encyclopaedic' Narrative in MS AM 764 4to', *Sagas, Saints and Settlements*, ed. Gareth Williams and Paul Bibire (Leiden: Brill, 2004), pp. 1–11.

31 The order of the seven planets and the durations of their orbits are shown in a diagram that accompanies both of the Icelandic hemispherical world maps, AM 736 I 4to f. 1v and AM 732b 4 f. 3r. See Chapter 1 and figure 9. The term used for heaven on this Latin diagram is *caelo*, which has been translated into lopt in this Old Norse text.

32 *AI* II, pp. 246–47. Simek has identified an Old Norse translation of Honorius's *Imago mundi* in the fifteenth-century manuscript AM 685d 4to into which these lines have been inserted. These lines themselves, however, are not from Honorius. Simek, *Altnordische Kosmographie*, p. 399.

that begins on folio 10v has been fragmented so that the sections on the relevant planets have been interspersed between the two folios that contain the planetary diagrams, on folios 10v and 11r. At the top of folio 10v, the orbits of the moon, Mercury, and Venus are described above the diagram that shows their orbits.

This short text introduces a section in the encyclopaedia on planetary kinematics. The version of this text on folio 2v contains a coda (absent from the 10v–11r version) that makes explicit the distinction between the planets and the stars, and therefore the distinction between this and the previous section of the encyclopaedia: 'þessar eru allar reikendr, en adrar stiornr snuaz með festingu himens' ('these are all wanderers; the other stars turn with the firmament'). The 'others that turn with the firmament' are the fixed stars, which as a category comprise lone stars (*stellae*), star clusters (*sidera*), and constellations (*astra*) (Isidore, *Etymologies* 3.60). These stars turn with the celestial sphere, the concavity of the night sky against which the sun, moon, and other planets move. The planets (*planetes*, from the Ancient Greek 'to wander') are also a variety of star, so-called because they are not fixed in their positions in relation to other stars but range across the cosmos (3.67). Information about the fixed stars is concentrated in the first half of this encyclopaedia, which yields to a study of the movements of the planets on these folios. Isidore's definition of a planet calls to attention the peculiarities of their orbits that are the subject of the texts and diagrams assembled in this division of the encyclopaedia:

> Vnde pro eo, quod errant, retrograda dicuntur, uel anomala efficiuntur, id est, quando particulas addunt et detrahunt. Ceterum quando tantum detrahunt, retrograda dicuntur; stationem autem faciunt, quando stant.
>
> (It is because of their wandering that they are called retrograde, or are rendered irregular when they add or subtract orbital degrees. When they pull back only they are called retrograde, and they make a 'station' (*statio*) when they 'stand still' (*stare*).)[33]

Thus the former section of the encyclopaedia concerns the constellations and their mythologies; the latter concerns the wandering planets. The motions of the planets are the focus of the texts and diagrams that follow in 1812 I.

[33] Isidore, *Etymologies* 3.67. All translations are from Isidore of Seville, *The Etymologies of Isidore of Seville*, trans. Stephen A. Barney, W. J. Lewis, J. A. Beach, and Oliver Berghof (Cambridge: Cambridge University Press, 2006), p. 104.

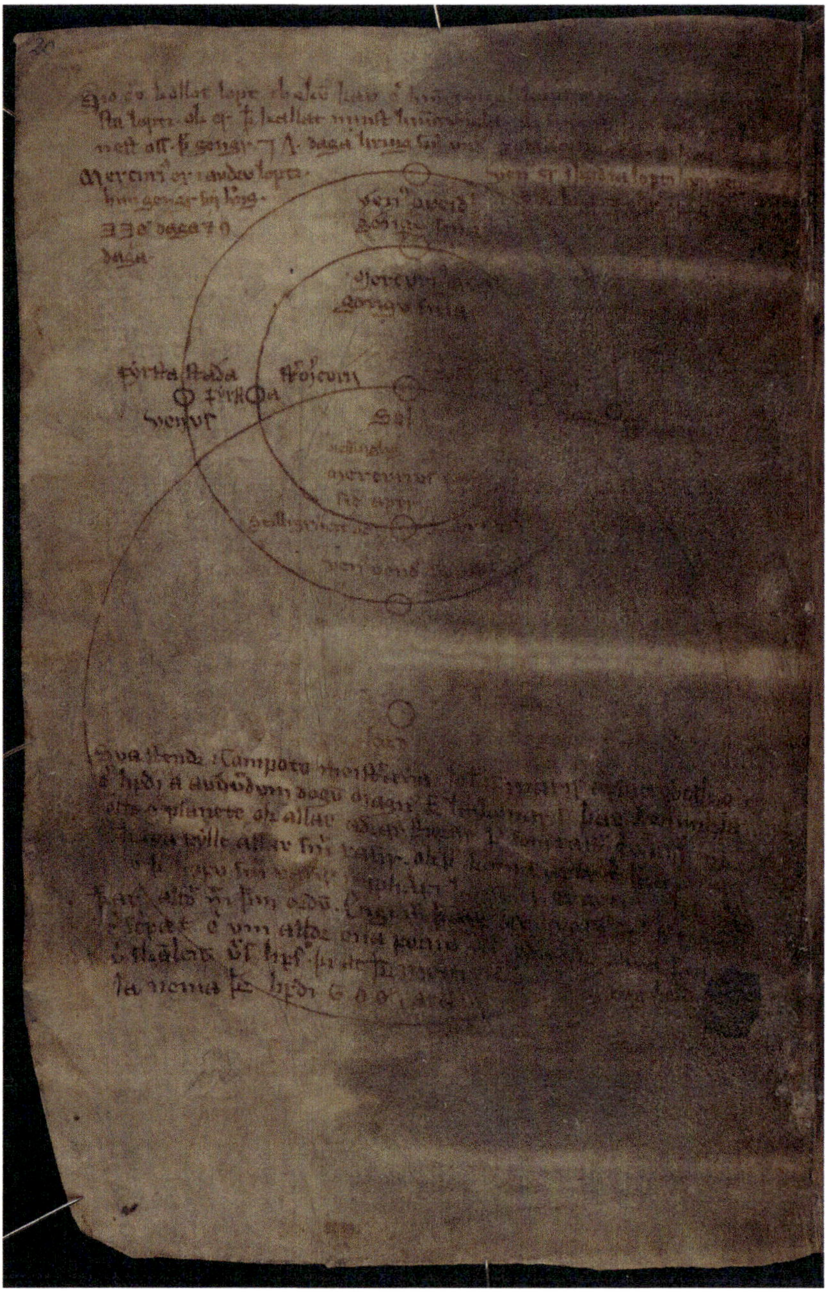

FIGURE 10: Folio 10v begins an exposition on planetary kinematics, and contains an introduction to the seven planets and the heavens through which they move, a diagram showing the heliocentric orbits of Mercury and Venus, and a text attributed to the English-born Parisian scholar Johannes de Sacrabosco, translated into Old Norse. Reykjavík, Stofnun Árna Magnússonar, GkS 1812 I 4to, f. 10v, 1315–*c.*1400.

The Motions of the Inferior Planets

This diagram (figure 10), preserved on folio 10v, describes and explains a pattern in the motions of the inner planets, Mercury and Venus. It accompanies an Old Norse text (at the top of the folio) that details the time it takes for these planets to complete their orbits around the Earth, and an astronomical note (at the bottom of the folio) from the *computus* of Johannes de Sacrobosco (d. 1256), the English-born Parisian scholar 'who lived during the early days of King Magnús Hákonarson' ('computo meistaranna Iohannis i Paris af Sacrobosko, er lifði a avnðverdvm dogvm Magnus konungs hakonar sonar'). The text that begins 'sio eru kollut lopt i bokum, þau er himin tungl hverfi um' ('seven heavens, on which celestial bodies turn, are mentioned in books') introduces the section on the wandering planets, and is woven, through its fragmentation across two folios, into the two diagrams that illustrate their motions. The diagram is paired with another on folio 11r, also with 'Iord' ('Earth') at its centre, that shows the orbits of the remaining outer planets: Mars, Jupiter, and Saturn.

The diagram accounts for two observations in the motions of the inner planets, Mercury and Venus.[34] Firstly, ancient astronomers saw that these planets did not stray far from the sun in the course of their orbits around the Earth. This had led Cicero in his *Dream of Scipio* to call Mercury and Venus the *comites solis* ('the sun's companions').[35] Secondly, they saw that while the planets usually moved eastwards across the night sky, they did not appear to turn in their orbits around the Earth uniformly. Rather, the planets appeared periodically to stall and then reverse direction, moving westwards relative to the stars instead of their usual eastwards. This phenomenon was described by ancient astronomers, whose observations were summarised by Pliny the Elder (*Historia Naturalis* II.12), Seneca (*Naturales questions* VII.xxv.6–7) and Isidore of Seville (*Etymologies* 3.67–69). The diagram accounts for these two observations by introducing two cosmological principles: the eccentric epicycle, and the geo-heliocentric orbits of Mercury and Venus.

The heliocentric orbits of the inner planets
(Reykjavík, Stofnun Árna Magnússonar, GkS 1812 I 4to, f. 10v) 1315–c.1400.

Iord	Earth
Sol	Sun
messinglig	brass
Fyrsta staða Venus	First station of Venus
Ven*us* greid*ir* gongu sina	Venus proceeds in its course

34 Its inscriptions have been transcribed in a footnote in *AÍ* II, p. ccxiii.
35 Macrobius, *Commentary* 1.19.4.

Aun*n*r st*að*a Ven*us*	Second station of Venus
Ven*us* vend*ir* sic apt*r*	Venus turns itself around
Stilligr heit oc varringligr	Calming, hot and coppery [Mercury]
Fyrsta Mercurius	First [station] of Mercury
Mercuri*us* greid*ir* gongu sina	Mercury proceeds in its course
Aun*n*r st*að*a Mercuri*us*	Second station of Mercury
Mercurius vend*ir* sic apt*r*	Mercury turns itself around

In modern terms, we know that the planets appear to move eastwards relative to the stars because of the daily rotation of the Earth on its axis, and that they appear periodically to change direction when they overtake, or are overtaken by, the Earth in the course of their yearly orbits around the sun.[36] The planets also appear to orbit the Earth at different speeds at different times in their orbits.[37] For medieval natural philosophers, however, this phenomenon, known as apparent retrograde motion, troubled the Aristotelian model of the nested planetary spheres that guided their regular and symmetrical motions around the Earth.

This diagram shows the system developed by Claudius Ptolemy (*c.* 100 – *c.* 170 CE) to explain these planetary motions: the eccentric epicycle.[38] Between around 1150 and 1250, two cosmological systems were introduced into Western Europe: those of Aristotle and Ptolemy.[39] The Aristotelian system was known through translations of Aristotle's natural books, and commentaries on them. Ptolemy described the eccentric epicycle in a treatise entitled *Hypotheses of the Planets*. This treatise was not known directly in the Latin Middle Ages, but was known in abstract

36 The question as to whether or not the Earth turned on an axis was contentious among medieval natural philosophers and their antique forebears. See Edward Grant, *Planets, Stars, and Orbs: The Medieval Cosmos, 1200–1687* (Cambridge: Cambridge University Press, 1994), pp. 494–510.

37 This is because of the elliptical nature of their orbits, so that they appear to move faster when their angular distance from the Earth is smaller (when they are at their apogee), and slower when it is more oblique (when they are at their perigee).

38 On the origins of epicycles and deferents in the Hellenic world see Duhem, *Le Système du Monde*, vol. 1, p. 427.

39 Edward Grant, 'Eccentrics and Epicycles in Medieval Cosmology', in *Mathematics and its Application to Science and Natural Philosophy in the Middle Ages, Essays in Honor of Marshall Clagett*, ed. Edward Grant and John E. Murdoch (Cambridge: Cambridge University Press, 1987), pp. 189–214, p. 189; Grant, *Planets*, p. 275; Grant, *Foundations*, p. 104. On Aristotelian and Ptolemaic astronomy in the Hellenistic world see Duhem, *Le Système du Monde*, vol. 1, pp. 488–97.

through Arabic treatises, which became assimilated into the European canon in the twelfth and thirteenth centuries.[40]

Both the Aristotelian and Ptolemaic systems held that the cosmos comprised seven planets located in a variable number of spheres, which accounted for variations in their movements.[41] The main difference between the two systems was whether or not these spheres were concentric with the Earth. Aristotle described the cosmos as a series of concentric orbs with the Earth at its absolute centre. However, medieval natural philosophers observed that this system could not account for the complexity of the planets' motions.[42] Ptolemy, on the other hand, proposed that the planetary spheres were eccentric, which is to say that they did not have the Earth at their absolute centres, and that the planets' orbits were epicyclic, that is, that the planets turned on smaller circles (*epicycles*) as they proceeded in their wider orbits (*deferents*) around the Earth.[43] This solution was known in the central Middle Ages through Sacrobosco's *Tractatus de sphaera* (*c.* 1230), which drew heavily from Ptolemy's *Almagest* and its Arabic commentaries.[44] Ptolemy's theory better represented planetary motions, and could account in particular for the periods in which their orbits appeared to reverse direction. When the planet turned in its epicycle, it would appear momentarily to stand still, then resume its course in the opposite direction.[45]

The second principle illustrated by this diagram pertains to Mercury and Venus's heliocentric orbits. Again, this theory is rooted in observation. In modern terms, we know that Mercury and Venus orbit between the Earth and the sun, and that they cannot appear, from an earthly

40 Duhem, *Le Système du Monde*, vol. 1, p. 429. Many of the avenues by which newly translated cosmographical treatises were assimilated into Western Europe have yet to be identified or analysed.

41 Grant, *Foundations*, p. 104.

42 Edward Grant, 'Cosmology', p. 280; Grant, *Foundations*, p. 105.

43 A modern analogy might be an orbit within an orbit: the moon circles around a point (the Earth) that in turn circles around the sun. The inferior planets (Mercury and Venus) were thought to orbit the sun which orbited the Earth. However, the superior planets (Mars, Jupiter, and Saturn) orbited imaginary points that orbited the Earth.

44 The channels by which Sacrobosco became familiar with these texts are mostly unknown. Pedersen, 'Astronomy', p. 318.

45 Grant, 'Eccentrics and Epicycles', p. 195. Medieval natural philosophers sought a compromise that would make the epicycle and deferent compatible with Aristotle's theory of the nested concentric spheres, in order to demonstrate that 'eccentric and epicyclic orbs did not imply consequences that were subversive and destructive of Aristotelian cosmography and physics', on which the canon was based. The solution assigned each planet three orbs, a compromise adapted in Sacrobosco's *De tractatus sphaera*. Thorndike, ed and trans., Johannes de Sacrobosco, *The Sphere of Sacrobosco and its Commentators*, (London: Cambridge University Press, 1949), p. 114.

perspective, to stray far from the sun's path. This led ancient astronomers to theorise that Mercury and Venus circled the sun, centuries before Copernicus and Tycho Brahe developed their heliocentric and geo-heliocentric models of the solar system.[46] Ancient astronomers proposed that these planets turned on their epicycles around the sun, as it in turn drew a larger orbit around the Earth.

The Icelandic diagrams resemble others that illustrate widely known astronomical works, such as Sacrobosco's *Tractatus de sphaera*, William de Conches's *Dragmaticon philosophiae*, and Daniel of Morley's *Liber de naturis inferiorum et superiorum*.[47] A diagram in a twelfth-century manuscript of William de Conches' *Dragmaticon* (IV.4.6–7) has been identified by Obrist as a borrowing from the eighth-century Persian Jewish astronomer Māshā'allāh's *Liber de orbe* (Ch. 32), an abundantly illustrated cosmographical treatise that drew from the Ptolemaic treatises in which the theory of the eccentric epicycle was developed.[48] It shows the Earth amid the sun's course, with Mercury and Venus at four positions in its epicycle: in forward and retrograde motion, and in its two stations (Vatican City, Bibliotheca Apostolica Vaticana, MS Pal. lat. 1042, f. 23r).

The Icelandic diagram's inscriptions describe the forward and retrograde motions of the planets. The inscription 'Ven*us* greid*ir* gongu sina' ('Venus proceeds in its course') describes the usual course Venus takes in its orbit from west to east across the night sky. The 'fyrsta staða Venus' ('first station of Venus') is the point at which the planet appears to stall in its course before reversing direction. Thus 'Ven*us* vend*ir* sic apt*r*' ('Venus turns itself around') describes the retrograde motion of the planet when it appears to move east to west across the night sky, and 'aun*n*r st*aða* Ven*us*' ('second station of Venus') the point at which it stalls before resuming

46 For an overview of heliocentric astronomy in the Hellenic world see Duhem, *Le Système du Monde*, vol. 1, pp. 339–426. On its intersection with the epicycle and deferents and its mathematical basis, see especially pp. 441–68.

47 Examples of the circumsolar diagrams are described in Duhem, *Le Système du Monde*, vol. 3, pp. 153–55; Rudolf Simek, *Heaven and Earth in the Middle Ages: The Physical World before Columbus*, trans. Angela Hall (Woodbridge: Boydell, 1996), p. 15; and especially Bruce S. Eastwood and Gerd Grasshoff, *Planetary Diagrams for Roman Astronomy in Medieval Europe, ca. 800–1500* (Philadelphia: American Philosophical Society, 2004), for Calcidian diagrams see pp. 75–86; for Capellan diagrams see pp. 133–35. Sacrobosco's *Tractatus de sphaera* also circulated with diagrams derived ultimately from Ptolemy's *Almagest*, though by which channels is unknown. The diagrams that accompany these treatises are rarely reproduced in textual editions, and when they are, manuscript shelf marks are seldom supplied.

48 The influence of Arabic cosmographical treatises on William's encyclopaedia are evaluated in Barbara Obrist, 'William of Conches, Māshā'allāh, and Twelfth-Century Cosmology,' *Archives d'Historie Doctrinale et Littérature du Moyen Âge* 76 (2009), 29–87. His sources were also considered in Duhem, *Le Système du Monde*, vol. 3, pp. 87–101.

its usual west to east motion. This is described in diagrammatic terms in Sacrobosco's *Tractatus de sphaera*.

> Si igitur due line ducantur a centro terre ita quod includant epiciclum, una ex parte orientis, reliqua ex parte occidentis, punctus contactus ex parte orientis dicitur statio prima, punctus vero contactus ex parte occidentis dicitur statio secunda. Et quando planeta est in alteruta illarum stationum, dicitur stationarius. Arcus autem epicicli superior inter duas stationes interceptus dicitur directio, et quando planeta est in illo, dicitur directus. Arcus vero epicicli inferior inter duas stationes dicitur retrogradatio, et planeta ibi existens dicitur retrogrades.
>
> (If, then, two lines are drawn from the centre of the earth to include an epicycle, one on the east and the other on the west, the point of contact on the east is called the 'first station' (*statio prima*), while the point of contact to the west is called the 'second station' (*statio secunda*). And when a planet is in either of those stations it is called 'stationary' (*stationarius*). The upper arc of the epicycle intercepted between these two stations is called 'direction' (*directio*), and when a planet is there it is called 'direct' (*directus*). But the lower arc of the epicycle between the two stations is called 'retrogradation' (*retrogradatio*), and a planet existing there is called 'retrograde' (*retrogrades*).)[49]

On the diagram, Mercury and Venus are shown in four stages of their epicycle: in the stationary position in which the epicycle begins; in direct motion; in its second stationary position; and in retrograde motion, after which it reappears on the other side of the sun and begins its epicycle anew. The Old Norse *fyrsta* and *aunnr staða* are loan translations of the Latin *statio prima* and *statio secunda* (cf. Old Norse *solstaða*, 'solstice'). The Latin *directio* and *retrogradatio* are not adopted as loanwords but have been allocated more descriptive terms: the planet proceeds in its course (*greiðir göngu sína*), and then turns itself around (*vendir sik aptr*) as it turns in its epicycle. The accompanying texts do not contribute to the diagram's description of these planets' circumsolar orbits; the diagram does not explain facts previously established by the texts, but operates for the most part independently of them.[50] More detailed explanations of the eccentric epicycles are not to be found in the Icelandic encyclopaedic literature that survives. These translation moments are especially valuable, therefore, for what they tell us about the transmission of these ideas into medieval Iceland.

49 Sacrobosco, *Sphere*, pp. 114–15.

50 Barbara Obrist examines William de Conches's appropriation of cosmographical diagrams from the *Liber de orbe*, and observes that they were 'quite easy to assimilate for, although they are ultimately derived from figures used in geometrical demonstrations, they are now endowed with a purely illustrative function'. Obrist, 'William of Conches', 53.

Sacrobosco in the Days of King Magnús Hákonarson

Beneath the diagram showing the motions of the inferior planets on folio 10v there is an Old Norse astronomical note attributed to Sacrobosco.[51] This short text begins 'sva stendr i computo meistaranna Iohannis i Paris af Sacrobosko' ('thus it says in the *Computus* of the master John of Sacrobosco in Paris') and connects this exposition of planetary motions to the compilation's prevailing interest, the *computus*, reminding the reader of the connection between the determination of the calendar, predicated on the motions of the sun and moon, and the wider study of cosmology.

Little is known about John of Sacrobosco (Sacrobosco a Latinised form of his hometown, literally 'holy wood', which may be Holywood or Halifax in West Yorkshire) other than he was an English astronomer who taught at the University of Paris from *c.* 1230 until his death in *c.* 1256.[52] Indeed, we can glean most of this from his introduction in this Icelandic text, which describes 'computo meistaranna Iohannis i Paris af Sacrobosko, er lifði a avnðverdvm dogvm Magnus konungs hakonar sonar' ('the *Computus* of master John of Sacrobosco in Paris, who lived during the days of King Magnús Hákonarson').[53] Sacrobosco wrote three of the most widely-read scientific treatises of the medieval period. The first, the *Algorismus vulgaris*, describes simple arithmetical functions, such as the extraction of square and cubic roots. The second, the aforementioned *Tractatus de sphaera*, is an exposition on spherical astronomy and cosmography that often contained diagrams similar to those found in 1812 I. The third is a *computus* manual known as *De anni ratione*, which can be identified as the 'computo' paraphrased and translated into Old Norse here. These works are elementary treatises aimed at the instruction of relative beginners, or non-specialists, in the sciences of arithmetic, astronomy, and *computus*. The *Tractatus de sphaera* numbers among the clearest and most popular textbooks in astronomy and cosmography from the thirteenth to the seventeenth century, but is also of an elementary level.

The link between Sacrobosco and the reign of King Magnús Hákonarson (Magnús VI of Norway, 1238–80) conforms to a general tendency in Icelandic encyclopaedias, as we saw in Chapter 1, to synchronise Icelandic and wider European perspectives on the world and time. This endeavour surfaces in the mention of local saints, including Magnús Erlendsson, Sunniva, and Bishop Guðmundr, alongside other European saints in the calendrical texts near the beginning of this manuscript. Icelanders were keen to demonstrate that their own local enquiries into

51 *AÍ* II, pp. 257–58.

52 Olaf Pedersen, 'In Quest of Sacrobosco', *Journal for the History of Astronomy* 16 (1985), 175–220.

53 *AÍ* II, p. 257.

natural philosophical matters were contiguous with, and could enrich, the Latin learning they had inherited with the advent of bookish culture that accompanied their conversion to Christianity centuries earlier. As well as translating these texts into the vernacular, they supplemented them with their own local terms of reference; placing Latinate place-names into apposition with their Icelandic equivalents, measuring astronomical phenomena with reference to local saints' days, or, as here, locating the canonical works treating planetary astronomy within their own regional chronologies.

This short translation from Sacrobosco describes how long it takes for the planets and the fixed stars to turn in their orbits around the Earth, and cites the Roman historian Josephus (*c.* 37–100 CE).

> Þar er en mikla ölld, planete ok allar aðrar stiornr, þær sem fástar erv a himni, hafa fyllt sinar rasir ok þær koma aftr til þeira stada, er þær hofv sinar rasir i vpphafi heimsins. Oc getr Iosephus þessar alldar met þessum orðvm: Engi maðr þarf at hugsa, at þat se rangt, er skrifat er um alldr ena fornv manna eða iafna þeira efi við skamleic vars lifs, þi at þeir metti ei skyra gang himin tvngla, nema þeir lifdi 600 ara.)

> (Over a great period of time the planets and all the other stars, those that are fixed in the heavens, have completed their courses and come back to their places, which they had at the beginning of the world. Josephus says about these ages these words: No one must think that it is wrong, that which is written in the time of the ancients, or compare their age with the shortness of our lives, because they will not understand the passages of the planets, unless he lives for six hundred years.)[54]

The citation emphasises the difficulties in observing and thinking about the stars and planets' revolutions around the Earth within human time-scales. While the text on the seven planets gives the orbit of Saturn as thirty years, the fixed heavens turn at a speed not perceptible over the course of a person's life. In the *Tractatus de sphaera*, Sacrobosco states that the firmament moves by just one degree in one hundred years, and that what the antique literature says on the revolution of the firmament must be accepted. The Old Norse translation, however, diverges from Sacrobosco's original.

> De quo Iosephus sic meminit, Nullus ad vitam modernam et annorum nostrorum breuitatem uitam comparans antiquorum, falsa putet eorum scripta, que nunc ediscere posset, si sexcentos uiueret annos.

> (Concerning this, Josephus advises us thus: no one, comparing our modern life and the brief course of our years with those of the ancients,

54 *AÍ* II, p. 258. My translation.

could ever think their writings false, if he were able to study them thoroughly, even if he should live for six hundred years.)[55]

By design or accident, the Old Norse substitutes 'eorum scripta' ('their writings') for 'gang himin tungla' (the 'motions of the planets'), moving the focus from the interpretation of ancient writings to the courses of the planets themselves. Indeed, the diagrams derive from far more detailed expositions on planetary kinematics than their companion texts, which do not mention such complexities as retrograde motion, circumsolar orbits, or the eccentric epicycles. This citation from Sacrobosco's *De anni ratione*, slightly altered in its Old Norse translation, appears to eschew engagement with the mathematical astronomy that would be needed to fully contextualise and explain these planetary motions.

The Motions of the Superior Planets

On folio 11r begins a second stage in this visual exposition on planetary kinematics. At the top of the folio there is a continuation of the text on the seven planets that began on the facing verso. This describes the position and orbit of Saturn. Further fragments from this text, describing the positions and orbits of Jupiter, the sun, and Mars, are interspersed into the body of the diagram next to their relevant diagrammatic depictions.

This diagram (figure 11) explains the apparent retrograde motion of Mars, Jupiter, and Saturn: the naked-eye planets that appear to stop in their orbits of the Earth and appear periodically to move in the opposite direction.[56] As explained above, the retrograde planets move in small circles (or epicycles) on their eccentric orbits (deferents) around the Earth. These circles are parallel to the sun's plane, the ecliptic, which is shown on the diagram. The three planets are shown in their apogees, the point in their orbits farthest away from the Earth. The planets' apogees and perigees (the points in their orbits closest to the Earth) were situated in different signs of the zodiac.[57] Martianus Capella, and Pliny the Elder, whom he follows, placed the apogee of Jupiter in Virgo, Saturn in Scorpio, and Mars in Leo, as shown in the diagram.[58] The diagram contains inscriptions that relate

55 Johannes de Sacrobosco, *Libellus Iohannis de Sacrobosco, de Anni ratione, seu ut vocantur vulgo, Computus Ecclesiasticus* (Paris: Guillaume Cauellat, 1550), p. 18.

56 Bradley Carrol and Dale Ostlie, *An Introduction to Modern Astrophysics* (San Francisco: Addison-Wesley, 2007), p. 4.

57 Pedersen, 'Astronomy', p. 307. The planets in their absides (their near and far points, apogees and perigees) were sometimes represented in circular diagrams in manuscripts of Pliny's *Historia Naturalis*. See Eastwood and Grasshoff, *Planetary Diagrams*, pp. 31–32.

58 Bruce S. Eastwood, 'Invention and Reform in Latin Planetary Astrology', in *Latin Culture in the Eleventh Century. Proceedings of the Third International*

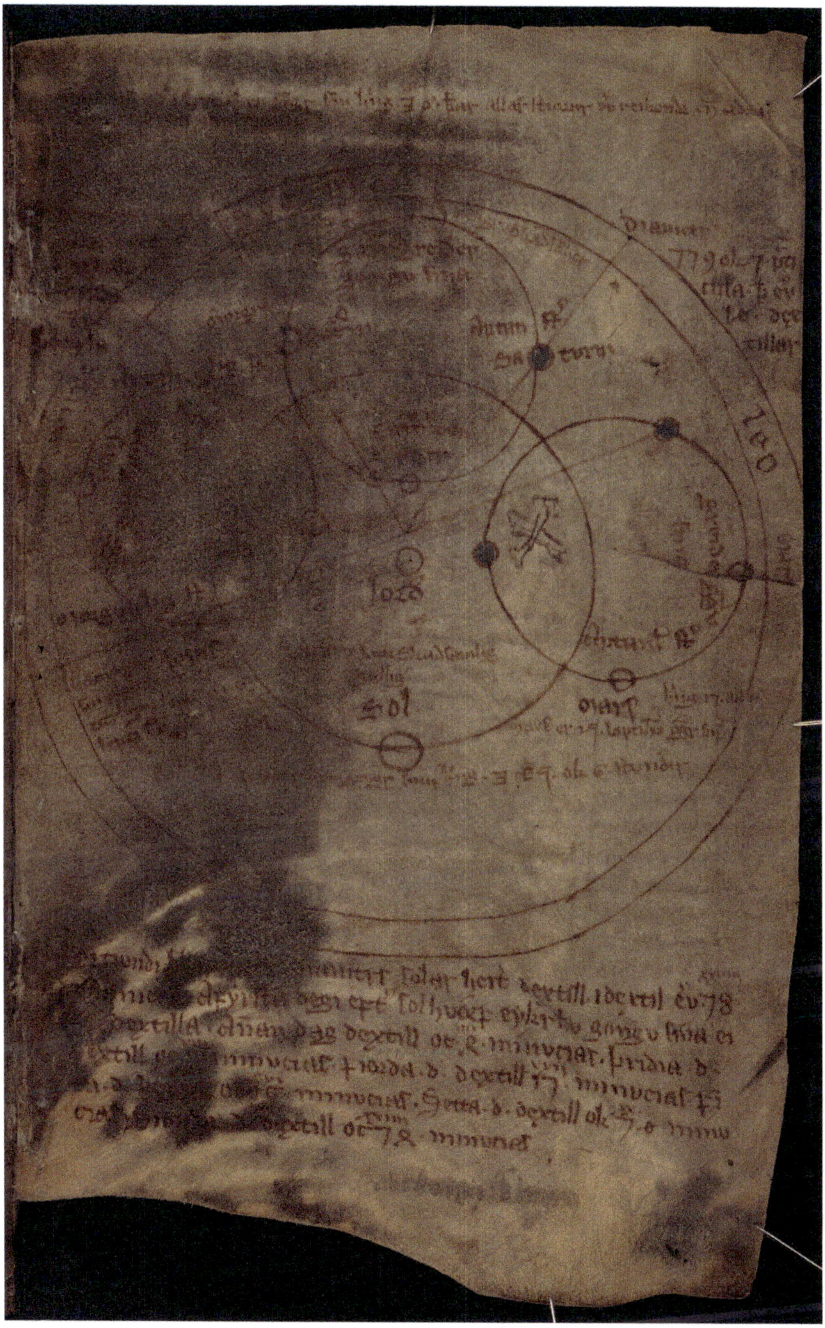

FIGURE 11: Folio 11r bears a continuation of the description of the seven planets, a diagram showing the revolutionary motions of the outer planets, and three short notes on their dimensions. The compass hole at the centre of the diagram is reused for the zonal map overleaf. Reykjavík, Stofnun Árna Magnússonar, GkS 1812 I 4to, f.11r, 1315–*c.*1400.

to the position of the retrograde planets in the sky in the mornings and evenings. A compass hole at the centre of the Earth is reused on the verso overleaf for another circular diagram, the Icelandic zonal map.[59] Like the diagram on folio 10v, this one marks the positions of the planets in two stations, and shows the forward and retrograde motions between them.

The motions of the outer planets
(Reykjavík, Stofnun Árna Magnússonar, GkS 1812 I 4to, f.11r) 1315–c.1400.

Iord	Earth
Heit oc þur skadsamlig	Hot and dry harmful
gullig	golden
Sol	Sun
hamingjusamligr heitr	Happy hot [Jupiter]
oc vatr eitrligr	and wet poisonous
morgunlig staða Jovis	Station of Jupiter in the morning
Jovis greid*ir* go*n*gu sina	Jupiter proceeds in its course
Aptanlig st*aða* Jovis	Station of Jupiter in the evening
...[60]	[Jupiter turns itself around]
Þvr oc kalldr saturnus	Dry and cold Saturn
blyligr skadsamligr	leaden harmful
Morgin staða Saturni	Station of Saturn in the morning
Saturn*us* greid*ir* gongu sina	Saturn proceeds in its course
Aptan st*aða* Saturn*us*	Station of Saturn in the evening
Saturn*us* vend*ir* sic apt*r*	Saturn turns itself around
...	[Station of Mars in the morning]
Mars greid*ir* gongu sina	Mars proceeds in its course
Aptanlig st*aða* Mars	Station of Mars in the evening
...	[Mars turns itself around]

We see the planets in two stations, the points at which they appear to stall and then change direction and become prograde or retrograde in

Conference on Medieval Latin Studies, Cambridge, September 9–12 1998, vol. 1, ed. Michael W. Herren *et al.* (Turnhout: Brepols, 2002), pp. 264–97, p. 273.

59 *AÍ* II, p. ccxiv.

60 A number of legends are missing. There is enough room for these inscriptions, and no space central to the diagram has been destroyed by trimming. Such omissions might imply that the diagram's scheme was only partially understood or valued.

their movements. The morning and evening stations have to do with whether or not the planet is in opposition to the sun. Because the orbits of Mercury and Venus are closer to the sun than the Earth they can never be in opposition, that is, Mercury and Venus can never be opposite the sun in the sky, or have the Earth come between them. When the superior planets are in opposition to the sun, they rise before it and are seen in the sky at evenings; when they are not in opposition, they are seen in the sky at mornings.[61]

Another feature of this and the previous diagram are the qualitative characteristics attributed to the planetary bodies. The most fundamental mark of Aristotelian inheritance on medieval cosmology is the distinction between the terrestrial and celestial realms, which were described in terms of stark opposition.[62] The terrestrial realm extends from the centre of the Earth to the lunar sphere, and is characterised by change and impermanence. The terrestrial world was composed of the four sublunary elements, in ascending order earth, water, air, and fire. In contrast, the celestial realm begins at the lunar sphere and extends to the eighth sphere in which the stars are fixed. The celestial realm is fundamentally unchangeable because no contrary qualities – hot or cold, wet or dry – exist there. These two realms had radically different natures, as Grant puts it, 'if incessant change was basic to the terrestrial region, then lack of change had to characterise the celestial region'.[63] The four sublunary elements – earth, water, air, and fire – whose interactions mobilise change in the terrestrial realm have no place in the celestial realm.

The planets, however, were nonetheless endowed with qualities, as can be seen on the two planetary diagrams in this manuscript fragment. The diagram on folio 10v describes Mercury as calming, hot, and copper, and the diagram on folio 11r describes Jupiter as happy (that is, *jovial*), hot, wet, and poisonous; the sun as hot, dry and harmful; and Saturn as dry, cold, leaden, and harmful. The qualities attributed to the planets had the potential to be problematic because, in Aristotelian thought, generation and corruption will take hold wherever contrary qualities exist.[64] Planets are regularly assigned qualities that are characteristic of the sublunary world.[65] The planets are characterised as they are on the Icelandic diagrams in the fifteenth-century *Ymago mundi* of Pierre d'Ailly, which describes Saturn as cold and dry, pale and evil in disposition; Jupiter as hot and wet, clear and pure, and thus tempering the maliciousness of Saturn; Mars is hot and dry, fiery and radiant, and therefore provokes war. Venus and Mercury are both radiant, defined by their relationship with the sun,

61 Simek, *Heaven and Earth*, p. 17.
62 Lewis, *Discarded Image*, pp. 104–09; Grant, *Foundations*, p. 55.
63 Grant, *Foundations*, p. 63.
64 Grant, 'Cosmology', p. 287; Grant, *Foundations*, p. 109.
65 Grant, *Planets*, pp. 467–69.

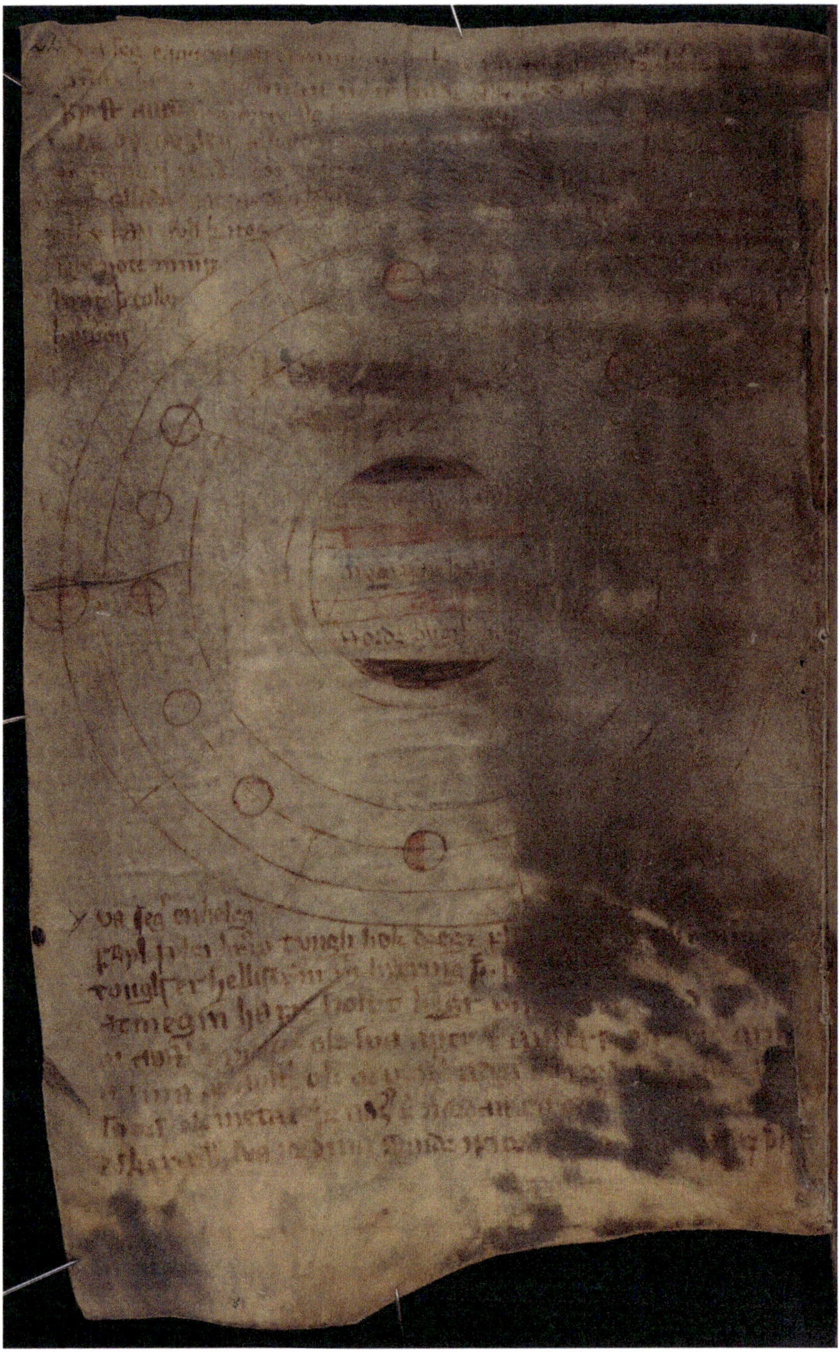

FIGURE 12: Folio 11v contains extracts from Helpericus's *De computo* and William of Conches's *De philosophia mundi*, paraphrased and translated into Old Norse. The Icelandic zonal map features amid a lunar diagram. Reykjavík, Stofnun Árna Magnússonar, GkS 1812 I 4to, f.11v, 1315–*c*.1400.

from whose course they never stray far.[66] Natural philosophers described these qualities as associative, not formal: the planets did not exhibit these qualities in their own natures, but had the capacity to produce them in the sublunary world. The sun, therefore, was not warm itself, but had the capacity to warm; as Robertus Anglicus, a thirteenth-century English commentator on Sacrobosco's *Sphere* puts it, 'non omne illud quod nigrificat est nigrum' ('not everything that blackens is black').[67] It is notable that the scheme presented on the Icelandic diagrams is partial, with only a few planets assigned their commonly held characteristics, it being likely that they, or the exemplars from which they were copied, are incomplete.

The Sizes of the Earth and Sun

Three short texts on the diameter of the Earth and sun accompany the diagram on folio 11r. Near the upper-left corner of the diagram there is an Old Norse note on the Earth's diameter, with parallels in other Old Norse astronomical texts; near the upper-right is an Old Norse note on the diameter of the sun, which has similar parallels.[68] These two notes border the central diagram of the motions of the superior planets. Below the diagram are six lines in Old Norse on the diameters of the sun and its orbit of the Earth in the days following the equinox.[69]

These texts are preserved elsewhere in the Icelandic encyclopaedic manuscripts AM 415 4to (early-fourteenth century) and AM 685 4to (fifteenth century). In these manuscripts, these notes are integrated into a highly technical geometrical treatise on the geometry of circles and the calculation of the circumference.[70] No such text accompanies the statements presented here. This section of 1812 I is concerned with the structure of the Earth and its dimensions. However, the diameters are provided without the means to calculate them. Cosmographical doctrines are presented here without the apparatus needed to reproduce them, suggesting that this is a more rudimentary handbook.

The Computus

The Icelandic zonal map (figure 12) appears on folio 11v. Having examined thus far the visual argument constructed through the encyclopaedia's combination of diagrams, we can see that the zonal map appears amid an exposition of the size and structure of the cosmos, and

66 Edward Grant, ed., *A Source Book in Medieval Science* (Cambridge, MA: Harvard University Press, 1974), p. 632.

67 Sacrobosco, *Sphere*, p. 155.

68 *AÍ* II, p. 233.

69 *AÍ* II, p. 91.

70 *AÍ* II, pp. 231–35.

its physical operations. The folio is headed with a short text attributed to the computist Helpricus, or Helpericus. The text is written in Old Norse and describes the establishment of the Julian calendar by Julius Caesar, and the subsequent need for intercalary days.[71] The text explains that in order to achieve parity between the lunar and solar calendars, a system of 'golden numbers' ('*aureum numerum*') was established. The numbers one to nineteen could be assigned to dates in the calendar to number the years in the nineteen-year Easter cycle.[72] The *saltus lune* (Old Norse *tunglhoppun*) happens when the year accumulates an extra lunar day, because the average lunation (or lunar month) is slightly longer than the notional lunar month of 29.5 days. Therefore, the calculated age of the moon, or *epact*, must 'jump' one day every nineteen years, in the final year of the nineteen-year Easter cycle. This will ensure that the lunar year and the solar year begin on the same day, that is, on January 1st, once every nineteen years. This inserted lunar day is called an *embolism*, and compensates for the fact that the moon loses eleven days each year (that is to say, the moon is eleven days younger on January 1st each year) so that the solar and lunar years achieve parity. In this Old Norse paraphrase, Latin terminology is not translated. Thus we have: 'aureum numerum', 'epactiss', and 'embolissmis'. The exception is 'saltus lune', to which has been added: 'þat kaullum ver tunglhoppun' ('which we call the leap moon').

This text is an Old Norse paraphrase from one of the most popular textbooks of *computus* of the Middle Ages.[73] The date at which Helpericus's *De computo* was composed is uncertain; it was common for teachers and students to copy *computus* handbooks and 'modernise' the worked examples, using the current date for the *annus praesens* ('present year'). Faith Wallis observes that Helpericus's readers seemed to regard *De computo* as a condensed and simplified version of Bede's *De temporum ratione*, Helpericus's treatise often serving as an introduction to Bede's more detailed composition.[74] Such appears to be the case here, the excerpt from *De computo* in 1812 I immediately preceding a text attributed to Bede. Simek has shown that this text attributed to Bede is actually an Old Norse paraphrase and translation of William de Conches's *De philosophia mundi* (see below), though the attribution might

71 *AÍ* II, p. 92.

72 An Old Norse table of lunations with Golden Numbers is preserved on 1812 II f. 23r. *AÍ* II, pp. 259–60.

73 Heplericus's *Liber de computo* is edited in *PL*, vol. 137, cols. 17–48.

74 This is the case in Oxford, St John's College MS 17, and several other manuscripts of English provenance. Faith Wallis, '10. Helpericus, *De Computo*, fols. 123r–135v: Overview', *The Calendar and the Cloister: Oxford, St John's College MS17*, McGill University Library, Digital Collections Program, 2007. http://digital.library.mcgill.ca/ms-17 (accessed 26th December 2019).

signal the compiler's awareness of the thematic similarities between Helpericus's and Bede's works.[75]

This note links cosmology to *computus* through a reminder of the disparity between the lunar and solar calendars. It is again a reminder of one of the fundamental rationales for astronomical enquiry, located in the disparity in the lunar and solar calendars. This context is similar to the Icelandic hemispherical world maps, which similarly accompany a note on the error in the Julian calendar and in the case of 732b other *computus* texts.

The Icelandic Zonal Map

On examination of its companion texts and diagrams, the Icelandic zonal map emerges as one stage in a complex visual exposition on the structure of the physical universe and its operations. The zonal map has much in common with the two planetary diagrams that accompany it, all showing the Earth in its cosmological contexts. The difference between them is one of scale and focus; while the Earth features as only a point in the two planetary diagrams, it is enlarged and shown in greater detail on the zonal map. While the two planetary diagrams concern circumsolar Mercury and Venus (f. 10v) and the superior planets (f. 11r), the zonal map concerns the Earth and moon. Further, the Icelandic zonal map and the planetary diagrams are linked by the material circumstances of their production: the zonal map shares a compass hole with the diagram of superior planetary motions overleaf. These diagrams were drawn by their scribe at the same time, and with the same apparatus, so it is appropriate that their connection is acknowledged and interpreted. When the zonal map is restored to its manuscript context, its role as a continuation of this scheme of planetary diagrams is obvious.

The zonal map also complements the encyclopaedia's computistic interests, expressed most fully in its initial folios. The map features after an extract from Sacrobosco's *On the reckoning of years* (*De anni ratione*), and immediately below an extract from Helpericus's *On computus* (*De computo*), which, like the note that accompanies the Icelandic hemispherical world maps, concerns the establishment of the Julian calendar. The abbreviated lunar cycle that surrounds the map is incomplete and poorly implemented, which might again indicate the handbook's rudimentary nature. However, world maps and *computus* texts both concern the immediate structure of the cosmos, and the implications of the sun and moon's motions for both the Earth's climate and the calendar.

75 Sharpe has noted that the 'very instability' of notions of medieval authorship 'can be revealing about the text's reception'. Richard Sharpe, *Titulus: Identifying Medieval Latin Texts: An Evidence-Based Approach* (Turnhout: Brepols, 2003), p. 8.

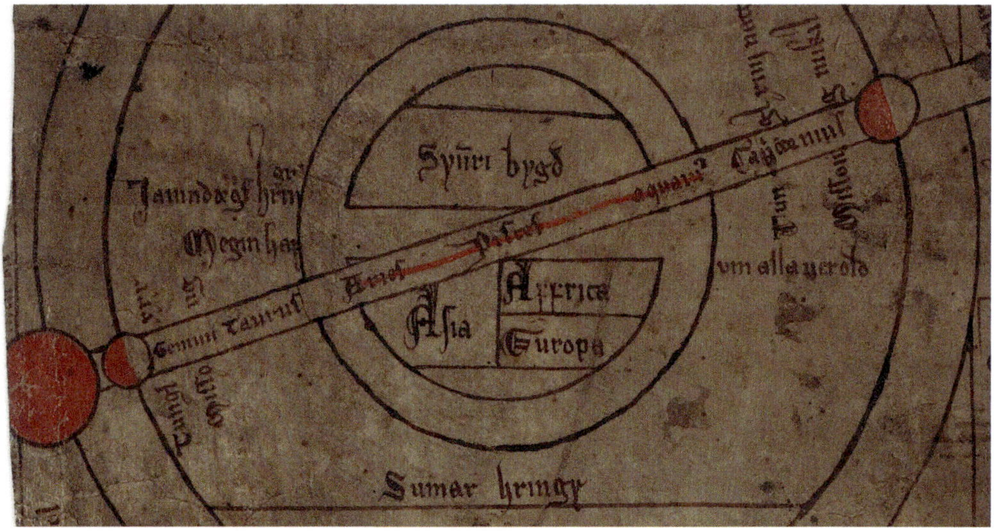

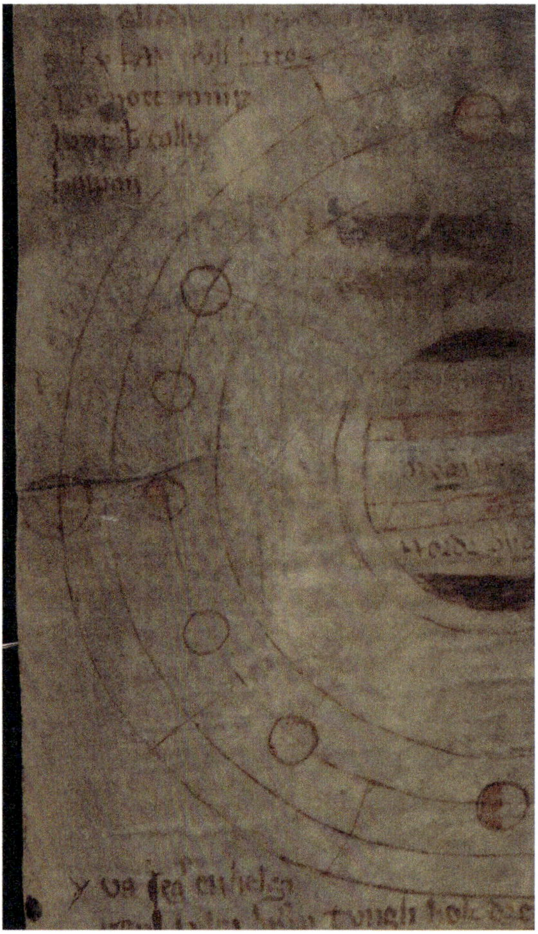

FIGURE 13: Details from the Icelandic zonal and hemispherical world maps show the sun and the part of the moon that faces it coloured red. The new moon is shown in both examples on the map's left, where the part of the moon that faces the Earth is not shaded. The full moon is faded and no longer visible on the zonal map, which, however, adds representations of the moon at approximately five and eleven nights (top and bottom). These are the times at which the *meðaldagar* ('neap tides') occur. There are no inscriptions on the zonal map to identify these features. Reykjavík, Stofnun Árna Magnússonar, GkS 1812 I 4to, f.11v, 1315–*c.*1400 and Copenhagen, Arnamagnæan Collection, AM 736 I 4to, fol. 1v, *c.* 1300. Photograph Suzanne Reit. Reproduced with permission from the Arnamagnæan Institute.

Previous facsimiles have reproduced only the central zonal map, removing from view the surrounding lunar scheme. However incomplete or abbreviated, this lunar scheme connects the map to a wider visual programme: that of the structure of the cosmos and the movement of its parts. The diagram clearly shows the movement of the moon around the Earth and in relation to the sun (figure 13). The sun (coloured red) is shown on the outermost circle on the left-hand-side of the diagram. This can be identified as the sun, despite its lack of an inscription, because the part of the moon at the bottom of the diagram that faces it is coloured red. This represents the half-moon at around five nights into the lunation. The lunar diagram might have been left incomplete because the scribe lacked the ability to complete it. Alternatively, the diagram might have been purposefully left incomplete when the structure of the encyclopaedia became apparent to the scribe as they produced the last diagram in this section of the manuscript, a diagram dedicated to the representation of the lunar month on folio 12v. It might be suggested that the scheme around the Icelandic zonal map was left incomplete to reduce overlap between the encyclopaedia's numerous diagrams; in their illustration of the lunar phases, the zonal map and diagram of the eclipses (described below) are equivalent, and would overlap considerably if the zonal map were complete.

The Philosophy of the World

The map's function emerges further through examination of the text written beneath it. This is an Old Norse text about the world's winds and oceans, attributed in its first line to the Venerable Bede. Kålund supported this attribution and said that it derived from *De natura rerum*, Ch. 39.[76] While references to Bede and his works abound in Icelandic encyclopaedics, this particular text has been identified by Simek as an Old Norse paraphrase of the French scholastic philosopher William de Conches's *De philosophia mundi* (III.xiv–xv, 'On the Ocean Currents' ('De refluxionibus Oceani') and 'On the Origin of the Winds' ('De ortu ventorum').[77] This text appears to have been influential in Iceland and has parallels in other Icelandic encyclopaedic manuscripts.[78] This text extends from the bottom of folio 11v to the top of folio 12v, where it meets a diagram of the eclipses.

The text describes the Cratesian quadripartition of the world by intersection of the equatorial and meridional oceans, described in Chapter 1.

76 *AÍ* II, p. 85.

77 This misattribution has been well documented: William's treatise is edited under the name of Bede in *Patrologia Latina* vol. 90 and under the name of Honorius Augustodunensis in vol. 172. Simek, *Altnordische Kosmographie*, pp. 63–69.

78 *AÍ* II, pp. 85–86.

Megin hafit sialft ligur um midia iordina, sem lindi ur austri i vestur ok svo aptr til austurs, enn retter arma sina ur austri ok vestri, adra til sudurs, en adra til nordurs, ok mætaz þeir adrir fyri nordan, en adrer fyri sunan iordina, ok skera þeir svo iordina i sundr i fiordunga.

(The ocean itself lies around the middle of the world and encircles it from east to west and back to the east, and reaches its arms out eastwards and westwards, and another to the south, and another to the north, and they meet each other in the north and in the south of the world, and thus divide the world into its four parts.)[79]

These two arms of the ocean divide the world horizontally ('retter arma sina ur austri ok vestri') into a northern and southern hemisphere, and vertically ('adra til sudurs, en adra til nordurs') into an eastern and western hemisphere. This is shown pictorially on the Icelandic hemispherical and zonal maps. On the zonal map, the central inscription *megin haf* identifies the equatorial ocean, while the circle that encloses the map is the meridional ocean. The text continues to describe further the lay of these oceanic branches, with supplementary information on the effects of the conjunctions of the sun and moon on the tides. The Old Norse text appears to be losely based on the parallel chapter in *De philosophia mundi*, a free and independent translation that does not adhere strictly to the original.

Two maps are associated with William's *De philosophia mundi*. Of the 120 or so extant manuscripts of this encyclopaedia, sixty-seven contain at least one map.[80] The map, after Macrobius, accompanies the chapters on the ocean currents and the winds and shows a spherical Earth with a small geographical nomenclature in the northern habitable region. This map bears a strong resemblance to the Icelandic zonal map: the oceans feature prominently on both, and both show the world divided by the equatorial ocean into northern and southern hemispheres. We see more of the map's origins in the text of William's *De philosophia*.

Qualiter vero ascendat et descendat si quis scire desiderat, et quae nomina, ex quibus regionibus sumpta mappam mundi consulat. Sed quia facilius illabitur animo subjecta descriptio, id quod diximus oculis subjiciamus: in posita sphaerula occidentalis refluxio ad septentrionem vergens, ex Atlante monte adjacente, Atlanticum mare vocatur, infra quam est Anglia, et vincinae insulae. Ex orientali refluxione ad septentrionem vergente nascitur Indicum mare. Similiter ex aliis refluxionibus ad austrum se vergentibus credendum est, maria nasci diversa. Sed hoc nostra attestione describi non debet, quia propter torridam zonam interjectam situs illarum nobis incognitus perseverat.

(If anyone desires to know in what manner [the Mediterranean Sea] ascends and descends, and which names it takes from such regions,

79 *AÍ* II, p. 85.

80 Destombes, *Mappemondes*, p. 97.

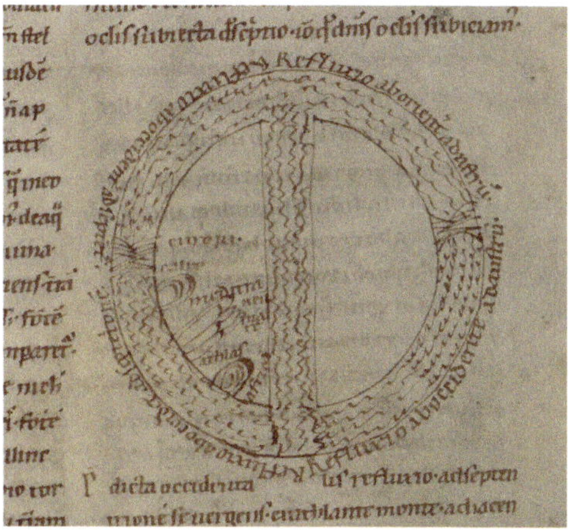

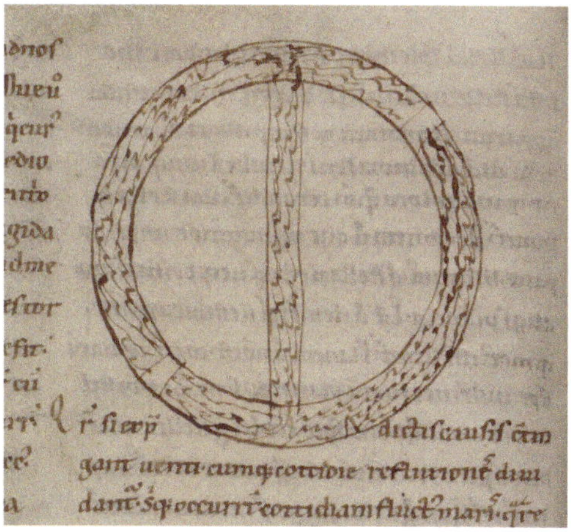

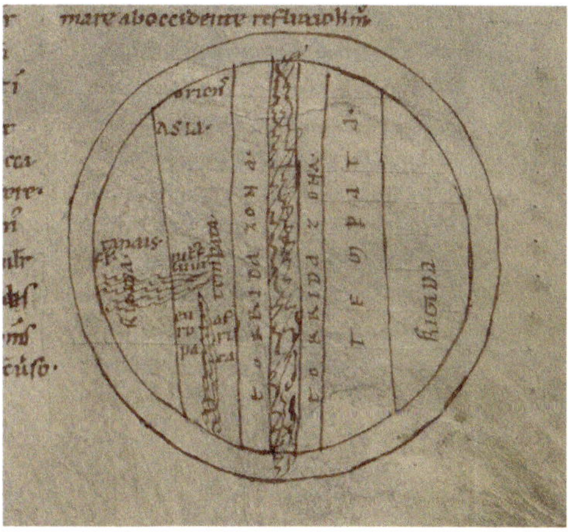

FIGURE 14: Three maps in a manuscript of William of Conches's *De philosophia mundi*. The map on the left accompanies III.xiv on folio 13r. The map bottom left appears at the end of III. xx overleaf on folio 13v. The zonal map with geographical nomenclature below appears on folio 15r. Philadelphia, University of Penn, LJS 384, *c.* 1150 (Lawrence J Schoenberg Collection of Manuscripts, Kislak Cetner for Special Collections, Rare Books and Manuscripts, University of Pennsylvania).

> consult a *mappa mundi*. And because the accompanying diagram sinks into the mind more easily, let us place under the eye what we have said: into the given sphere the western current inclining towards the north, adjacent to the Atlas Mountains, is called the Atlantic Ocean, in which is England and its neighbouring islands. The Indian Ocean is born from the eastern current that inclines towards the north. Similarly, we must believe that from the other currents inclining south that other seas are born. But this ought not to be described in our testimony, because their locations remain unknown on account of the intermediate torrid zone.)[81]

William, like Macrobius, acknowledges that a diagram can overcome the difficulties inherent to written geographical descriptions and explicitly enjoins the inspection of a *mappa mundi*, the purpose of which was to illustrate the lay and disposition of the world's oceans.

81 William of Conches, *De philosophia* III.xiv, *PL* 172, col. 80.

The Icelandic zonal map emanates from this tradition, and although its companion Old Norse text does not include the Latin original's call for a world map, the map nevertheless illustrates the quadripartition of the globe and the relative positions of the oceans. The map therefore sustains relationships with both its companion texts on folio 11v, depicting the lunar month, described in the extract from Helpericus's *De computo*, and showing the quadripartition of the Earth, described in this paraphrase from William's *De philosophia mundi*.

The Eclipses

The Old Norse paraphrase of William's *De philosophia mundi* (III.xiv–xv) spans folios 11v–12r, and connects two circular diagrams: the Icelandic zonal map on folio 11v and a diagram of the lunar eclipses on folio 12v (figure 15). At the top of folio 12v begins an Old Norse text about the monthly variation in the tides, with reference to the spring tides (*missong*) and neap tides (*meðaldagar*).[82] The text is similar to the tidal note that accompanies the Icelandic hemispherical world maps. The most striking similarity between the tidal note and this text is the directive on which it ends: 'þetta ma giorr skilia i þessi figurv' ('you can understand this in this diagram'), which formalises its connection with the diagram below.

Unlike the others in 1812 I, this diagram has been reproduced in facsimile and its legends transcribed.[83] A number of verbal echoes exist between the diagram's Old Norse inscriptions and the accompanying text. The diagram shows the configurations of the sun and moon in relation to the Earth described in the accompanying text as the causes for the spring and neap tides. The outer circle is inscribed with the names of the twelve signs of the zodiac and thus represents the ecliptic, the path on which the sun moves around the Earth. Beneath theses are written the names of the signs' ruling planets, the planetary domiciles; in astrology, the planet was said to exert a stronger influence when it appeared in these signs. Into this circle has been drawn a large sphere representing the sun, which is drawn at the bottom of the diagram. The middle circle represents the course of the moon around the Earth. Into this circle are drawn four representations of the moon at different points in the lunar month: at the top, the sun and moon are in opposition (they are opposite one another in the sky, and the moon is full); at the bottom, the sun and moon are in conjunction (they are in the same part of the sky, and the moon is new); on the left the half-moon is waxing; on the right the half-moon is waning. The dark cone shows the shadow cast upon the Earth by the moon, and upon the moon by the Earth, during the eclipses, when the sun and moon pass the ecliptic at the same time. Inscriptions on the diagram indicate

82 *AÍ* II, p. 89.
83 *AÍ* II, p. 89.

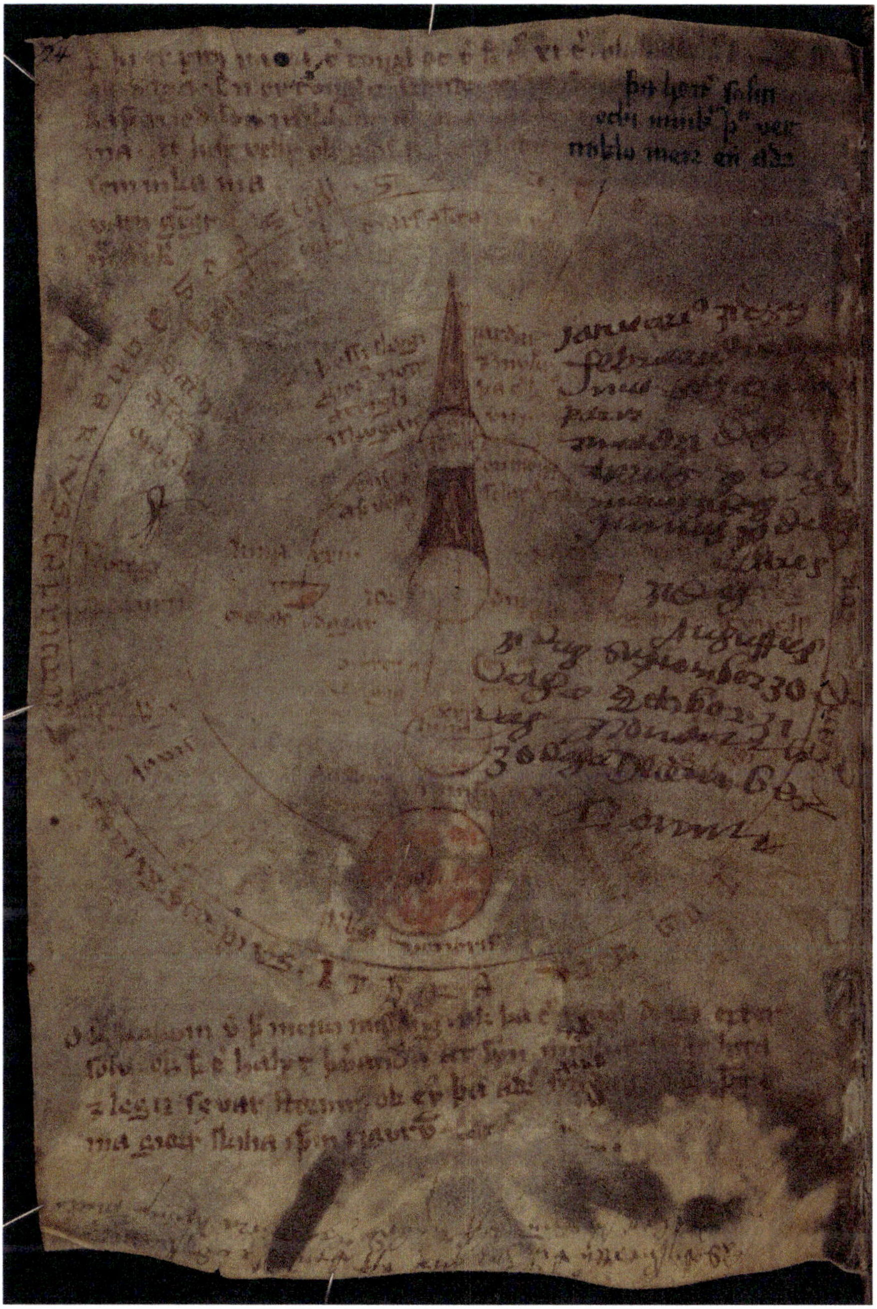

FIGURE 15: Folio 12v contains the conclusion of the Old Norse translation of William's *De philosophia mundi,* and a diagram of the eclipses and tides. Reykjavík, Stofnun Árna Magnússonar, GkS 1812 I 4to, f.12v, 1315–*c.*1400.

how old the moon is when it is in these positions, as well as the occurrence of the neap tides at nine and twenty-two nights, and the springs and high springs at thirty and zero nights.

The diagram and the zonal map both illustrate tidal processes, but it is the hemispherical world map that shares in the greatest degree of functional similarity with the eclipse diagram. Both are employed in order to explain the generation of the tides and the influences exerted on them by the moon and sun. A number of inscriptions are common to both diagrams, they depict the same celestial arrangements in different ways, and illustrate functionally equivalent texts.

The presence of the eclipse diagram might retrospectively explain why the zonal map's lunar phase diagram has been abbreviated or left incomplete: this function has been transferred to the latter diagram. The compiler of the manuscript has not duplicated entirely two overlapping lunar schemes. The functional complementarity of the eclipse diagram and the zonal map presents a strong argument against the examination of these different diagrams in generic isolation. The zonal map must be seen in the wider context of 1812 I as a visual exposition on the structure of the cosmos. Furthermore, similairites between these diagrams and the Icelandic hemispherical world maps urge that consideration be given to diagrams of different types across different manuscripts, and that careful attention be directed towards the circumstances of their use.

The eclipses and tides
(Reykjavík, Stofnun Árna Magnússonar, GkS 1812 I 4to, f.12v) 1315–c.1400

ARIES	Aries
Mars a þetta [mar]k	Mars in this sign
[TAV]RO	Taurus
[Venus a þetta mark][84]	[Venus in this sign]
[GEMINE]	[Gemini]
[Mercurivs a þetta mark]	[Mercury in this sign]
CANCER	Cancer
[herbergi] tungls	[domicile of the moon]
LEO	Leo
hergergi solar	domicile of the sun

84 Some of the diagram's texts are now faded or overwritten, but were apparently visible to, or were reconstructed by, Kålund. See *AÍ* II, pp. 89–90, fn. 2. These readings are visible in AM 252 fol., copied from GkS 1812 4to by the Icelandic scribe Ásgeir Jónsson in *c.* 1686–1707 (f. 60v). The notes written on the right side of the diagram are, according to Kålund, in a hand dated *c.* 1600.

VIRGO	Virgo
herbergi mercurius	domicile of Mercury
LIBRA	Libra
h*er*bergi veneris	domicile of Venus
SCORPIVS	Scorpio
[h*er*bergi Mars]	[domicile of Mars]
SAGITARIVS	Sagitarius
herb*er*gi Iovis	domicile of Jupiter
CAPRICORNVS	Capricorn
h*er*bergi [satur]n*us*	domicile of Saturn
AQUARIVS	Aquarius
Sat*ur*nus a þetta mark	Saturn in this sign
PISCES	Pisces
Iovis a þetta marc	Jupiter in this sign
þessi skugi iarðar	This shadow of the Earth
gior*ir* nott *ok* myrkr[85]	makes night and darkness
a tvngli þa er þat verdr	on the moon when it passes
i skvgan vm	into shadow
misgong en meiri	High springs
af vellu solar hita[86]	from the boiling heat of the sun
Myrkva solu af	Darkening of the sun
skuta tungls[87]	from the passage of the moon
Misgong en minni	Low springs
[af voco tungls][88]	From the influence of the moon

85 Kålund has 'mylir' for 'myrkr', which does, in 1812, appear to be the correct reading (a confusion with *mylin*, 'moon', 'luminary'?). The copy in AM 252 fol., however, has 'myrkr'.

86 Compare the line from the tidal note accompanying the hemispherical world map: 'Enn tunglit stendr gegnt sol e*r*u missaung af vellu solar hita. þviat solin er þa iþeim lut zodiaco er iafnt stendr yfir megin hafino' ('when the moon stands opposite the sun, spring-tides result from the boiling heat of the sun, because the sun is then in the part of the zodiac which stands over the ocean'). *AÍ* II, pp. 117–18.

87 Kålund omits this legend.

88 Again this legend can be elaborated upon with the tidal note accompanying the hemispherical world map: '7 þa er ny verðr i þessum maurkum. skytur tunglit meire sinn er vauku ahafit en vant er. þviat þa stendr tungl gengr

luna xxii	moon [at] twenty-two [nights]
medals dag*ar*[89]	neap tides
xiii	moon [at] thirteen [nights]
luna	
luna vi natta	moon [at] six nights
medals dag*ar*	neap tides
xxx	moon [at] thirty [nights]
luna	
ior din	the Earth
sol	sun

Conclusion

This chapter has demonstrated that the Icelandic zonal map generates a productive suite of relationships with its companion items. The map belongs to a sequence of texts and diagrams that thematise planetary kinematics, providing a background to the origins of calendrical computation covered in the preceding folios. The zonal map is one step in a staged exposition of the structure of the physical universe and its clocked processes, following on from two diagrams showing the Earth amid the eccentric and epicyclic orbits of the inferior (f. 10v) and superior (f. 11r) planets. The zonal map and diagram of the eclipses narrow the exposition's focus, and show how the revolutionary motions of these planetary bodies necessitated the medieval science of *computus*. Relatedly, the zonal map shows the quadripartite division of the globe by the intersection of the equatorial and meridional oceans, and its partially realised lunar scheme implies the effects of the moon upon them.

This section of GkS 1812 4[to] can be characterised as an elementary introduction to the structure of the physical universe accompanied by practical verses that aid memorisation of basic calendrical principles. Despite the technical nature of its sources, it is not an advanced treatise on cosmology or mathematical astronomy, and its texts and diagrams

yfir hafino' ('and when it becomes new in these signs, the moon exerts its influence more than usual, because then the moon's course stands above the ocean'). *AÍ* II, p. 118.

89 Icelandic tidal treatises describe the cycle of *missong* ('spring tides') and *meðaldagar* ('neap tides') caused by the motions of the sun and moon. The neap tides are those that occur at around five and eleven nights into a lunation, at the midpoints between the full and new moons. At these times the tidal forces are weakest and produce the least tidal range. See discussion of the tides in relation to the hemispherical world map in Chapter 1.

do not provide the apparatus needed to establish or verify the cosmological principles it relates. Rather, it appears to have been intended for a non-specialist readership, for whom it provides access to literature translated into Latin from Arabic in the twelfth and thirteenth centuries. Its diagrams derive from popular astronomical handbooks in wide circulation. It is not clear whether or not their assembly here preserves the structure and contents of its exemplar. A compiler may have adapted their sources in an independent way, perhaps following the general structure of an original compilation but not every detail. The introductory text preserved across folios 10v and 11r, for instance, appears twice in this fragment, its duplication implying the combination and recombination of texts to produce different local effects through their interactions with their companion texts and diagrams.

It is possible to understand these diagrams in terms more complete than those available to those who copied them. The planetary diagrams in particular are not accompanied by the prose explanations, in works by John of Sacrobosco or William of Conches, that would have elucidated their visual arguments. It is not obvious how much information was available to the copyist who produced them, or how far readers were familiar with the traditions from which they draw. The translated legends describing forward (*greiðir göngu sína*) and retrograde (*vendir sik aptr*) motion provide the only evidence we have for Icelanders' acquaintance with the theory of the eccentric epicycle, and these translation moments provide valuable insights into astronomical learning in medieval Iceland.

The Icelandic zonal map grants visual expression to cosmological concepts that had been theorised in antiquity – the latitudinal distinction of the climatic zones, and the oceanic globe. In Iceland, however, these global concepts were not only theoretical, but related to the Icelanders' own experience of the geography of their regions. Icelanders possessed their own native awareness of the northern polar regions, and may have examined speculative literature about the world's northern regions to locate the world of their experience within the world described by Latin natural philosophy. The map shows how the northern regions, extending almost to the polar circle, are a stable part of the contemplated world.

The zonal map is oriented with the northern hemisphere at the bottom, perhaps so that its Icelandic viewer was positioned bodily to look from their own northern regions southwards towards the wider world. The map calls the space inhabited by medieval Europeans the *norðr bygilig hálfa* ('northern inhabitable region'), describing a temperate region between the cold pole and hot equator on the Earth's northern hemisphere. This term was sometimes, as we saw in the note on the ocean's tides that accompanies the Icelandic hemispherical world maps, interchangeable with *vorri byggð* ('our inhabited region'). 'Our', here, means 'everyone's'; this tricontinental space encompassed all human civilization and history. Icelanders, however, sometimes referred to their own northern regions, the Scandinavian world and the islands of the North Atlantic, as the *norður hálfur* ('northern regions').[90] The Icelandic zonal map's *norðr bygilig hálfa*, then, may have conveyed to Icelanders a double meaning, describing at once the global concept of the northern hemisphere – the space Icelanders shared with the rest of humanity – and their own northern regions more narrowly.

The Icelandic zonal and hemispherical world maps are representations of global space, and represent cartographic varieties that are studied less than those maps that show the dispersion of places and peoples within the known world. These maps, however, were no less productive than their counterparts in informing the Icelanders' sense of the physical world and its geography. In the three chapters that follow, our attention turns to two maps of a very different type. These maps are principally maps of the known world, a space whose representation on the zonal and hemispherical world maps is schematic. These maps take for granted the Earth's sphericity, its climatic zones and oceans, but are related forms of expression which described the world, and located the Icelanders' experience within it.

90 For example *Nítíða saga*, pp. 140–41.

CHAPTER 3

The Two Maps from Viðey

The manuscript in Reykjavík's Stofnun Árna Magnússonar with the shelf mark GkS 1812 4to, sometimes called the Viðey book, contains between its covers three of the world maps that come down to us from medieval Iceland. This manuscript is not made up of regular gatherings but is a compilation of fragments written between *c.* 1200 and the fourteenth century. The zonal map on folio 11v, in the section of the manuscript dated 1315–*c.* 1400, belongs to a complement of cosmological diagrams that describe the Earth and its cosmic position. In a section of the manuscript dated *c.* 1225–50 are a further two world maps, drawn on two leaves that have been inserted into the abundantly illustrated section of the book that accommodates the zonal map. The larger map spans an entire manuscript opening on folios 5v–6r, and shows the geographic dispersion of places and people within the *kringla heimsins*. Its companion on folio 6v is a small schematic T-O map, showing the division of the world into its continental thirds – Africa, Asia and Europe. These two maps – one complex and variegated; the other simple and iconic – are complementary visions of world geography encompassing all miniaturised Creation.

These two maps are unlike the hemispherical and zonal maps in that, taking for granted the Earth's sphericity, they show only the known world. This space had been known, since Herodotus's *Histories*, as the *ecumene* (or *oikoumene*) and compassed the civilised world known to the Greeks and Romans, in contradistinction to those regions at its fringes that were unknown or inhabited by barbarians.[1] This space, which contained all human civilization and history, was called in Latin the *orbis terrarum*, and in Old Norse the *kringla heimsins*.

1 Peter Fibiger Bang and Dariusz Kołodziejczyk, '"Elephant of India': Universal Empire through Time and across Cultures', in *Universal Empire: A Comparative Approach to Imperial Culture and Representation in Eurasian History*, ed. Peter Fibiger Bang and Dariusz Kołodziejczyk (Cambridge: Cambridge University Press, 2012), pp. 1–40, pp. 30–33.

The larger map is one of few maps found in medieval European books to span a full manuscript opening.[2] The manuscript seam divides the map's representations of Asia in the east (f. 5v) and Europe and Northern Africa in the west (f. 6r). This seam has been inscribed, on folio 5v, with the names of the two rivers, 'Tanakvisl fluvi*us* maxim*us*' ('Tanakvísl, the greatest river') and 'Nilus flum*en* egipti' ('Nile, the river of Egypt'), which separate Europe from Asia, and Asia from Africa. The 'Mediterraneum mare' ('Mediterranean Sea', at the middle, *medius*, of lands, *terrae*) completes the map's T-O framework, marking the boundary between Europe and Africa. The map contains 104 legends that show the dispersion of places, peoples, rivers and mountains, in addition to thirty-two names designating cosmological concepts. The smaller map overleaf presents a minimalist depiction of terrestrial space, a simple T-O map that bears only the names of the three continents.

These two maps, drawn *c.* 1225–50, are the earliest witnesses to the cartographic culture of medieval Iceland. They are broadly contemporary with the two English Psalter maps (London, British Library, Add. 28681, ff. 9r and 9v, *c.* 1250), which are similarly drawn on the recto and verso of the same manuscript folio.[3] The Viðey maps are timely with the appearance of the large monumental wall-mounted maps from Ebstorf (*c.* 1250) and Hereford (*c.* 1300), and the considerable cartographic output of the Benedictine monk and chronicler Matthew Paris.[4] Icelanders were not latecomers to cartographic production, but fared with their English and Continental contemporaries in thinking about the wider world and their place within it.

Both maps derive their geometrical basis from the natural phenomena that are grouped in their conceptual frames, each map ensconced within

2 Simek, 'Cosmology', p. 110. The most common double-page maps in medieval books are those associated with the second book of Beatus of Lièbana's *Commentary on the Apocalypse* (776–86), global visions of the known world and subequatorial continent that illustrate the apostles' mission to bring the Christian religion to the whole world. Chekin, *Northern Eurasia*, p. 171. Double-page maps are also associated with Ranulf Higden's *Polychronicon* from the second half of the fourteenth century. Destombes, *Mappemondes*, pp. 149–53.

3 Bettina Schöller, *Wissen speichern, Wissen ordnen, Wissen übertragen: schriftliche und bildliche Aufzeichnungen der Welt im Umfeld der Londoner Psalterkarte* (Zürich: Chronos, 2015). These paired maps, now on folios 9r and 9v, were once placed at the opening of the psalter. See also Chet van Duzer, 'The Psalter Map (c. 1262)', in *A Critical Companion to English Mappae Mundi of the Twelfth and Thirteenth Centuries*, ed. Dan Terkla and Nick Millea (Woodbridge: Boydell, 2019), pp. 179–96, p. 179. Van Duzer asserts that the pairing of maps on the recto and verso of the same manuscript folio is 'highly unusual', but the Viðey maps add to the number of attested examples.

4 Naomi Reed Kline, *Maps of Medieval Thought: The Hereford Paradigm* (Woodbridge: Boydell, 2001), p. 81; Birkholz, *Two Maps*, p. iii.

a circle that correlates the four cardinal directions – north, south, east and west – with other fours appearing in nature. Alongside the names of the twelve winds that blow inwards from their points around the map's horizon, the map-maker groups the four cardinal directions, the four seasons, the four ages of man (*aetates hominum*), and the components that make up the human microcosm. The smaller T-O map is situated within a diagram that shows, in addition to the natural fours disposed around the larger map, the twelve months of the year and the twelve signs of the zodiac.

There is considerable overlap in what these two maps, on the recto and verso of the same manuscript folios, show. Both maps show the division of the known world into the three continents – Africa, Asia and Europe – but this structure can be more easily discerned on the smaller map, which depicts the three waters dividing the continents – the Rivers Tanais and Nile and the Mediterranean Sea – as double lines. Both also show the order of nature through aligning the cardinal directions with their associated natural fours. This geometrical basis is depicted on the smaller map, the conceptual frame divided into fours by radials extending outward from the central T-O map. Its simple and minimalist depiction of geographic space may have done an exegetical service to the larger map, enabling the viewer to discern more easily its T-O framework and the quadripartite structure of its conceptual frame.

No proper names have developed to distinguish between these maps from others in the Icelandic corpus. The hemispherical and zonal maps are examples of comparatively distinct cartographic genres that appear with some frequency in works associated with Macrobius and William of Conches. The maps in 1812 III are not *sui generis*, but do not have such precise parallels or associations with known authors. The larger map has been known by as many names as there are studies of it, appearing in its slender literature as the 'planisphère', 'veraldar kringlan', the 'mappa-mundi', the 'Icelandic Mappa Mundi', the 'Große Mappa mundi', and the 'Icelandic Map'.[5] These names do not helpfully distinguish this map from other maps from Iceland, or convey its relationship with its companion map on the verso overleaf. Because this section of the manuscript appears to be associated with the Augustinian monastery at Viðey, these paired

5 Rafn, *Antiquités Russes*, pp. 392–94; *DI* I, p. 182; *AÍ* III, p. 72; Pritsak, *The Origin of Rus'*, vol. 1, pp. 514–16; Simek, *Altnordische Kosmographie*, pp. 419–23; Rudolf Simek, 'Skandinavische Mappae Mundi in der europäischen Tradition', in *Ein Weltbild vor Columbus: Die Ebstorfer Weltkarte*, ed. Hartmut Kugler *et al.* (Weinheim: VCH, 1991), pp. 167–84, p. 178; Chekin, *Northern Eurasia*, p. 70; and subsequently Alfred Hiatt, '*Beowulf* off the map', *Anglo-Saxon England* 38 (2009), 11–40, 22.

maps are refered to throughout this book as the larger (ff. 5v–6r) and smaller (f. 6v) Viðey Maps.[6]

Earlier work on these maps focussed on their representation of regions assumed to be familiar to their Icelandic maker. The maps make their earliest appearance in the second volume of Rafn's *Antiquités Russes*, for the larger map's inclusion of Russia, Kiev, and, probably, the nomads of the Steppe. Rafn thought this map was contemporary with the oldest part of this manuscript (GkS 1812 IV 4to) and dated it *c.* 1150.[7] Kålund revised this earlier dating and assigned the map to the section GkS 1812 III 4to, which he dates on palaeographic grounds to *c.* 1225–50.[8] This date has been accepted by all modern commentators except for Omeljan Pritsak, who, in his *The Origin of Rus'*, simply calls it 'The Icelandic Mappa Mundi of about 1150 or 1250'.[9] Destombes dates the map more broadly than has been customary in the Scandinavian literature to the thirteenth century.[10] Haraldur Sigurðsson's *Kortasaga Íslands* reproduces just half of the map in facsimile, limiting our perspective to the depiction of Iceland and Europe on folio 6r.[11] Chekin's is the most comprehensive commentary on the map, but covers only the legends pertaining to Northern Eurasia.[12]

Despite the long interest in its larger companion, little has been written on the smaller Viðey Map. Facsimile reproductions are partial, presenting only the central medallion that bears the names of the three continents. The map is usually presented as evidence of thirteenth-century Icelanders' familiarity with the ubiquitous understanding that the world was divided into three continents. Rafn provides a facsimile of the T-O at its centre but omits the greater part of the conceptual wheel that encloses it, erroneously describing this apparatus as a wind rose. The conceptual frame is mischaracterised and omitted from the facsimile, but all its legends are transcribed.[13] Simek again references this map alongside its

6 The practice of naming maps for the centres that produced or held them has precedent: for example, the map in Oxford, St John's College, MS 17, on f. 6r produced at Thorney Abbey is commonly known as the Thorney Map.

7 Rafn, *Antiquités Russes*, p. 392.

8 *AÍ* II, pp. ccxi–ccxii.

9 Pritsak, *Origins of Rus'*, vol. 1, pp. 514–15.

10 Destombes, *Mappemondes*, p. 175. The attendant description is perfunctory. Destombes states that the map features a nomenclature of seventy names (there are 104) presented in columns (true only for the map's African legends). He states that the map features amid a perimeter of winds, but makes no mention of the other fours assembled around its perimeter.

11 Haraldur Sigurðsson, *Kortasaga Íslands*, p. 54. The map has been reoriented to place north at the top, so that the inscriptions Thule and Iceland can be read the right way up.

12 Chekin, *Northern Eurasia*, pp. 69–71.

13 Rafn, *Antiquités Russes*, p. 391.

copy in AM 252 fol., but again excerpts the T-O from its conceptual frame and transcribes only the names of the three continents.[14]

Interest in these maps can be seen as early as the late seventeenth century, when an assortment of items from the Viðey book were copied by the Icelandic scribe Ásgeir Jónsson in *c.* 1686–1707.[15] Ásgeir did not copy the book in its entirety, or preserve the structure of his exemplar as we know it, but reorganises its material into a new logical structure. This new composition (Reykjavík, Stofnun Árna Magnússonar, AM 252 fol.) excerpts the book's visually interesting contents and recombines them into a new complement of circular diagrams. The two Viðey Maps bookend this series, the smaller map copied on folio 58r and the larger map on folio 62r. Between them, Ásgeir has arranged the diagram showing the motions of the inner planets (f. 58v), the motions of the superior planets (f. 59r), the zonal map (f. 59v), and the diagram of the eclipses (f. 60v). To these has been added on folio 61r a geometrical diagram drawn from the Old Norse *Algorismus* (copied from GkS 1812 II, 4[to], fourteenth century, Norwegian?, f. 13v). This diagram illustrated an arithmetical treatise comprising an Icelandic prose translation of the Latin hexametrical *Carmen de Algorismo*, written by the French priest Alexander de Villa Dei (1170–1250), and elements from Sacrobosco's *Algorismus Vulgaris*.[16] The larger Viðey Map is not drawn on a manuscript opening but on a fold-out sheet of paper. When Ásgeir Jónsson produced his copy of the Viðey book in the seventeenth century, therefore, he reordered its maps and diagrams into a new logical structure, pulling the paired Viðey Maps apart so that they opened and closed a graphic study of the world and its cosmic position. This process of selection and rearrangement may well resemble the circumstances in which the Icelandic encyclopaedias themselves were compiled in the thirteenth and fourteenth centuries, their compositors free to reorganise their material into new logical structures with each new copying. The early modern copy begins with a glossary –

14 The same map is reproduced in facsimile earlier in the same volume with different foliation. Simek, *Altnordische Kosmographie*, p. 99. The folio number provided in the caption to the earlier facsimile is the correct one. Simek, *Altnordische Kosmographie*, pp. 508–09.

15 Simek dates this manuscript to c. 1500. Simek, *Altnordische Kosmographie*, p. 424. There are monochrome facsimiles of the 1812 III map and its apograph in AM 252.

16 The Icelandic treatise describes the ten Arabic numerals (including the *cipher*, or zero, introduced to European thought by the *Carmen*), the place-value number system, odd and even numbers, the seven arts of computing (addition, subtraction, doubling, halving, multiplication, division, and the extraction of square and cubic roots), and the elements (fire, air, water, and earth) and the numbers with which they are associated. The treatise is also preserved in Hauksbók (ff. 89v–93r). See Kristín Bjarnadóttir, 'Mathematical Education', p. 43.

written in Árni Magnússon's hand – of Old Norse astronomical terms with their Latin translations, headed with the title 'Ex computo Antiquo Regis', which may have been an early modern name for this composition, AM 252 fol., or its exemplum, GkS 1812 4to, which Bishop Brynjólfur Sveinsson had presented to the Danish King Frederick III in 1662.

The relationship between the two Viðey Maps has remained unseen; the maps have, from their earliest study, been examined as individual objects of attention.[17] The present chapter introduces the two maps together, and is a touchstone for the two that follow. The maps were drawn to be consulted in sequence, both with each other, and, from at least the sixteenth century, with the Icelandic zonal map bound into the same compilation.

Genre and Sources

The question of map genre is vexed by the relatively small number of cartographic survivals. Map-makers were free to elaborate upon their exemplars with each new copy, incorporating new information about the regions known to them or, as modern transcriptions of medieval maps attest, misplace, omit, or double their names. The larger map has been characterised as belonging to the 'non-schematic T-O' or 'non-schematic tripartite' type, a group comprising those that show the world divided

17 In Simek's *Altnordische Kosmographie,* more than ninety pages separate facsimile reproductions of these two maps. The maps' relation both to one another and to the items preserved with them is poorly understood. In the only article-length study on the Icelandic maps, the first and last maps described are the smaller and larger Viðey Maps respectively, without it being mentioned that these maps were drawn on the recto and verso of the same folio. Simek, 'Skandinavische Mappae Mundi', pp. 170, 178. On the contrary, Simek wrongly attributes the map in his discussion (though not in the caption that accompanies the map facsimile) to folio 11v. Folio 11v actually accommodates the zonal map. Similar foliation errors in Simek's *Altnordische Kosmographie* confused the map's most recent commentator, Leonid S. Chekin, who questioned whether the small T-O map described by Simek was indeed one map or two. Chekin, *Northern Eurasia,* pp. 69–70. Chekin did not examine the manuscript in which these maps are bound, and there are a number of infelicities in his description of 1812 III that can be traced to errors in earlier accounts. Kålund erroneously identified the second fragment of 1812 III as folios 34r–36v. *AÍ* II, p. ccx. However, this second fragment of 1812 III is a bifolium that spans folios 35r–36v and does not include folios 34r–34v. Chekin's reported 'astronomical fragments' actually belong to 1812 IV. Additionally, Chekin states that 'a list of Icelandic bishops' is preserved on folio 5r. However, this list of bishops is an addition to the folio dated to the 1480s; the folio originally contained only the list of forty highborn Icelandic priests.

into continental thirds and contain a geographic nomenclature.[18] The map does not derive its inscribed names from a single source, like Lucan's *Pharsalia* or Sallust's *Jugurthine War*.[19] Chekin categorises the map among others whose place-names are arranged in a geographically informed rather than abstract way.[20]

Medieval map-makers drew the preponderance of their information from ancient authoritative authors like Herodotus, Pliny the Elder, and the Roman historians whose works had been canonised for the High Middle Ages by scholars such as Orosius (*c.* 385–*c.* 418 CE) and Isidore of Seville (*c.* 560–636). The geographical descriptions of Orosius's *Histories Against the Pagans* (*Historiae Adversus Paganos*) (book one) and Isidore's *Etymologies* (book fourteen), which themselves borrowed from multiple and overlapping written traditions, provided a foundation of names in geographical sequence that medieval map-makers were sometimes able to enlarge with their own local or topical information. For the lands venerated by medieval Christianity, map-makers may have consulted the Bible and the various exegetical works available to them. The place-names appearing in Scripture were compiled in Eusebius of Caesarea's (*c.* 260–340 CE) *Onomasticon* (or *On the place-names in Holy Scripture*), translated into Latin by Jerome, which comprised an alphabetically-arranged gazetteer of scriptural place-names, with information on their geographical locations.

Maps may themselves have been used as the basis for new studies in world geography, and may have been copied, translated, turned towards new regional emphases, or rendered into written descriptions. Detailed descriptions of medieval maps, compiled for instructional or educational purposes, illuminate cartographic traditions that are poorly attested in the surviving record. The *Descriptio mappa mundi*, attributed to the Parisian theologian and schoolmaster Hugh of St Victor (*c.* 1096–1141), is

18 Woodward, 'Medieval *Mappaemundi*', p. 295. The term 'tripartite' can be misleading when used of these maps; there were many ways in which the world could be ordered and anatomised and, as the Viðey maps show, these were not mutually exclusive. In medieval geographical writings the world is more frequently described as quadripartite, the world divided between four points on the horizon: where the sun rises and sets on the summer solstice (NE and NW) and winter solstices (SE and SW). On the division of the world into its parts in the Middle Ages, see Suzanne Conklin Akbari, *Idols in the East: European Representations of Islam and the Orient, 1100–1450* (Ithaca: Cornell University Press, 2009), pp. 19–27.

19 Destombes, *Mappemondes*, p. 175; Chekin, *Northern Eurasia*, p. 9.

20 The Viðey Map bears little resemblance to the other eight maps in this group, especially in terms of the density of its inscriptions. It bears nearly twice as many legends as the second-most inscribed map in this group, the Thorney map, which contains around fifty-five legends. Chekin, *Northern Eurasia*, pp. 59, 63.

a written description of a medieval world map that provides an incomplete but scrupulously detailed account of the names inscribed onto its surface.[21] The *Expositio mappe mundi* (*c.* 1190), attributed by Patrick Gautier Dalché to the Yorkshire scholar Roger of Howden, and surviving in two manuscripts, is likewise a meticulous description of a map written in enough detail for a copy to be drawn by its example.[22] It lists place-names according to their locations relative, usually 'opposite', to each other, and in sequence on coastlines and rivers. It is likely that medieval cartographers consulted such works in compiling their maps.[23]

The medieval world map was not so much an image of the world as it was thought to be as an archive of geographic information drawn from multiple and overlapping textual traditions. The map-maker may have drawn from these selectively, and without a view to divulging the totality of their geographical awareness. A map's sources can thus seldom be singled out with certainty, but the broad traditions from which they emanate can sometimes be described. It is likely that the Viðey Maps are the product of a long process of combination and recombination of information from various sources, combined mostly before the map was drawn in Iceland. To these, the Icelandic map-maker certainly made additions in the regions known to them. The map contains, as we shall see in Chapter 4, a number of vernacular place-names in its depiction of the Scandinavian countries, demonstrating that its Icelandic maker was able to assemble place-names from multiple sources and organise them within a T-O framework.

The traditions from which the map's names emanate are valuable not only as 'sources' underlying the map's composition, but for understanding how the map's medieval viewers may have understood it. Access to these sources and their explanations of obscure names conditioned not only the

21 Hugh of St Victor, *La "Descriptio mappe mundi" de Hugues de Saint-Victor: texte inédit avec introduction et commentaire*, ed. Patrick Gautier Dalché (Paris: Études Augustiniennes, 1988). On Hugh's *Descriptio* and its relation to the so-called Munich Map see Nathalie Bouloux, 'The Munich Map (c. 1130): Description, Meanings and Uses', in *A Critical Companion to English Mappae Mundi of the Twelfth and Thirteenth Centuries*, ed. Dan Terkla and Nick Millea (Woodbridge: Boydell, 2019), pp. 92–111, pp. 99–102.

22 Roger de Howden, *Du Yorkshire à l'Inde: une géographie urbaine et maritime de la fin du XIIe siècle (Roger de Howden?)*, ed. Patrick Gautier Dalché (Geneva: Droz, 2005).

23 In his transcription of the Hereford Map's legends, Scott D. Westrem has shown that the *Expositio* was likely a source or analogue. On Roger of Howden and geographical knowledge in the north of England see Dan Terkla, 'Books and Maps: Anglo-Norman Durham and Geospatial Awareness', in *A Critical Companion to English Mappae Mundi of the Twelfth and Thirteenth Centuries*, ed. Dan Terkla and Nick Millea (Woodbridge: Boydell, 2019), pp. 68–91, pp. 87–91.

map's original composition, but also how far it may have been understood by its medieval readers. The less familiar of the map's legends may have been desemanticised exotica, empty displays of erudition that conveyed beguiling texture in the depiction of the world's distant places, but carried no associations with any specific place. Some of the map's legends appear in forms so corrupted (e.g. 'caspies', 'Cori*n*gagena') that it is only really possible to know what might have been intended through comparison with other maps and geographical treatises. In the texts and translations provided in this book, the transcriptions of names are the truest indication of what the map's readers saw. Comparison with other maps and written descriptions, however, illuminate how readers with access to the written descriptions on which it was ultimately based may have interpreted it.

Maps did not develop in isolation from other species of medieval diagram, and the Viðey Maps can be likened to other diagrammatic traditions through their conceptual frames. As Naomi Reed Kline has shown, maps relate to other diagrams that grant visual expression to patterns discerned in nature. Natural phenomena – from the twelve winds and four seasons, to understandings of the human microcosm – found diverse diagrammatic expression throughout the High Middle Ages.[24] As we shall see in Chapter 5, the Viðey Maps relate to other diagrams that thematised the order of nature and the relatedness of its component systems.

The *kringla heimsins*

The larger Viðey Map teems with legends that show the geographic dispersion of places and people. This map is the only large-scale visualisation of world geography that survives from medieval Iceland, its 104 geographic legends marking the locations of lands and regions, cities, rivers, mountains, and peoples. The names of the three continents are written outside the *kringla heimsins* in positions at the head of each continent on the map's conceptual frame. The three waters that separate them, 'Tanakvisl fluvi*us* maxim*us*' ('Tanakvísl, the greatest river'), 'Nilus flum*en* egipti' ('Nile, the river of Egypt'), and 'Mediterraneum mare' ('Mediterranean Sea') converge in the map's middle, and lend the map its T-O framework.

The map's depiction of Africa draws verbatim from Isidore's *Etymologies*, and is prevailingly limited to the North African regions – Libya, Mauritania, and Numidia – that had been subdued by the Roman Empire. Icelanders may have been familiar with these regions from the Roman historian Sallust's account of the Jugurthine war, which

24 Kline, *Medieval Thought*, p. 12.

had been known in Iceland, and incorporated into *Rómverja saga*, the Old Norse saga of the Romans.[25] The map's eleven African legends are lengthier than those elsewhere on the map, which tend to comprise a single word, and relate a miscellaneous assembly of remarkable facts about its lands and peoples, from 'Getulia ibi infants ludunt serpenti*bus*' ('Getulia, where children play with snakes'), to 'Trogita prouincia ibi uenitur carbunculus igneus et alter exacontalitus lx coloribus micans' ('Province of the Troglodytes, from that place come fiery carbuncles and also 'sixtystone', which sparkles in sixty colours'). These legends are arranged with no obvious sense of their geographical locations relative to one another, excepting perhaps Ethiopia and its surrounding deserts, which are said to border one another in Isidore's *Etymologies*.[26] Their arrangement into columns is a bookish format that visually echoes their origin in ancient writings.

The map's depiction of Asia comprises fifty-four geographical legends – more than Europe and Africa combined – and extends from India in the east to the River Tanais, which marks its conventional border with Europe, in the west. As with Africa, the map's depiction of the Asian regions derives from bookish sources, but conveys a deeper awareness of their relative geographical locations. The names are disposed along broad arcs within which several geographic areas can be discerned. The map is particularly conscientious in its arrangement of the Roman

25 *Rómverja saga* combines Old Norse translations of Sallust's *Bellum Jugurthinum* and *Bellum Catilinae* and a prose paraphrase of Lucan's *Pharsalia* into a saga about the Romans. See Þorbjörg Helgadóttir, ed. *Rómverjasaga*, 2 vols. (Reykjavík: Stofnun Árna Magnússonar í Íslenskum Fræðum, 2010), vol. 1, p. xiii; and Þorbjörg Helgadóttir, 'On the Sallust Translation in *Rómverja saga*', *Saga-Book* 22:3–4 (1987–88), 263–77. The so-called 'African excursus' (17.1–18.12) is the first of three formal digressions in the *Bellum Jugurthinum*, two chapters on African geography that correspond with the latter half of *Rómverja saga* Ch. 7. On the African excursus and its alignment with Sallust's social themes see Ronald Syme, 'Military Geography at Rome', *Classical Antiquity* 7 (1988), 227–51, 227; Thomas Wiedemann, 'Sallust's 'Jugurtha': Concord, Discord, and the Digressions', *Greece & Rome* 40 (1993), 48–57, esp. 49–52; and Robert Morstein-Marx, 'The Myth of Numidian Origins in Sallust's African Excursus (*Iugurtha* 17.7–18.12)', *The American Journal of Philology* 122 (2001), 179–200, esp. 179–80.

26 Evelyn Edson describes those maps comprising a basic cartographic frame onto which a geographical nomenclature has been added, sometimes in a way that is geographically suggestive, as list maps. Evelyn Edson, *Mapping Time and Space: How Medieval Mapmakers Viewed Their World* (London: British Library, 1998), pp. 5–6; Evelyn Edson, 'World Maps and Easter Tables: Medieval Maps in Context', *Imago Mundi* 48 (1996), 25–42, 30. One such example, in Lambert of St Omer's Liber Floridus, lists African place-names under the continental heading 'Affrica habet provincias' ('the provinces in Africa'). See Chekin, *Northern Eurasia*, p. 479.

provinces of Asia Minor (an area coextensive with modern-day Turkey), where Phrygia and Troy are located, the Parthian silk road kingdoms, and Mesopotamia. In these regions, names are arranged in sequence with their description in bookish sources like Isidore's *Etymologies*. In Mesopotamia, the map-maker has written the river name Euphrates so that it runs through the name Babylon, the ancient city that written descriptions placed on its banks. The map's emphasis, however, is on Palestine and its surrounding regions. These lands were commonly accorded disproportionate space on medieval world maps, and were also depicted in detail in some of the earliest surviving regional maps.[27] The map shows Jerusalem and the neighbouring towns Jaffa and Hebron, where the Biblical Adam is buried; the Arabian Peninsula, where Mount Sinai rises; and the Mesopotamian city of Carrhae, where the Biblical Abraham settled. The map may have been an exercise in visualising the places described in the Bible.

The map's thirty-five European legends include medieval place-names unknown to ancient authors, among them the names of the Scandinavian regions. The larger Viðey Map shows Iceland in its north-western corner, but it was not the first map to do so. As we shall see in Chapter 4, that distinction belongs to the eleventh-century Cotton Map, produced *c.* 1050 at Canterbury, England. Iceland and Thule – which may already have been incorporated into the map's exemplar before it came to Iceland – are prominently depicted beneath the legend *Europa*, which sits outside the world circle on the map's conceptual frame. The Scandinavian Peninsula reaches out from Iceland towards the map's centre, and comprises Norway, the two provinces of Sweden, 'Gautl*and*' ('Götaland') and 'Suiþjoð' ('Svealand'), and Russia. The depiction of the Scandinavian regions will be examined closely in Chapter 4. Beyond Russia lies the map's other Northern Eurasian legends, Scythia, Kiev, and a legend probably indicating the nomads of the Steppe. The map is timely with the Mongol predations on the Rus' from 1237, which precipitated in the fall of Kiev in 1240, and the map-maker's attention to these names may demonstrate a topical awareness of the events that took place there.[28]

Among the map's medieval legends is one marking the lands inhabited by the *Bjarmar*, a people whose presence in the White Sea region was reported to King Alfred the Great by the Norwegian traveller Ohthere, and incorporated into the Old English translation of Orosius's *Histories*

27 Some eight regional maps of the Holy Lands, extant in twenty-three copies, come down to us from between the late twelfth and mid-fourteenth century. P. D. A. Harvey, *Medieval Maps of the Holy Land* (London: British Library, 2012), p. 1.

28 Chekin, *Northern Eurasia*, p. 224. See Peter Jackson, 'The Mongols and Europe', in *The New Cambridge Medieval History: Volume V c. 1198–c. 1300*, ed. David Abulafia (Cambridge: Cambridge University Press, 1999), pp. 703–19, p. 706.

Against the Pagans.[29] Ohthere reports that he encountered the *Beormas*, who spoke a Finnic language, near the mouth of a large river that empties into the White Sea. The Bjarms and their native region, Bjarmaland, are mentioned with some frequency in Scandinavian historical writings and, in a vague and fictionalised way, in the Icelandic *fornaldarsögur*.[30] Their placement on the map accords with their north-westerly location in these writings. *Bjarmar* is the only ethnonym appearing in the map's depiction of Europe. In Africa, the map-maker notes the places where the Troglodytes and the children of Getulia live, while several among the map's Asian legends likely originated as ethnonyms, but may not have been recognised as such by the map's Icelandic audiences.

The regions depicted on Europe's western seaboard – Normandy, Brittany, and Gascony – are those lands that until 1204 had comprised the Continental reaches of the Angevin Empire. The extent of the French kingdom, and its border with the Holy Roman Empire, was carefully marked on Matthew Paris's near contemporary (*c.* 1250) itinerary map, with the Anglo-Norman inscription 'ci part lempire e le regne de france' ('here are distinguished the Empire and the kingdom of France').[31] The French regions are delineated on the larger Viðey Map by the placement of the British Isles between the Angevin's Continental possessions and the kingdom of France, represented on the map as Francia and Gallia.[32] Their depiction on the map may be a remnant of a tradition that emphasised the distinctiveness of the French and English realms, and the Plantagenet and Capetian overlordship of these areas some decades earlier. The map's depiction of Europe extends in a sequence of names along the Mediterranean Sea as far as its depiction of the Byzantine world,

29 Alan S. C. Ross, *The 'Terfinnas' and 'Beormas' of Ohthere* (London: Viking Society for Northern Research, 1981).

30 Mervi Koskela Vasaru, *Bjarmaland* (Oulu: University of Oulu, 2016); and Tatiana Jackson, 'Bjarmaland Revisited', *Acta Borealia* 19:2 (2002), 165–79.

31 Miller, *Mappaemundi*, vol. 3, p. 88; Katharine Breen, 'Returning Home from Jerusalem: Matthew Paris's First Map of Britain in Its Manuscript Context', *Representations* 89 (2005), 59–93, 69.

32 We might compare this arrangement of Continental European names with those presented in a geographical vision in the fourteenth-century *Nitiða saga*, pp. 140–41. Nítíða uses a magical stone to reveal a vision of the Southern European lands, described as 'Frakkland, Provintiam, Ravenam, Spaniam, Hallitiam, Friisland, Flandren, Norðmandiam, Skottland, Grikkland, og allar þær þjóðir þar byggja' ('France, Provence, Ravenna, Spain, Hallitia, Frisia, Flanders, Normandy, Scotland, Greece, and all the people living there'). The appearance of Scotland in this list may reveal the saga author's awareness of the Franco-Scottish alliance (1295–1560), established to resist the territorial ambitions of the English Edward I. This, and the example of the Angevin territories above, may demonstrate that Icelandic thinkers were able to manipulate cartographic space to show contemporary geo-political situations.

comprising Constantinople, Thrace, and Greece.[33] In the vicinity of these legends, the map-maker has placed Sparta and Moesia, a Roman province situated on the Danube that extended as far as Macedonia. The map's sequence of European names branches into two around Germania, with one sequence – Saxonia, Da*n*morc, and Frisia – tending northwards up towards the British Isles, and another – Francia and Gallia – tending southwards to the 'parmo montes', likely the Pyrenees, that mark their boundary with Spain. A pen line, one of few outlines drawn on the map, separates Francia and Germania, the commencement of Matthew's 'Empire'.

Earlier scholarship has focussed on the map's European place-names, primarily for what they can tell us about the Icelanders' geographical awareness in the thirteenth century. However, the map is of uneven historical value for what it can tell us about the realities of the territories it depicts. That it was not drawn to document the totality of its maker's geographic awareness can be seen in its many obvious omissions of places that would certainly have been known to its maker. Orkney, Shetland, the Faroe Islands, Greenland, and places associated with the Vínland voyages were without doubt well-known to Icelanders in this period, and yet find no place on the maps they drew in these centuries. The map is concise in its depiction of Europe; the preponderance of its European names are traditional, and their arrangement conveys, for the most part, little specifically updated knowledge or interest in these regions.

The fallacy that early maps represent stages in our developing geographical awareness, or are partially realised views of a world we today know completely, can mislead, especially when the names of ancient territories or peoples coincide with terms in modern political usage.[34] *Germania,* for instance, is not coextensive or continuous with modern Germany, and *Francia,* at the time the map was drawn, likely described an area that extended little beyond the Île-de-France, and was certainly to be contrasted with areas such as Brittany and Normandy.[35] So too are the boundaries between continents not quite the same as we would have them; Asia, as on the Icelandic Viðey Map, conventionally includes Egypt and its environs, extending from its boundary with Europe at the Tanais (Don) up to the Nile.[36] How we understand

33 David Jacoby, 'Latin Empire of Constantinople and Frankish States in Greece', in *The New Cambridge Medieval History: Volume V c. 1198–c. 1300,* ed. David Abulafia (Cambridge: Cambridge University Press, 1999), pp. 525–42, p. 528.

34 On related assumptions in the reading of medieval literature, see Kenneth Hodges, 'Introduction: Places of Romance', in *Mapping Malory: Regional Identities and National Geographies in Le Morte D'arthur,* ed. Dorsey Armstrong and Kenneth Hodges (New York: Palgrave Macmillan, 2014), pp. 11, 14.

35 I am grateful to Erika Graham-Goering for her thoughtful comments on the map's depiction of the French regions, and to Rosa María Rodríguez Porto for her insight into the map's Hispanic territories.

36 On the continents as cultural rather than natural constructs see Asa Simon

a legend depends largely on which map it is that interests us: the map copied by an Icelandic scholar at Viðey *c.* 1225–50; the earlier map from which it was copied; or our own world map. The legend 'massagete' may be taken as an example. The name 'massagete' appears nowhere else in Old Norse writings and may have been, for the Icelandic map-maker, an exotic term representing no region in particular. For Isidore of Seville, however, the Massagetes (e) are 'a nation descended from the Scythians that inhabits the region between the source of the Tigris and the City of Carrhae' (*Etymologies* 9.2.63, from Herodotus, *Histories* 1.215–6).[37] In modern ethnographic terms, the Massagetes are an ancient eastern Iranian nomadic confederation that inhabited the steppes of Central Asia (modern Turkmenistan, western Uzbekistan, and southern Kazakhstan). These three perspectives on the world and its places converge in the map image, and care must be taken, in describing what the map shows, to acknowledge their distinctiveness.

The map's image of the world does not conform to any one historical moment but is an asynchronous amalgam of names both ancient and contemporary.[38] Medieval world maps were visualisations of history as well as geography, and depicted events in time as well as space. The Hereford Map, for example, shows the expulsion of Adam and Eve from Paradise, the Israelites' crossing of the Red Sea, and the crumbling of the city walls at Troy. A map's geographical contents may have proceeded in sequence with the events of time, depicting the commencement of human history in the earthly Paradise, located in the east, to scenes from the Last Judgement and resurrection, sometimes enacted by the map's European audiences who contemplated the map's holy places in anticipation of their own judgement and salvation. As Suzanne Conklin Akbari puts it, 'the orient was a place of both geographical and temporal origins, with the earthly paradise located at once in the region furthest east and in the remotest past'.[39] The Viðey Map does not show the earthly Paradise, but three of the four rivers that flowed from Eden – the Tigris, the Euphrates, and the Nile – course through the map's depictions of the ancient Mesopotamian and Egyptian kingdoms. The map's west

Mittman, 'The Vercelli Map (c. 1217)', in *A Critical Companion to English Mappae Mundi of the Twelfth and Thirteenth Centuries*, ed. Dan Terkla and Nick Millea (Woodbridge: Boydell, 2019), pp. 127–46, pp. 144–45.

37 Reference to the *Histories* follow Herodotus, *The Landmark Herodotus: The Histories*, trans. Andrea L. Purvis, ed. Robert B. Strassler and introd. Rosalind Thomas (London: Quercus, 2008).

38 On the combination of ancient and contemporary geographical information on the English Sawley Map see Alfred Hiatt, 'The Sawley Map (c. 1190)', in *A Critical Companion to English Mappae Mundi of the Twelfth and Thirteenth Centuries*, ed. Dan Terkla and Nick Millea (Woodbridge: Boydell, 2019), pp. 112–26, p. 114.

39 Akbari, *Idols*, p. 3.

contains a greater incidence of place-names derived from contemporary medieval sources, not least in its depiction of the Scandinavian regions and Iceland. These regions, northernmost in Europe, stand at the other end of the map's historical scale, their comparatively recent discovery and Christianisation representing an 'essential precursor to the end of the world – the diffusion of the Church over the entire world'.[40] The larger Viðey Map does not present a view of the world as it really was in the thirteenth century, but is an historical proposition written into a geographical framework. We shall explore the map's historical argument in Chapter 4.

Orientation

The two Viðey Maps – as well as the Icelandic zonal map, and the two hemispherical world maps – are all oriented with south at the top. While examples of *mappae mundi* with all four cardinal points at the top survive, an eastern *orientation* (from Latin *oriens*, 'east') was most common, then, in descending order, northern, southern and western.[41] The northern and less common southern orientations are most frequently encountered on maps whose geometries were centred on the Earth's axis and the climatic zones, but T-O maps were usually drawn with Asia at the top, to venerate the holy places of Christianity in the east.

While the larger Viðey Map opens with south at the top, there is no dominant textual orientation or clear way up from which to read its legends; its reader must turn the map medium to read its inscribed names.[42] The orientation of the map's legends partly undermines some editorial strategies that separate groups of legends along continental fault lines. If the map is turned so that the Mediterranean sequence of European place-names can be read the right way up, the Holy Lands are also brought into view and made readable by their shared orientation. When the reader holds the map with east at the top, its Continental European names are aligned into a sequence that traces the route, through Rome and Constantinople, into the Holy Lands, compelling the viewer's gaze from their place in western Europe towards the places contemplated and intensely desired by medieval Christianity.[43]

40 Alessandro Scafi, 'Defining *mappaemundi*', in *The Hereford World Map: Medieval World Maps and their Context*, ed. P. D. A. Harvey (London: British Library, 2006), pp. 345–54, p. 348.

41 Woodward, 'Medieval *Mappaemundi*', p. 336.

42 Anna-Dorothee von den Brincken, *Finnes Terrae: Die Enden der Erde und der vierte Kontunent auf mittelalterlichen Weltkarten* (Hanover: Hahnsche Buchhandlung, 1992), p. 129; Chekin, *Northern Eurasia*, p. 69.

43 In certain contexts of usage, the larger Viðey Map can resemble Matthew's itinerary map, and enact a similar function in guiding the reader's eye onwards towards important sites of pilgrimage. For similar uses of medieval

That all the Icelandic world maps are oriented with south at the top is remarkable for the fact that they are representative of different cartographic genres. This was noticed by Rafn, who sought to understand this peculiarity in terms of the models that might have been available to Icelandic copyists. Rafn was aware that a south orientation was common among contemporary maps from the Islamic world, but suggests that the larger Viðey Map was more likely influenced by a map of the zonal type.[44] This suggestion has been developed by Chekin, who notes that the larger Viðey Map is similar in form and orientation to the double-page Pseudo-Isidorian Vatican Map (Vatican City, Biblioteca Apostolica Vaticana, Vat. Lat. 6018, ff. 63v–64r) probably produced in Italy between 762 and 777.[45] He observes that this double-page map is also oriented with south at the top, and suggests that one of the models used in its composition was a hemispherical world map depicting the latitudinal division of the world's climatic regions. The larger Viðey Map bears two legends reading *frigida* ('cold'), one positioned in the map's depiction of Central Europe and another in Asia. The European *frigida* has traditionally been paired with the adjacent place-name *Scythia* to form *Scythia frigida* ('Scythia the Cold'), but Chekin favours an interpretation that sees *frigida* as a 'rudiment of zonal structure', a term designating one of the globe's two polar regions that has found its way onto the map in two places.[46] The Asian *frigida*, however, is almost certainly a rendering of the place-name *Phrygia*, an ancient region in eastern or central Anatolia among the provinces of Asia Minor mentioned in Isidore's *Etymologies* (13.3.41). The European

maps see Edson, *Time and Space*, p. 14; Daniel K. Connolly, *The Maps of Matthew Paris: Medieval Journeys through Space, Time, and Liturgy* (Woodbridge: Boydell, 2009), p. 2; Breen, 'Returning Home', p. 75.

44 Rafn, *Antiquités Russes*, p. 390. On the south orientation common to Islamic maps see Ahmet T. Karamustafa, 'Introduction to Islamic Maps,' in *The History of Cartography* vol. 2, book 1: *Cartography in the Traditional Islamic and South Asian Societies*, ed. J. B. Harley and David Woodward (London: Chicago University Press, 1992), pp. 1–9. Rafn observed that only two maps in the *Atlas do Visconde de Santarém*, a compendium of lithographic reproductions of medieval European maps and navigational charts (the third edition of which was published in 1849) exhibited this orientation. The two maps Rafn identified were found in a Latin manuscript of the *Cosmographia* of Asaph the Jew from the fifteenth century (Paris, Bibliothèque Nationale de Paris, MS no. 4744) and a twelfth-century manuscript of Honorius Augustodunensis's *Imago mundi*. On the influence of Honorius's *Imago Mundi* on Old Norse texts see Peter Springborg, 'Weltbild mit Löwe Die *Imago Mundi* von Honorius Augustodunensis in der altwestnordischen Textüberlieferung', in *Cultura classica e cultura germanica septentrionale*, ed. Pietro Janni, Diego Poli, and Carlo Santini (Rome: Herder, 1988), pp. 167–219.

45 Chekin, *Northern Eurasia*, p. 126. See also Leonid S. Chekin, 'Easter Tables and the Pseudo-Isidorean Vatican Map', *Imago Mundi* 51 (1999), 13–23, 15.

46 Chekin, *Northern Eurasia*, p. 70

frigida is also probably one of the several Phrygias mentioned in antique geographical descriptions. This region appears on the Hereford Map as *Frigia*, close to the legend *Scythia*.[47]

The southern orientation may, however, have been driven by the map-maker's familiarity with other map types. The two maps in 1812 III (*c.* 1225–50), on the recto and verso of the same manuscript folio, are the only Icelandic maps that we know for certain were drawn with knowledge of one another. These maps may have been known, however, to the compositor of the younger fragment 1812 I (1315–*c.* 1400), which contains the zonal map. The compiler of this younger fragment may have drawn the zonal map with south instead of the more usual north at the top for consistency with the two older maps in 1812 III. The maps can be compared with those drawn by the cardinal and theologian Pierre d'Ailly, a fifteenth-century chancellor of the University of Paris, whose *Imago Mundi* sometimes included a complement of eight diagrams that show the Earth's parts anatomised in multiple ways: into two hemispheres, seven climatic zones, and three continents.[48] All these different maps were oriented with north at the top, so that they could be seen together and compared.[49] The maps in GkS 1812 4[to] may similarly have been turned towards a common orientation to harmonise their perspectives on world geography in a way that enabled their viewers to compare, transpose and mentally sort a diversity of cartographic images into a single image of the world.

It is conceivable that the south orientation common to the Icelandic maps emanates from the Icelanders' sense of their geographical position on Europe's northernmost periphery. Icelanders may have reoriented their maps to enact their own perspectives on world geography, turning them so that the world became aligned with their vision, its places reaching southwards before them. It is improbable, however, that this trope would have occurred to all the maps' compilers independently; it is more likely that their shared orientation stems from their compilation, analogy with other cartographic genres, and the models available to their makers.

The Viðey book as it comes down to us today, a compilation of fragments composed between *c.* 1200 and *c.* 1400 and combined as late as the sixteenth century, is an ensemble that tasks its reader with mentally sorting these maps and diagrams into a single coherent image of the world. The Viðey book does not contain three isolated maps, but rather a sequence of maps through which its compilers seek to epitomise ways of visualising of global space from multiple perspectives. The zonal map and its associated diagrams (ff. 10r–12v) focalise multiple perspectives

47 Westrem, *Hereford Map*, pp. 142–43.

48 Miller, *Mappaemundi*, vol. 3, pp. 127–28.

49 Akbari, *Idols*, p. 25.

on the Earth and its cosmic position: from a point vanishingly small amid the orbits of the much larger planets, to a schematic depiction of its surface. The two Viðey Maps (ff. 5v–6v) are also conjoined into a single cartographic statement: one presenting a vision of the world teeming with names and detail, the other a simple statement of its basic order. The maps, oriented to a common perspective, enable the world to be seen across a range of scales and perspectives that can be compared and analysed together. The book thematises the world and the medieval Icelanders' imaginative attachments to it, and we shall see in the two following chapters how these maps come to produce meaning through their relationships with their companion items.

CHAPTER 4

Iceland in Europe

The larger Viðey Map is the only map in the Icelandic corpus that shows Iceland, and, as such, is a unique instance of Icelandic self-portraiture. The name 'Island' ('Iceland') is one of the map's few medieval place-names. The opening chapters of the Icelandic *Landnámabók* (*Book of Settlements*), a history of the Icelandic settlement written in the twelfth century, describe the succession of names by which Iceland was known to Norse seafarers in the early decades of the ninth century. It attributes Iceland's discovery to Naddoðr, a Norwegian Viking who sights Iceland on his way to the Faroe Islands. On his return to Norway in the autumn, he sees Iceland's mountains white with snow, and so names it *Snæland* ('Snowland'). The next Scandinavian to reach Icelandic shores, however, named it a second time. Garðarr Svavarsson, we are told, dwelt for a short time in the north of the island, which he named *Garðarshólmi* ('Garðarr's Island'). Iceland's third discoverer, Hrafna-Flóki ('Raven-Flóki'), sought Iceland with the assistance of three ravens, which, like the Biblical Noah, he let fly in search of land. Flóki made land at Vatnsfjörður in the northwest, and on seeing the fjord to the north frozen over, called the island *Ísland* ('Iceland').

Such is the history of the name *Iceland* presented by the anonymous compilers of *Landnámabók*. The name's earliest written attestation, however, appears at Canterbury in England in *c.* 1050. The name's earliest witness is the English Cotton Map (London, British Library, MS. Cotton Tiberius B.V., f. 56v, *c.* 1050), which also has the distinction of being the earliest map to show the island. The Cotton Map (depicted on this book's cover) was drawn sometime between the two conquests of England – the first Danish, the second Norman – that took place in the eleventh century. When the Danish prince Cnut, 'the Great', ascended the English throne in 1016, the Danish throne in 1018, and the Norwegian throne in 1028, England became part of an expansive maritime realm. The unusual detail with which the map-maker depicts North Atlantic coastlines may, as Michelle P. Brown suggests, draw upon Anglo-Scandinavian awareness of these regions, and the wider mercantile and ecclesiastical networks of which they were a part.[1] Although reportedly a Norse coinage, the name that appears on this map antedates any alphabetic writing in a Scandinavian language, and there is nothing in the name 'Island' that can be used to distinguish between an Old Norse or Old English provenance at the time the map was made. The precise channels by which the name came to the attention of a map-maker at Canterbury are unknown, but its appearance on an eleventh-century English map demonstrates that the 'Island' of the larger Viðey Map need not be the original work of an Icelandic map-maker. The name may well have been copied from an earlier map.

The map thematises Iceland's place in a wider world, with the island shown, in the company of its elusive double, Thule, at the *caput Europae*, immediately beneath the legend *Europa* on the map's frame. Iceland is separated from the map's representation of the Scandinavian Peninsula and the Baltic littorals – Norway, Götaland, Svealand, and Russia – by the waters of the North Atlantic, while mainland Denmark assumes its place nearer the map's Continental European names, where it borders Saxony.

In its depiction of Scandinavia, the larger Viðey Map may bear witness to a number of cartographic firsts, through additions original to the Icelandic map-maker in the thirteenth century. On medieval world maps, Scandinavia was most commonly known by the conventional names *Norwegie* ('Norway') and *Scanzia* (a name that appears in Pliny's *Historia Naturalis*, probably related to the name *Skåne*, Old Norse *Skáney*). These were traditional Latinised names that could stand for the entire Scandinavian region.[2] These traditional names do not appear on the Viðey Map, its maker instead drawing on more particular regional terms available in the vernacular to convey the distinctiveness of these regions as they were known to thirteenth-century Icelanders. The map replaces the traditional *Scanzia* with two Old Norse names, *Suiþioð* ('Svealand') and *Gautland* ('Götaland'), following Adam of Bremen's distinction of the Swedish territitories into *Sueonia* and *Gothia*.[3] The map may be the

1 Michelle P. Brown, 'Making Manuscripts and Mappae Mundi', in *A Critical Companion to English Mappae Mundi of the Twelfth and Thirteenth Centuries*, ed. Dan Terkla and Nick Millea (Woodbridge: Boydell, 2019), pp. 20–43, pp. 26–27. On the geographical awareness that underlies this map see Dan Terkla, 'Books and Maps: Anglo-Saxon Glastonbury and Geospatial Awareness', in *A Critical Companion to English Mappae Mundi of the Twelfth and Thirteenth Centuries*, ed. Dan Terkla and Nick Millea (Woodbridge: Boydell, 2019), pp. 44–67.

2 Leonid S. Chekin, 'Mappae Mundi and Scandinavia', *Scandinavian Studies* 65 (1993), 487–520.

3 These regions would not become unified until the thirteenth century. See Stefan Brink, 'Naming the Land', in *The Viking World*, ed. Stefan Brink and Neil Price (London: Routledge, 2008), pp. 57–66, p. 60; Judith Jesch, *The Viking Diaspora* (London: Routledge, 2015), p. 21.

earliest to show Sweden unambiguously, being broadly contemporary with the map drawn by Matthew Paris (d. 1259), a Benedictine monk from St Albans, England, who visited the court of the Norwegian King Hákon IV in 1248. The map illustrating his *Chronica majora* (Cambridge, Corpus Christi College, MS 26 284) marks Sweden with the name *suenica*.[4] The Viðey Map retains the conventional *Norwegie*, but its sense is narrowed from the entirety of Scandinavia to Norway.

The Viðey Map may also be the earliest to show Denmark. While the ambiguous name *Dacia*, the name of a central European Roman province sometimes used as an archaic name for Denmark, appears quite commonly on medieval maps, it is never certain that Denmark was intended.[5] Denmark has been placed apart from the lands of the Scandinavian Peninusla among the map's Continental European names, where it borders Saxonia ('Saxony'). Its placement has a number of parallels on other world maps and in contemporary historical writings. In his *Gesta*, Adam of Bremen describes Denmark as the commencement of the Scandinavian countries, and that 'transeuntibus insulas Danorum alter mundus aperitur in Sueoniam vel Nordmanniam' ('in going beyond the islands of the Danes there opens up another world in the direction of Sweden and Norway').[6] Adam conceptualises Denmark as the frontier between the European Continent and the *alter mundus* ('other world') that is insular Scandinavia. Denmark is likewise imagined as a border space in the *Leiðarvísir*, an Icelandic itinerary to Rome probably compiled in the fourteenth century, where Denmark is perceived as the Scandinavian region beyond which lies the *austrvegr* ('eastern way'), the Old Norse term compassing the Baltic Sea region and Russia.[7] The placement of Denmark next to Saxony on the Viðey Map has a parallel in the English Hereford Map, which emphasises the distinction between the territories

4 Chekin, 'Scandinavia', 487–520, 514. On this map, see Daniel K. Connolly, 'In the Company of Matthew Paris: Mapping the World at St. Albans Abbey', in *A Critical Companion to English Mappae Mundi of the Twelfth and Thirteenth Centuries*, ed. Dan Terkla and Nick Millea (Woodbridge: Boydell, 2019), pp. 147–78.

5 Chekin, 'Scandinavia', p. 497, fn. 12. An insular *Dacia* appears in the vicinity of Iceland and Thule, but also continental Westphalia, on a map in Ranulf Higden's fourteenth-century *Polychronicon* (Cambridge, Corpus Christi College, MS 21, f. 9r). This island does seem to be Denmark, its presentation consistent with medieval rumours of an insular Scandinavia.

6 Citations from Adam's *Gesta* are drawn from Adam of Bremen, *Gesta Hammaburgensis ecclesiae pontificum*, MGH SRG 2, ed. B. Schmeidler (Leipzig: Hahnsche Buchhandlung, 1917) IV.21, p. 250. Translations follow Adam of Bremen, *History of the Archbishops of HamburgBremen*, trans. Francis J. Tschan (New York: Columbia University Press, 2002), p. 202.

7 'I gegnum Danmork gengr sior i austr-veg' ('through Denmark the sea extends towards the eastern way'). *AÍ* I., p. 11. For arguments about its dating, see Vídalín, 'Óláfr Ormsson's Leiðarvísir', 212–34.

of the Danes and Saxons. The legend 'Terminus Danorum et Saxonum' ('the frontier between the Danes and the Saxons') has been written along the course of the River Elbe, which demarcated these territories.[8] Legends marking *termina* ('frontiers' or 'limits') are few on the Hereford Map, which, besides the one between Denmark and Saxony, shows only the 'terminus Francie et Burgundie' ('the frontier between France and Burgundy'), the 'terminus Europe' ('edge of Europe'), and the 'terminus Affrice' ('edge of Africa').[9] Two of these frontiers, the edges of Europe and Africa, mark fundamental categorical distinctions between the world's continental thirds, while those between France and the Duchy of Burgundy, and the Danish Kingdom and Saxony, are important regional distinctions in Europe at the time the map, or its exemplar, was drawn.[10]

The distinction between the Saxons and the Danes was both political and diocesan, marking the boundary between the Holy Roman Empire and the Danish possessions north of the River Elbe and, following the establishment of the Nordic archdiocese at Lund in 1103 and at Niðaróss (Trondheim) in 1152, the limits of the metropolitan authority of the archdiocese at Hamburg-Bremen.[11]

'Da*n*morc' ('Denmark') is one of the map's few place-names written in the vernacular Old Norse. The preponderance of the larger Viðey Map's legends are written in Latin. The map tends to show places by their Latin names – 'Constantinopolis', 'Russia', 'Grecia' – even when Old Norse equivalents – Miklagarðr ('Constantinople'), Garðaríki ('Russia'), Gríkkland ('Greece') – are well attested in Old Icelandic literature. Unlike the Icelandic hemispherical and zonal maps, whose legends have been mostly translated or calqued into Old Norse, the larger Viðey Map presents no especially technical vocabulary, like the names of the parallels of latitude, that might have compelled an Icelandic map-maker to translate their exemplar into more easily understandable language. But there was nonetheless an established vernacular onomasticon that the map-maker might have used, but chose not to. The distinction between the map's Latin and Old Norse contributions is not the distinction between unfamiliar and familiar; many of map's Latin names represent places that were stable parts of the Icelanders' geographic imaginary. The map contains only four place-names written entirely in Old Norse:

8 Westrem, *Hereford Map*, pp. 222–23.

9 Westrem, *Hereford Map*, pp. 286–87; 334–35; 370–71.

10 On the 'terminus Danorum et Saxonum' on the English Sawley Map, *c.* 1190, see Hiatt, 'The Sawley Map', p. 118. As Hiatt notes, this boundary 'reflects a division within Carolingian, rather than Roman, Europe'.

11 See Kurt Villands Jensen, 'The Blue Baltic Border of Denmark in the High Middle Ages', in *Medieval Frontiers: Concepts and Practices*, ed. David Abulafia and Nora Berend (Aldershot: Ashgate, 2002), pp. 173–83, at pp. 177–80 for an account of the relationships between the Danish kings and German emperors in the thirteenth century.

'Island' ('Iceland'), 'Gautl*and*' ('Götaland'), 'Suiþioð' ('Svealand'), and 'Da*n*morc' ('Denmark'). To these may be added two macaronic legends, written in a combination of Old Norse and Latin: 'Bja*r*ma*r* habitauit hic' ('Bjarms live here') and 'Tanakvisl fluvi*us* maxim*us*' ('Tanakvísl, the greatest river').[12] The map's remaining vernacular contributions are the names of the four cardinal directions in the map's conceptual frame, which are placed in apposition with their Latin equivalents: 'suðr meridies' ('south'), 'occidens vestr' ('west'), 'norðr septentrio' ('north'), and 'austr oriens' ('east').

Medieval maps may have derived their written contents from a rich ferment of written and oral sources, sometimes incorporating place-names from vernacular languages. Changes in language could also be used to distinguish between geographical and cultural regions. Matthew Paris's itinerary maps, for example, contain legends written in English, Insular French, and Latin. The itinerary's legends begin in England with the names of London's city gates written in English, then advance through Europe in French, the more mobile language of the aristocracy, before culminating in the Holy Lands in Latin.[13] Katharine Breen argues that the use of English in naming local sites in England is determined by popular usage, while the movement into Latin better enables Matthew to elicit a name's spiritual or historical relevance.[14] More generally, Alfred Hiatt observes a tendency for 'geographical description within the area(s) familiar to the makers and audience of the map to be linguistically mixed', as place-names were known in local languages as well as in Latin.[15] For example, while the majority of the Hereford Map's 1091 legends are in Latin, around half of its eighty-one British inscriptions are in English, with, perhaps, a slight inclination to enlist Latin in the names

12 These legends are macaronic in that they combine Old Norse and Latin vocabulary with inflectional endings in both languages. Thus the *Bjarmar*, the Old English *Beormas*, retain an Old Norse plural ending though the rest of the legend is written in Latin. The term *Tanakvísl*, combines *Tanais*, which has been adapted to the inflectional system of Old Norse, with the Old Norse noun *kvísl* ('river branch', 'tributary').

13 Breen discusses Matthew's own views on the three languages he uses through examination of the references to them scattered throughout his works. Breen, 'Returning Home', 64, 90, fn. 27.

14 Breen, 'Returning Home', 64.

15 Alfred Hiatt, 'From Hulle to Carthage: Maps, England, and the Sea', in *The Sea and Englishness in the Middle Ages: Maritime Narratives, Identity, and Culture*, ed. Sebastian I. Sobecki (Cambride: D.S. Brewer, 2011), pp. 133–57, p. 143. The English Cotton Map contains, in addition to the linguistically ambiguous *Island*, four vernacular place-names: *Suðbryttas* ('Brittany'), *Neronorroen* ('Norway'), *Sleswic* ('Sleswick'), and *Scridefinnas* (from the Old English *Orosius*, 'Scridefinne'). Hiatt numbers *Tylen* also among the map's vernacular inscriptions, perhaps because of its later association with *Island*. Hiatt, 'Hulle to Carthage', pp. 135–36.

of those places with particular political or ecclesiastical significance.[16] Insular French, the elite language of the map's patrons, is limited to the map's pentagonal frame, and in all instances relate to the contemplative or salvational uses to which the map might have been put by its elite audiences.[17] The map's donor inscription, written in Insular French, enjoins the map's communities of owners, listeners, readers, and viewers to pray for the soul of its patron, the clergyman Richard de Bello:

> Tuz ki cest estoire ont
> ou oyront ou lirront ou veront
> prient a Jhesu en deyte.
> De Richard de Haldingham e de Lafford eyt pite
> Ki la fet e compasse
> Ke ioie en cel li seit donc.

(Let all those who have this history – or who shall hear, or read, or see it – pray to Jesus in his divinity, to have pity on Richard of Haldingham and Lafford, who has made and planned it, to whom joy in heaven may be granted.)[18]

The Hereford Map's Insular French legends are all accommodated by the map's frame, and speak directly to its audiences in the elite language of its patrons, the language in which the map may have been commissioned, and its contents discussed, lectured on, or sermonised about ('or who shall hear, or read, or see it'). Dan Terkla describes a similar distribution of Latin and Insular French on the fragmentary Duchy of Cornwall Map (*c.* 1286). While the preponderance of its texts are written in Latin, the language of its frame is Insular French. At the base of the map is a gallery of paired roundels in which personifications of the *aetates hominum* (the

16 Hiatt, 'Hulle to Carthage', pp. 142–43.

17 In addition to the donor inscription, there are four Insular French texts in this space. Two are spoken by angels on either side of the Judgement scene depicted at the map's apex. On the left an angel guides the faithful to heaven: 'Levez! Si vendres a joie pardurable' ('Arise! You shall come to joy everlasting'); while on the right an angel guides the sinful to hell: 'Levez! Si alez au fin de enfer estable' ('Arise! You are going to the fire established in hell'). The third Insular French inscription appears near the top of the map's frame next to the image of the *Virgo lactans*: 'Veici, beu fiz, de deinz la quele chare priestes' ('See, dear son, my bosom, in which you took on flesh'). A fourth legend speaks to the map's reader directly, in an inscription that has been taken as the map's motto: 'Passe avant' ('Go ahead'); an invitation to the map's reader, perhaps, to explore its contents. Westrem, *Hereford Map*, p. 11. On the donor inscription, see Marcia Kupfer, 'The Hereford Map (c. 1300)', in *A Critical Companion to English Mappae Mundi of the Twelfth and Thirteenth Centuries*, ed. Dan Terkla and Nick Millea (Woodbridge: Boydell, 2019), pp. 227–52, p. 233.

18 Westrem, *Hereford Map*, p. 11.

ages of man) respond to questions put to them about their mortality; the map's personification of 'decrepitude' intoning that 'La mort me haste p<our> sa rente' ('Death hurries me along to the Day of Reckoning').[19] The map uses Insular French to speak to its audiences about their mortality and hope for salvation; as Terkla puts it, 'Anglo-Norman is the banderole language – and thus the language of heaven'.[20]

The influence of other European vernaculars, notably Old Norse, has also been detected in the Hereford Map's place-names. The map renders Norway in the vernacularised form 'Noreya', which seems to combine the vernacular *Nóregr* with the Old Norse toponymic element *–ey*, a place-name generic denoting 'island'. The name's peculiar form may be an attempt to imitate Scandinavian toponymic conventions, like the map's 'Fareie' ('Faroe Islands'), or is perhaps expressive of persistent rumours of an insular Scandinavia inherited from antiquity, and not reliably dispelled until the later Middle Ages.[21]

On the larger Viðey Map, Old Norse is mostly restricted to the Scandinavian regions – 'Island', 'Gautl*and*', 'Suiþioð', and 'Da*n*morc' – across which varieties of the language were spoken. The term for the Old Norse language in the medieval period was the *dansk tunga* or *Dacisca lingua*, 'the Danish language', a term appearing in the early eleventh century in both Old Norse and Latin.[22] The term was frequently used to describe the Scandinavian origins of its speakers. In *Knýtlinga saga* (1250s), for example, an Icelandic history of the Danish kings from the tenth to the thirteenth century, King Cnut (*c.* 985 or 995–1035) established

19 Dan Terkla, 'The Duchy of Cornwall Map Fragment (c. 1286)', in *A Critical Companion to English Mappae Mundi of the Twelfth and Thirteenth Centuries*, ed. Dan Terkla and Nick Millea (Woodbridge: Boydell, 2019), pp. 197–226, pp. 211–15. On the ages of man on the larger Viðey map, see Chapter 5.

20 Terkla, 'The Duchy of Cornwall Map', p. 216.

21 On the uncertainty about the topography of the Scandinavian peninsula see Chekin, 'Scandinavia'; and Westrem, *Hereford Map*, pp. 192–93. Other place-names on the map that include the *–ey* place-name element are *Fareie* ('Faroe Islands') and *Lindeseya* ('Lindsey', though here ultimately from its Old English cognate *–īg* or *īge*).

22 The earliest witness to the term *dansk tunga* appears in strophe 15 of the *Víkingavísur* composed by the Icelandic skald Sighvatr Þórðarson in praise of King Ólafr Haraldson, *c.* 1015. See Håkon Melberg, *Origin of the Scandinavian Nations and Languages. An Introduction, I–II* (Halden: Eget Forlag,1953), p. 17. This term is contemporary with the earliest appearance of the term *Dacisca lingua* ('Danish language') in Latin, in the *Historia Normannorum* (Ch. 11) of the Norman historian Dudo of Saint-Quentin (*c.* 960–1026) in the first quarter of the eleventh century. See Stephen Pax Leonard, *Language, Society and Identity in Early Iceland* (Oxford: Wiley-Blackwell, 2012), pp. 121–22, p. 128. Less common, but sometimes seen in texts from the thirteenth century is the term *norræna* ('Norse'). Stefán Karlsson, *The Icelandic Language* (London: Viking Society for Northern Research, 2004), p. 9.

a hostel along the pilgrimage route to Rome for the use of those 'er þar kæmi af danskri tungu' ('who came of the Danish language'), while King Eírikr Sveinsson the Good (r. 1095–1103) in the same saga established a hostel on the road to the pilgrimage site at Borgo San Donnio so that 'allir pílagrímar, þeir er danska tungu mælti, skyldi ókeypis nógt vín drekka' ('all pilgrims who spoke the Danish language should have enough free wine to drink').[23] As much as the language the term denotes the geographical origins and a sense of common cultural belonging among its speakers.[24] In *Óláfs saga Tryggvasonar,* for example, Snorri Sturluson describes the eponymous Óláfr as 'frægstr maðr var a dansca tungu' ('the most famous man of the Danish language').[25] The language may have been associated with Denmark because it was the most populous Scandinavian country for much of the Middle Ages, and geographically marked the commencement of the northern lands where the language was spoken and its speakers understood.[26] This may have been the implication in the *Leiðarvísir,* which punctuates the description of the pilgrims' journey between the towns of Minden and Paderborn in Saxony with the observation 'nu skiptazt tungur' ('now the languages change').[27] Kålund suggests that the itinerary has been corrupted in transmission and that originally this linguistic discontinuity would have marked the entry into Saxony from Denmark.[28] The Danish border, as we have seen, marked a frontier that was at once political, diocesan, and linguistic, and its position on the Icelandic Viðey Map – away from the map's other Scandinavian regions and yet likened to them through the common language of their inscription – evokes its singular status as a frontier.

The Icelandic map-maker appears to use Old Norse to relate cartographically distant legends. Its distribution is coextensive with the map's depiction of the Scandinavian regions, the territorial and linguistic space sometimes called the *dansk tunga.* Denmark ('Da*n*morc') has been placed away from the Scandinavian Peninsula among the map's Continental European place-names, but is tied to the other Scandinavian regions – 'Island', 'Gautl*and*', 'Suiþioð' – through the common language in which they are written. Names inscribed onto medieval maps in local vernaculars

23 *Danakonunga sögur,* ed. Bjarni Guðnason, Íslenzk fornrit 35 (Reykjavík: Hið Íslenska Fornrit Félag, 1968), pp. 112 and 213. See Ian McDougall, 'Foreigners and Foreign Languages in Medieval Iceland', *Saga-Book* 22 (1986–89), 180–233.

24 McDougall, 'Foreign Languages', 212–13; Leonard, *Language,* p. 128.

25 Snorri Sturluson, *Heimskringla,* 1:3.

26 McDougall, 'Foreign Languages', 212–213. On Denmark's importance in the Middle Ages, see Sverre Bagge, 'The Scandinavian Kingdoms', in *The New Cambridge Medieval History, vol. 5, c. 1198–c. 1300,* ed. David Abulafia (Cambridge: Cambridge University Press, 1999), pp. 720–42, p. 722.

27 *AÍ* I, p. 13.

28 Kristian Kålund, 'En Islandsk vejviser for pilgrimme fra 12. århundrede', Aarbøger for Nordisk Oldkyndighed og Historie, 3rd series, 3 (1913), 51–105, 66.

tend to represent regions that were familiar to the map-maker and their audiences, and were most commonly concentrated around the place where the map was produced. But as contemporary maps demonstrate, map-makers were alert to the possibilities of using language to shape their representations of the worlds' regions. Norway is the only Scandinavian land assigned a Latinate name, 'Norvegie', instead of the Old Norse *Nóregr*. The Latin incursion into a region otherwise described in Old Norse may have conveyed Norway's higher, supraregional status among the Scandinavian kingdoms, especially in Iceland. Norway was the seat of the Nordic archepiscopate at Niðaróss (Trondheim) (established in 1153) and, at the time the map was drawn, many prominent Icelanders were becoming vassals of the Norwegian King Hákon IV, who sought to bring the island under his sovereignty. The map uses language, as we shall see, to engage the Icelanders' history, and their relationship to the other Scandinavian regions.

Iceland and Thule

Iceland is not alone in the waters beyond Norway, but shares its position at the head of Europe with the elusive island of Thule. The name *Thule* has its origins in Classical European literature that relates the discovery of an island in the North Atlantic by the Greek navigator Pytheas Masilliensis (of Marseille) in the fourth century BCE. The names *Iceland* and *Thule* share a complex history.[29] In a well-known passage from the prologue to the Icelandic *Landnámabók*, Thule, which had been mentioned in the works of Bede, is presented as an earlier name for Iceland.

> Í aldarfarsbók þeirri er Beda prestr heilagr gerði er getit eylands þess er Thile heitir ok á bókum er sagt at liggi sex dægra sigling í norðr frá Bretlandi; þar sagði hann eigi koma dag á vetur og eigi nótt á sumar, þá er dagr er sem lengstr. Til þess ætla vitrir menn þat haft at Ísland sé Thile kallat at þat er víða á landinu, er sól skín um nætr, þá er dagr er sem lengstr, en þat er víða um daga, er sól sér eigi, þá er nótt er sem lengst.
>
> (In his book *De temporibus* the Venerable Priest Bede mentions an island called Thule, which in books is said to be six days' sail north of Britain. There, he said, there is no day in winter and no night in summer, when the day is at its longest. Therefore, wise men hold that Iceland is this Thule, because throughout the land the sun shines at night when the day is at its longest, and the sun is not seen by day when night is at its longest.)[30]

29 Dale Kedwards, 'Iceland, Thule, and the Tilensian Precedent in Medieval Historiography', *Arkiv för nordisk filologi* 130 (2015), 57–78. For the authoritative account of Thule's history since antiquity see see Monique Mund-Dopchie, *Ultima Thulé: Histoire d'un lieu et genèse d'un mythe* (Geneva: Droz, 2009).

30 Jakob Benediktsson, ed. *Íslendingabók / Landnámabók* vol. 1, Íslensk Fornrit 1

The prologue to *Landnámabók* paraphrases a description of Thule found both in Bede's *On Times* (*De temporibus*) (*c.* 703) and *The Reckoning of Time* (*De temporum ratione*) (725).[31] Bede situates this island six days' sail from Britain, noting that the sun does not rise there in the depths of winter or set at the height of summer. Bede himself was not among the 'wise men' who had associated Iceland with Thule; Iceland was not discovered and settled by the Norse until the ninth century, but the Icelandic author of *Landnámabók* appears keen to reference Bede in order to associate his history of the Icelandic settlement with a renowned authority.[32]

The convention of calling Iceland by the name *Thule*, however, was not original to *Landnámabók*. This historiographical motif traversed Northern European literatures, appearing in at least six histories written between the eleventh and thirteenth centuries. The works that explicitly associate Iceland with Thule are Adam of Bremen's *Gesta Hammaburgensis ecclesiae pontificum* (*Deeds of the Bishops of Hamburg-Bremen*) (*c.* 1075–1081), the anonymous *Historia Norwegiae* (*History of Norway*) (1160–1175), Theodoricus Monachus's *Historia de antiquitate regum Norwagiensium* (*The Ancient History of the Northern Kings*) (*c.* 1180), William of Newburgh's *Historia rerum Anglicarum* (*History of English Affairs*) (*c.* 1190s), Saxo Grammaticus's *Gesta Danorum* (*Deeds of the Danes*) (*c.* 1200), and *Breta sögur*, the Icelandic translation of Geoffrey of Monmouth's *Historia regum Britanniae* (*History of the Kings of Britain*).

On the larger Viðey Map, Iceland is presented as one of the focal points of Europe, its maker clearly interested in Iceland and its place in the wider world. Its appearance besides Thule, however, is ambiguous. The map may, like the prologue to *Landnámabók*, be a statement of these names' equivalence, suggesting that the two names, which no space or outlines separate, related to the same island. However, the Viðey Map is not unique in its double placement of Thule and Iceland, and maps including the English Cotton Map (*c.* 1050) and the Hereford Map (*c.* 1300) show both islands as separate entities located in the North Atlantic. Since the map makes sparse use of outlines, it is unclear whether the map-maker intended to show one island known by two names, or two

(Reykjavík: Hið Íslenska Fornrit Félag, 1968), p. 31.

31 Bede, *On the Nature of Things and On Times*, trans. Calvin B. Kendall and Faith Wallis (Liverpool: Liverpool University Press, 2010) 7, p. 111; and Bede, *Reckoning of Time* 31, p. 91. Bede in turn takes his information from Pliny the Elder's *Naturalis Historia* 2.77, ed. and trans. H. Rackham (Cambridge, MA: Harvard University Press, 1958), pp. 318–19; and 4.16, ed. H. Rackham (Cambridge MA: Harvard University Press, 1961), pp. 198–99; and Gaius Julius Solinus's *Collectanea rerum memorabilium* 22.9, ed. Th. Mommsen (Berlin: Weidmann, 1895), pp. 101–02.

32 On Bede's reputation in Iceland and the accessibility of his works see Donald K. Fry, 'Bede', in *Medieval Scandinavia: An Encyclopaedia*, ed. Philip Pulsiano (London: Garland, 1993), pp. 36–37.

separate islands. Their relationship on the larger Viðey Map, and in the traditions that may have informed it, have significant implications for the map's characterisation of Iceland that we shall explore in this chapter.

The name *Thule* originates in a now-lost treatise written by the Greek navigator Pytheas of Marseille, who allegedly circumnavigated Britain between 325 and 320 BCE.[33] At this time, the dominant power in the Western Mediterranean was Carthage, and a Punic blockade restricted access to the Atlantic Ocean and ensured that northern waters remained largely unknown to Mediterranean seafarers.[34] Pytheas's western explorations are contemporary with Alexander the Great's campaigns eastwards (356–323 BCE). Pytheas claimed that on entering the Atlantic Ocean he sailed northwards to Britain, following the coast as far as Orkney. From there he sighted but did not approach an island he named *Thule*. Although Pytheas's own record of his expedition, *On the Ocean*, no longer survives, his descriptions of the North Atlantic Islands and Thule were quoted and paraphrased in later works.[35] Thule became a staple if vague presence in the North Atlantic through the *Geographica* of the Greek historian Strabo (64/63 BCE–*c*. 24 CE) and Pliny's *Historia Naturalis* (*c*. 77–79 CE), whose accounts were canonised for later antiquity and the Middle Ages by Solinus's *Collectanea rerum memorabilium* (*Collection of Remarkable Facts*) (22.9) and Isidore's *Etymologies* (14.6.4).

Pytheas recorded geographical and astronomical information about the lands he visited, and is the earliest known author to relate systematically the length of the solsticial day to a place's latitude. What is important about Pytheas's description of Thule is that his observations place it on or near the formerly hypothetical Arctic Circle. As we saw on the Icelandic hemispherical world map, this circle was astronomically defined as the latitude above which there is at least one day in the

33 Pytheas's journey and his contributions to early medieval geographical knowledge are described in Þorvald Thoroddsen, *Landfræðissaga Íslands. Hugmyndir manna um Ísland, náttúruskoðun þess og rannsóknir, fyrr og síðar*, 4 vols. (Reykjavík: Ísafoldatprentsmiðja, 1892–1904), vol. 1, pp. 2–6; Vincent H. de P. Cassidy, 'Voyage of an Island', *Speculum* 38 (1963), 595–602, 595; and Haraldur Sigurðsson, *Kortasaga Íslands*, p. 23.

34 Cassidy, 'Voyage', 595.

35 Only one direct citation from it is known: Geminus of Rhodes' astronomical treatise *Introduction to the Phenomena* (first century BCE) cites Pytheas on the length of the solsticial day. 'It seems that Pytheas of Massilia came also to these [far northern] regions. At least he says in his treatise *On the Ocean*: 'The Barbarians showed us the place where the sun sets. For it happened that in these parts the night becomes extremely short, sometimes two, sometimes three hours long, so that the sun rises a short while after sunset". Geminus of Rhodes, *Geminos's Introduction to the Phenomena: A Translation and Study of a Hellenistic Survey of Astronomy*, trans. James Evans and J. Lennart Berggren (Oxford: Princeton University Press, 2006) VI.9, p. 162.

summer when the sun does not set (at the summer solstice), and one day in the winter when it does not rise (at the winter solstice).[36] Strabo, however, was sceptical of Pytheas's account of his travels, and cautions that an accomplished navigator and astronomer, like Pytheas, would have been able to fabricate observations of astronomical phenomena in the far north through his knowledge of the 'science of celestial phenomena and by mathematical theory'. That is to say, Pytheas may have theorised the occurrence of the midnight sun at higher latitudes having travelled only so far north as the British Isles.[37]

It is impossible to say how far north Pytheas sailed or what lands he might have seen.[38] It is commonly proposed that the earliest writer to apply the name Thule to Iceland was the Irish monk Dicuil in his geographical treatise *De mensura orbis terrae* (*c.* 825).[39] Dicuil takes his information from a community of Irish monks, who inform him of an island in the North Atlantic where they spend the lighter months of the year. According to their reports, the sun did not set on this island on the summer solstice, causing Dicuil to theorise, in accordance with what he understood about the sun's course, that it would not rise on the winter solstice either. He calls this island *Thile ultima*.[40] Some scholars have

36 The theoretical basis for the observation that the winter nights grow longer, and summer nights shorter at higher latitudes had arisen in the Hellenistic world out of the study of geometry. See Anjac, 'Foundations', pp. 136–42. Latitude was early associated with the length of the day, having been described by the late-antique geographer Pomponius Mela (d. 45 AD) in his *Pomponii melae de situ orbis* (Book 3). Bede's *De natura rerum* (Ch. 47) examines the relationship between a place's latitude and the length of the solsticial day more closely. He names dozens of place-names on eight parallels, one of which, the Scythian Circle, passes through the Rhiphaean Mountains, imagined mountains in the north of Scythia, and through Thule. Along this parallel 'the days are continued without interruption and alternately the nights'. This is from Pliny's *Historia Naturalis* (6.39). The imaginary Rhiphaei Mountains appear on the Hereford world map, see Westrem, *The Hereford Map*, pp. 136–37. On the association between latitude and the length of the day see Aubrey Diller, 'The Parallels on the Ptolemaic Maps', *Isis* 33 (1941), 4–7, 4; and Thoroddsen, *Landfræðissaga Íslands*, vol. 1, p. 7.

37 Strabo, *Geographica*, Loeb Classical Library, ed. and trans. H.L. Jones (Cambridge, MA: Harvard University Press, 1924) IV.5, p. 253.

38 Shetland, Iceland, and Norway are usually put forward as the most likely candidates. See Ian Whitaker, 'The Problem of Pytheas' Thule', *The Classical Journal* 77:2 (1982), 148–64, 159– 60; Mund-Dopchie, *Ultima Thulé*, pp. 24; 35–38.

39 This claim has been made by Thoroddsen, *Landfræðisaga Íslands, p.* 10; Fridtjof Nansen, *In Northern Mists: Arctic Exploration in Early Times* vol. 1, trans. Arthur G. Chater (London: William Heinemann, 1911), pp. 59–60; Cassidy, 'Voyage of an Island', 599; and Mund-Dopchie, *Ultima Thulé*, p. 114.

40 Dicuil names Pytheas, and derives his information about Thule from Pliny, Solinus, and Isidore, all of whom he cites. Dicuil, *Dicuili Liber de Mensura*

identified Dicuil's monks (*clerici*) with the hermits (*papar*) mentioned in the Icelandic *Íslendingabók*, the Sturlubók redaction of *Landnámabók*, and Theodoricus's *Historia*, whose presence was known to early Icelanders from the Irish books, bells, and crosiers they supposedly left behind.[41] Dicuil wrote his treatise, however, several decades before Iceland's discovery and settlement by the Norse in *c.* 870. Later medieval writers who called Iceland *Thule* did not attribute the ancient name to a remote island with no permanent inhabitants but to a contemporary European polity, and to interpret Dicuil's use of the name *Thule* in the same light as later appropriations is to overlook the nationalistic purposes to which it was put by these writers.

The medieval writings that associate these islands are not, as they may appear, earnest attempts to think about the historical reality of Pytheas's Thule, or to identify it with known islands in the North Atlantic. In its geographic sense, Thule was an elusive island situated somewhere in the North Atlantic, but in the Roman literature in which the name appears it figures primarily as a poetic synonym for the ends of the world, a *non plus ultra* that conveyed the limits of territorial reach and ambition. Thule's symbolic resonances are enshrined in a passage from Virgil's *Georgics* (*c.* 29 BCE) that was well-known to medieval authors. Virgil couples the name *Thule* with its companion adjective *ultima* ('outermost', 'farthest') when he extols the Emperor Augustus as a god, and wonders whether he will take earth, sea, or sky as his domain in death.[42]

Orbis Terrae, ed. C. A. Walckenaer (Paris: F. Didot, 1807) 7, pp. 38–39. On this passage see Michael Richter, 'Medieval Ireland and Iceland – worlds apart?', in *Saluting Aron Gurevich: Essays in History, Literature, and other Related Subjects*, ed. Yelena Mazour-Matusevich and Alexandra Korros (Liden: Brill, 2010), pp. 155–66, pp. 160–61.

41 Jakob Benediktsson, ed. *Íslendingabók / Landnámabók*, Íslensk Fornrit 1, 2 vols (Reykjavík: Hið Íslenska Fornrit Félag, 1968), pp. 4, 32–33; and Theodoricus, *Theodoricus Monachus: The Ancient History of the Northern Kings*, trans. David McDougall and Ian McDougall (London: Viking Society for Northern Research, 1998) 3.18–24, p. 6. On these hermits see Patrick Gautier Dalché, 'Comment penser 'l'Océan? Modes de connaissances des fines orbis terrarum du nord-ouest (de l'antiquité au XIIIe siècle)', *Actes des congrès de la Société des historiens médiévistes de l'enseignement supérieur public* 17 (1986), 217–33, 225. Margaret Clunies Ross suggests that reference to these artefacts in pre-settlement Iceland was intended to demonstrate that Iceland was from earliest times Christian. Margaret Clunies Ross, 'Land-taking and Text-making in Medieval Iceland', in *Text and Territory: Geographical Imagination in the European Middle Ages*, ed. Sylvia Tomasch and Sealy Gilles (Philadelphia: University of Pennsylvania Press, 1998), pp. 159–84, p. 174.

42 James S. Romm, *The Edges of the Earth in Ancient Thought: Geography, Exploration, and Fiction* (Princeton: Princeton University Press, 1992), p. 158.

> An deus immensi venias maris ac tua nautae
> Numina sola colant, tibi seruiat ultima Thule,
> Teque sibi generum Tethys emat omnibus undis.
>
> (Or whether you come as a god of the wide sea, and sailors pay homage to your divine presence alone, Ultima Thule obey you, and Tethys bequeath all her waters to you, as her daughter's new bridegroom.)[43]

These same symbolic resonances, of distance and the world's ends, are foregrounded in a similar passage from Seneca's *Medea,* which envisions Thule as the limit of the lands while looking ahead to a time it will no longer be so.

> Venient annis saecula seris, quibus Oceanus vincula rerum laxet, et ingens pateat tellus Tethysque novos detegat orbes nec sit terries ultima Thule.)
>
> (There will come an age in far-off years when the Ocean shall unloose the bonds of things, when the whole broad earth shall be revealed, when Tethys shall disclose new worlds and Ultima Thule not be the limit of the lands.)[44]

In its literary usage, Thule did not represent a distinct geographical area or European polity, but was conceptualised as a space lying outside the sphere of Roman political influence. In these literatures, Thule would be assimilated into the Roman Imperium only in mythic time, in the death of emperors or in the countless multitude of years.[45]

The earliest author to associate Iceland as a European polity with antique Thule is Adam of Bremen in the *Gesta Hammaburgensis ecclesiae pontificum* (*c.*1075–81). Adam presents *Thule* as an earlier name for Iceland in the fourth book of his history, which he devotes to the geography of the North Atlantic.

> Insula Thyle, quae per infinitum a ceteris secreta, longe in medio sita est occeano inquiunt, nota habetur. De qua tam a Romanis scriptoribus quam a barbaris multa referuntur digna praedicari. Ultima, inquiunt, omnium Thyle, in qua aestivo solsticio, sole cancri signum transeunte … Item Beda scribit in Britannia aestate lucidas noctes haut dubie repromittere, ut in solstitio continui dies habeantur senis mensibus, noctesque e diverso ad brumam sole remoto. Quod fieri in insula Thyle, Pytheas Massiliensis scribit sex dierum navigatione in septentrionem a

43 Virgil, *Georgics,* ed. and trans. Arthur S. Way (London: Macmillan, 1912) 1.29–31, pp. 2–4.

44 Seneca, *Seneca VIII, Tragedies,* vol. 1, The Loeb Classical Library, ed. and trans. John G. Fitch (Cambridge, MA: Harvard University Press, 2002) ll.375–379, p. 375.

45 Romm, *Edges,* p. 157; Mund-Dopchie, *Ultima Thulé,* pp. 67–82.

Britannia distante. Haec itaque Thyle nunc Island appellatur, a glacie quae occeanum astringit.

(About the island Thule, which is situated at an immense distance out in the Ocean far from all other islands, it is told that it is still rather unknown. However, both Roman writers as well as barbarians mention many things worth telling. Thule, they say, is the last of all islands in the Ocean. There is never night there at summer solstice when the sun passes through the Tropic of Cancer. Correspondingly, there is never daylight at midwinter solstice… Bede writes that the light nights in summer in Britain show that at summer solstice there will be continuous day for half a year and, by contrast, continuous night at midwinter solstice when the sun is far away. Pytheas from Marseille tells us that this is also the case in Thule, which lies six days' sailing from Britain. The said Thule is the island called Iceland because of its ice which makes the sea solid.)[46]

Like the anonymous author of *Landnámabók,* Adam ostensibly associates Iceland and Thule on the basis of their comparably high latitudes and geographical positions north of Britain. Adam mentions Iceland and its inhabitants four other times in the *Gesta,* but always as *Island,* without any reference to Thule.[47]

We can see Adam's motivations for appropriating the name *Thule* when we examine his rationale for describing the North Atlantic islands, which he says are 'non praetereundae sunt a nobis, quoniam Hammaburgensem parrochiam et ipsae respiciunt' ('not to be overlooked by us because they also belong to the diocese of Hamburg').[48] Adam was keen to credit the Hamburg diocese with the Christianisation of the far north, and so relies upon Thule's sense as a synonym for the ends of the world to exaggerate the scale of his diocese's authority. When Adam uses the name *Thule,* he envisions Iceland as the most distant region to which Christian evangelism

46 Adam of Bremen, *Gesta* IV.36, pp. 271–72; Adam of Bremen, *History,* p. 217. As the *Gesta*'s translators note, Adam obtained his information about Thule from Paulus Orosius, *Pauli Orosii Historiarum adversum Paganos Librii VII,* ed. Carolus Zangemeister (Vienna: Apud C. Geroldi Filium Bibliopolam Academiae, 1882) I.2.79, p. 29, Paulus Orosius, *Seven Books of History against the Pagans,* trans. A. T. Fear (Liverpool: Liverpool University Press, 2010), p. 45; and Bede, *De temporum ratione* 31, which he cites verbatim.

47 Firstly, Adam describes how the people of Iceland ('populis Islandorum') provide military aid to the royal St. Óláfr (*Gesta* II.61, p. 121); secondly, he exaggeratedly claims that King Haraldr Hardradi extended his empire as far as Iceland ('ille cruentum imperium usque ad Island extendit') (*Gesta* III.17, p. 159); thirdly, he mentions Icelanders alongside legates from the Orkney Islands and Greenland who had travelled to Bremen ('inter quos extreme venerant Islani') (*Gesta* III.24, p. 167); and fourthly, he mentions Iceland in reference to its Christianisation (*Gesta* III.72, p. 220).

48 Adam of Bremen, *Gesta* IV.34, pp. 271–72; Adam of Bremen, *History,* p. 215.

can aspire, and credits his foundation with having taken Christianity to the very ends of the Earth.[49] We can perceive a similar impulse in Adam's false claim that King Haraldr Hardradi had extended his empire as far as Iceland (III.37), which, reminiscent of Virgil's deployment of Ultima Thule in the *Georgics*, evokes the distant island to exaggerate the scale of his narrative and the achievements of its imperial protagonists.

Iceland is associated with Thule in two other works from medieval Scandinavia: the anonymous *Historia Norwegiae* (1160–75), and the *Gesta Danorum*, written by the Danish historian Saxo Grammaticus (*c.* 1200). The anonymous author of the *Historia Norwegiae* was certainly familiar with Adam's *Gesta*, where he would have encountered the claim for an Icelandic Thule. As Lars Boje Mortensen notes, the *Historia Norwegiae*'s geographical introduction elaborates upon and extends Adam's earlier description of the North Atlantic, updating it to serve the missionary mandate of the archdiocese at Trondheim (Niðaróss), which had been established in 1153/54.[50] Its author calls Iceland by two names. He explains that the island known to Norwegians as *Islandia* means 'the land of Ice', a Latinised name he combines with the calque *glaciei terra* ('que a Norwagensibus igitur Islandia, quod interpretatur glaciei terra, nuncupatur'). Similarly, Greenland is not given a vernacular name *per se*, but is named by the Latin calque *Virdis Terra*. The second name used in the *Historia Norwegiae* is the traditional name *Ultima Thule*, which its author presents as the island's earlier name.

> Que ab Italis *ultima Tile* dicta est, nunc quam magna frequencia colonum culta, quondam uasta solitudo et usque ad tempus Haraldi Comati hominibus incognita.
>
> That large island called by the Romans *Ultima Thule*, which today is inhabited by a great host of settlers, but which was once a vast wilderness and unknown to mankind right up to the days of Harald Fairhair.[51]

While this claim that ancient Thule is the island rediscovered as Iceland appears in chapter eight of the *Historia*, however, the author uses a demonym derived from the name *Thule* already in chapter one, when he credits the people of Thule, who are clearly the Icelanders, with the discovery of Greenland.

> Que patria a Telensibus reperta et inhabitata ac fide catholica roborata terminus est ad occasum Europa.

49 See Romm, *Edges*, p. 157. Christ twice enjoins his apostles to spread Christianity to the ends of the Earth (Matthew 28:19 and Mark 16:15).

50 *Historia Norwegie*, ed. Inger Ekrem and Lars Boje Mortensen and trans. Peter Fisher (Copenhagen: Museum Tusculanum Press, 2003), p. 17.

51 *Historia Norwegiae*, pp. 68–69.

(This country [Greenland], discovered, settled, and confirmed in the Catholic faith by Tilenses [Icelanders] marks the western boundary of Europe.)[52]

The demonym *Tilenses*, describing the people of Thule, appears without comment, before the author has introduced *Thule* as an alternative name for Iceland; the readers of the *Historia Norwegiae* are apparently expected to know who the *Tilenses* are.[53] Iceland and the Icelanders are similarly named *Tylen* and the *Tylenses* in Saxo Grammaticus's *Gesta Danorum* (*c.* 1200), where the *Tylenses*, the Icelanders, are likewise mentioned before Tylen. Saxo praises the *Tylenses* for their exceptional literary and historical achievements in the poetry and sagas he includes among his sources.[54] That neither the author of the *Historia Norwegiae* nor Saxo explain the who the Tilenses are implies that the motif of calling Iceland by the name *Thule* had become widespread among Scandinavian writers.

Not all writers, however, accepted the convention of using the name *Thule* for Iceland uncritically. The Norwegian monk Theodoricus in *The Ancient History of the Norwegian Kings* (*Historia de antiquitate regum Norwagiensium*) (*c.* 1180) acknowledges that the name *Thule* is sometimes attributed to Iceland when he describes its discovery and settlement by Norwegian seafarers centuries earlier. Theodoricus writes that some of its early discoverers believed that this island was Thule, but then equivocates: 'sed nos quia nescimus rei veritatem nec affirmamus nec negamus' ('but since I do not know I neither affirm nor deny the truth of this matter').[55] Theodoricus might make his view known, however, when he, like Saxo, names the Icelanders among his informants in the prologue to his *History*. Unlike Saxo and the author of the *Historia Norwegiae*, he does not call them *Tilenses* but *Islendinga*.[56]

Perhaps closest to the Viðey Map in their conception, and most suggestive of the relationship between its two mapped place-names, are the Icelandic works that associate Iceland with Thule. The best-known Icelandic attestation of this motif appears, as we have seen, in the prologue

52 *Historia Norwegiae*, pp. 54–55.

53 A number of terms for the inhabitants of Thule appear in the scholarship. Nansen, *Northern Mists*, uses *Telensias* (*passim*), McDougall and McDougall refer to the *Tilenses*, *Ancient History*, p. 55, fn. 3, while Mund-Dopchie, *Ultima Thulé*, uses *Thulites* (*passim*). Tilenses (or *Tylenses*) has the authority of being a loan from the Latin, and so has been used here.

54 Saxo Grammaticus, *Saxonis Gesta Danorum*, ed. J. Olrik and H. Ræder (Copenhagen: Levin & Munksgaard, 1931) 0.1.4, p. 5.

55 Theodoricus, 'Theodrici monachi Historia de antiquitate regum Norwagiensium', in *Monumenta Historica Norvegiæ: Latinske Kildeskrifter til Norges Historie i Middelalderen*, ed. Gustav Storm (Oslo: A. W. Brøgger, 1880), pp. 1–68, p. 8; Theodoricus, *Ancient History*, p. 6.

56 Theodoricus, *Ancient History*, p. 55, n. 3.

to *Landnámabók*, whose compositor cites Bede's 'aldarfarsbók' (either *De temporibus* 7 or *De temporum ratione* 31) for its information about Thule. The second Icelandic manifestation of this motif appears in *Breta sögur*, the Old Norse translation of Geoffrey of Monmouth's *History of the Kings of Britain* (*Historia regum Britanniae*). *Breta sögur* may have been translated into Icelandic from as early as *c.* 1200, but its earliest witness is preserved in a section of the manuscript known as Hauksbók compiled by the Icelandic statesman Haukr Erlendsson *c.* 1302–10.[57] Iceland is associated with Thule in a description of the fictional king of the Icelanders named Maluasius, who provides men to serve in the military campaigns of the mythical British King Arthur. In Geoffrey's Latin original, Maluasius is described as 'Maluasius rex Islandiae' ('Maluasius king of the Icelanders'). However, its Icelandic translator elaborates: 'Malvasius Tile konvngr. Þat heitir nv Island' ('King Maluasius of Thule, which is now called Iceland').[58] Geoffrey of Monmouth at no point associates his *Islandia* with Thule, using *Islandia* in all four of his references to Iceland, where, like Thule in Roman literature, the island is evoked in fictionalised contexts to emphasise the reputation and power of the legendary British kings.[59] The Hauksbók redaction of *Breta sögur* retains two of these references: when Iceland is mentioned for the second time, it numbers among the territories conquered by King Malgo, and is not associated with the name Thule as it has been earlier in the saga.[60]

The occurence of this motif in *Breta sögur* is not isolated. A version of *Landnámabók* based on the Sturlubók redaction and a now-lost version written by the lawspeaker and prior Styrmir Kárason is also preseved in Hauksbók and, like *Breta sögur*, is written in Haukr's own hand. It might be the case, then, that the incorporation of the name Thule into *Breta sögur* was motivated by the appearance of this motif in *Landnámabók* earlier in the Hauksbók compilation. This suggestion finds support in Marianne Kalinke's observation that the reference to an Icelandic Thule ruled by

57 Marianne Kalinke, 'Arthur, King of Iceland', *Scandinavian Studies* 87:1 (2015), 8–32, 9.

58 *Hauksbók*, p. 291.

59 Arthur takes his fleet to *Islandia*, and conquers the land (153.219); King Maluasius of *Islandia* pays homage to Arthur at Caerleon (156.345); Arthur leads an army made up of men from various lands including *Islandia* (162.524); and Malgo rules *Islandia* and other territories (183.119). See Geoffrey of Monmouth, *The History of the Kings of Britain: An Edition and Translation of De gestis Britonum (Historia regum Britanniae)*, ed. Michael D. Reeve and trans. Neil Wright (Woodbridge: Boydell, 2007). Reeve's edition does not supply any variant readings (e.g. *Thule* instead of *Islandia*) in any of these passages, suggesting that the case for their equivalence was original to the Icelandic translator.

60 See Geoffrey of Monmouth,*The History of the Kings of Britain* 183.119; *Hauksbók*, p. 295.

King Maluasius does not appear in the longer redaction of *Breta sögur*, preserved in a manuscript that does not contain *Landnámabók*.[61] In writing the motif into *Breta sögur*, Haukr invents a reference to Thule where none had existed in the Latin original, perhaps, earnestly, as a corrective to what he saw as the presumption that Iceland in its history had had kings, or, more subtly, to corroborate the association between these islands in the Icelandic *Landnámabók*.

This elaboration on Geoffrey's *Islandia* in the Icelandic *Breta sögur* can be compared with a similar passage in the *Historia rerum Anglicarum* (*History of English Affairs*), a history of post-conquest England written by the English historian William of Newburgh in the 1190s. In the proem to his *History*, William strives to undermine the credibility of Geoffrey's *History of the Kings of Britain* and denounce its fabulous tales of the British past. William cites the conquest of *Islandia* among the extravagant and clearly fabricated triumphs that Geoffrey attributes to King Arthur.

> His quoque addit Islandiam, quae ultima Tile secundum quosdam dicitur, ut ad Britonem illum in veritate spectare videatur, quod Augusto Romano a poeta nobili adulatorie dicitur: tibi serviet ultima Thule.
>
> (To these he [Geoffrey] adds as well Iceland, which some call furthest Thule, so that the flattering words addressed to the Roman Augustus by the celebrated poet [Virgil] seem in truth to refer to that Briton: 'Furthest Thule will be your slave'.)[62]

William is clearly familiar with the convention of calling Iceland by the name Thule, and applies it to his commentary on Geoffrey's *Islandia* in much the same way as the Icelandic translator of the *Breta sögur*. He does not do so, however, to claim their equivalence. Rather, drawing on Thule's symbolic resonances as a synonym for the ends of the Earth, which he exemplifies in a citation from Virgil's *Georgics* (1.30), he ridicules the extravagance of Geoffrey's claim that the British King Arthur's sovereignty exceeded that of the Roman Imperium.

The names *Iceland* and *Thule* have a complex relationship in writings that may have been known to the Icelandic map-maker. The authors of the *Historia Norwegiae* and *Gesta Danorum* use the name *Thule* to demonstrate that Scandinavian historical writings were contiguous with, and could enrich, Roman writings about the far north. Roman authors could only imagine a time when Ultima Thule would be assimilated into the Roman Imperium, but in these national histories of the Norwegian and

61 The longer version of *Breta sögur*, as yet unedited, is preserved in AM 573 4to, which does not contain *Landnámabók*. Kalinke, 'Arthur', 29, n. 18.

62 William of Newburgh, *The History of English Affairs*, book 1, ed. and trans. P. G. Walsh and M. J. Kennedy (Warminster: Aris and Phillips, 1988), pp. 32–33.

Danish peoples, Thule features as a stable part of the northern world. When the *Historia Norwegiae* records that Thule was unknown until the time of Harald Fairhair ('usque ad tempus Haraldi Comati hominibus incognita'), its author implies that the Norwegians have, in bringing human civilization to Thule, surpassed even the Romans' globalising achievement.

Since Iceland had only been settled in the ninth century, its history was short, and it had no authenticating presence in the seminal works of authoritative authors.[63] In claiming Thule as an ancient precursor to Iceland, however, Icelandic authors were able to claim for Iceland two things it sorely lacked: a human prehistory, and a presence in Roman literature. This Tilensian precedent enabled Icelandic authors to write their histories of the Icelandic people into a space shared with the historical keynotes of the High Middle Ages, as well as with Roman poets whose works formed the basis for the European curriculum.

Thule may have provided Icelanders with an opportunity to extend their history into antiquity, and in that sense resembles the Scandinavian origin story cultivated in the geographical introductions to *Snorra Edda* and *Ynglinga saga*. European ruling elites frequently traced their dynastic origins to the heroes who migrated westwards following the legendary Trojan war. The opening chapters of *Ynglinga saga* relate the establishment of the legendary Swedish Yngling dynasty at Uppsala by the god Freyr, the geographical description on which it opens showing the centralist origins of the Scandinavian royal families. The author of *Ynglinga saga* claims that the Norse gods were not true divinities, but exceptionally cunning people who migrated to Northern Europe from Central Eurasia and the regions surrounding the Tanais estuary. Óðinn, who was gifted with knowledge of the future, knew that he and his descendants would find posterity in the world's northern regions, and so travels through Garðaríki (Russia) and Saxland (Saxony) to establish himself on the island of Fyn in Denmark. His descendants, though his son Freyr, became the Ynglings, who are the legendary progenitors of the Scandinavian royal lines.

These transcontinental migrations into Northern Europe are described through a series of onomastic asides, through which the author provides etymological proofs of the Æsir's historical settlements in Central Eurasia. The name *Enea,* which *Ynglinga saga* and *Snorra Edda* pose as an alternative name for Europe ('en fyrir vestan kalla sumir Európa, en sumir Enea', 'and the west is called by some Europe, and others Enea'), appears to have been invented by an Icelandic author, perhaps Snorri Sturluson himself, to associate Europe with an eponymous founder in the Trojan

63 Diana Whaley, 'A Useful Past: Historical Writing in Medieval Iceland', in *Old Icelandic Literature and Society*, ed. Margaret Clunies Ross (Cambridge: Cambridge University Press, 2000), pp. 161–202, p. 163.

hero *Aeneas*.[64] Brutus, a legendary descendant of the hero Aeneas, was celebrated as the eponymous founder of Britain in Geoffrey of Monmouth's *History of the Kings of Britain* and its related vernacular literatures.[65] In the second part of his *Edda*, called *Skáldskaparmál* ('language of poetry'), Snorri Sturluson claims that the Norse 'god' Víðarr was in truth the Trojan Aeneas, demonstrating the Icelanders' connectedness to other European ruling elites who claimed descent from Troy.[66]

The alternative names for the River Tanais and its surrounding area provide the basis for a lengthier onomastic aside that authenticates a link between the Scandinavians and their historical settlements in Central Eurasia. The River called the Tanais, which formed the conventional boundary between Europe and Asia, is described as that river 'sú er at réttu heitir Tanais. hon var forðum kǫlluð Tanakvísl eða Vanakvísl' ('which is correctly called the Tanais. It was earlier called the Tana-river branch or Vana-river branch'). The region surrounding the Tanais River is also known by multiple names, as 'í Vanakvíslum var þá kallat Vanaland eða Vanaheimr' ('around the Vana-estuary was then called Vana-land or Vana-home').[67] These alternative names for the Tanais and the regions on its banks appear, like *Enea*, to have been fabricated by an Icelandic author to create an authenticating link between Central Eurasia and the divinities called the *Vanir*, whom the saga claims had their ancestral home there. Fittingly, the word *kvísl*, which is here used in its sense of 'river branch', 'tributary', 'an arm of a river', appears more commonly in its metaphorical sense as a 'branch of a family' or 'lineage'; the Tanakvísl, then, is both river branch, and the place where the Vanir branch of the Norse divinities had its origins. In *Gylfaginning*, Snorri similarly links the

64 *Heimskringla*, vol. 1, p. 10. See Simek, 'Snorri als Kosmograph', pp. 262–64.

65 On the myth of Trojan origins in medieval European literatures, see Sylvia Federico, *New Troy: Fantasies of Empire in the Late Middle Ages* (Minneapolis: University of Minnesota Press, 2003); and Emily Wingfield, *The Trojan Legend in Medieval Scottish Literature* (Cambridge: D.S. Brewer, 2014). On euhemerism and other mechanisms for a *translatio imperii* in Icelandic literature see Whaley, 'A Useful Past', p. 178; Sverrir Jakobsson, 'Hauksbók and the Construction of an Icelandic worldview', *SBVS* 31 (2007), 22–38, 27; and Jonas Wellendorf, 'Zoroaster, Saturn and Óðinn: The Loss of Language and the Rise of Idolatry', in *The Performance of Christian and Pagan Storyworlds*, eds. Lars Boje Mortensen and Thomas M. S. Lehtonen with Alexandra Bergholm (Turnhout: Brepols, 2013), pp. 143–70.

66 'Hann er Eneas, hann kom braut af Troju ok vann síðan stór verk' ('he is Æneas, he came out of Troy, and accomplished then great deeds'). Snorri Sturluson, *Snorra Edda: Skaldskáparmál*, ed. Anthony Faulkes (London: Viking Society for Northern Research, 2005) ll. 27–28, p. 6. Simek, 'Snorri als Kosmograph', pp. 262–64.

67 *Heimskringla*, vol. 1, p. 10.

Æsir to their ancestral home in Asia on confected etymological grounds, averring that the *Æsir* (singular *Ás*) were *Asiamenn* ('Asians').[68]

The legendary geography that accommodated the Trojan diaspora to European shores sometimes found cartographic expression. On the Hereford Map (*c.* 1300), Troy, called 'civitas bellicosa' ('the warring city'), is depicted in a ruinous state, its crumbling battlements representing the city's fall and the translation of its power to Rome through Aeneas's descendants.[69] The late medieval Gough Map of Great Britain (*c.* 1370, with revisions into the fifteenth century) bears an inscription on the Devonshire coast that reads 'hic Brutus applicuit cum Troianis' ('here Brutus landed with the Trojans'), evoking the legendary origins of the British people, who came to Britain's shores with its eponymous founder, the Trojan Brutus.[70] Matthew Paris (*c.* 1200–59) drew a map of Britain showing the ancient roads laid by Brutus's descendant, King Belinus, described in Geoffrey of Monmouth's *History*.[71]

The Viðey Maps are the only maps from medieval Iceland that were probably produced in Snorri Sturluson's lifetime, at the Augustinian monastery he himself founded, with Þorvaldur Gissurarson, in 1226. On the Viðey Map, the Troy legend appears initially to have been ommitted from the map's depiction of Asia Minor, but added later in a cramped hand above Phrygia, the province in which Isidore of Seville said it was located in his *Etymologies* (14.3.41). That its absence from the map was noticed and subsequently amended reveals, perhaps, the map-maker's awareness of Troy's centrality to European fantasies of national origin. On the border between Europe and Asia, the Icelandic map-maker has positioned the River Tanais, which served as the conventional boundary between these continental spaces. The name by which it is inscribed onto the map – 'Tanakvisl fluvi*us* maxim*us*' ('Tanakvísl, the greatest river') – is what *Ynglinga saga* describes as its archaic form ('var forðum kǫlluð'), a form notably distinct from its proper name *Tanais* ('sú er at réttu heitir Tanais'). As we have seen, the map names few places in Old Norse, but those it does are coextensive with areas of Scandinavian settlement. By enlisting the river's vernacular name *Tanakvísl*, the map-maker may gesture towards the mythic, archaic Scandinavian homelands in Central Eurasia, granting cartographic expression to the Trojans' migrations from Asia into Europe at the very place where these two continents meet. Through the map's two archaic names –Thule and Tanakvísl – the Icelandic map-maker extends Icelandic history into Classical antiquity

68 Snorri Sturluson, *Gylfaginning*, p. 6.

69 Diarmuid Scully, 'Augustus, Rome, Britain and Ireland on the Hereford mappa mundi: Imperium and Salvation', *Peregrinations: Journal of Medieval Art and Architecture* 4:1 (2013), 107–133, 110.

70 Birkholz, *King's Two Maps*, p. 66.

71 Delano-Smith and Kain, *English Maps*, p. 16.

to negotiate Icelandic anxieties about the island as a *terra nova*, and, like other Europeans, demonstrate their culture's centralist origins.

Mapping Medieval Iceland

The portrayal of Iceland on the larger Viðey Map is complicated by the double appearance of the names *Thule* and *Iceland*. As we have seen, medieval authors sometimes presented *Thule* as an alternative name for Iceland, and it is conceivable that the Icelandic map is a cartographic witness to a motif widespread in Icelandic and Scandinavian writings. Iceland, however, appears on four other medieval maps broadly contemporary with the historical writings that argued these names' equivalence: the Cotton Map (*c.* 1050), the Sawley Map (*c.* 1190), Gerald of Wales's map of Europe (*c.* 1200), and the Hereford Map (*c.* 1300).[72] Whether or not the discovery of Iceland was also the rediscovery of Pytheas's Thule, the advent of Iceland in the late ninth century did not displace Thule on contemporary maps. Contrary to the prevalence of narrative statements that associate these two names, Iceland became a double of Thule, with both islands sometimes drawn separately in the North Atlantic.

There are an additional two medieval maps, besides the larger Viðey Map, that show both Iceland and Thule in their depictions of the North Atlantic. The earliest of these, the exceptionally detailed Cotton Map (London, British Library, MS. Cotton Tiberius B.V., f. 56v, *c.* 1050), has the dual distinction of being the earliest map to show Iceland, and, as we have seen, of bearing the earliest written attestation of the name *Iceland*. The preponderance of its 146 legends derive from late antique geographical treatises as well, perhaps, as now-lost Roman administrative records and itineraries.[73] The map is, however, particularly notable for the detail in which it depicts North Atlantic coastlines, combining traditional names like *Orcades insulae* ('Orkney Islands') with newly available information about territories such as the Isle of Man, Shetland, the Faroe Islands, and, as we saw in the introduction to this chapter, Iceland.[74] Histories of

72 Haraldur Sigurðsson counts a further twenty-one maps and charts that show Iceland from the fourteenth and fifteenth centuries. Haraldur Sigurðsson, *Kortasaga Íslands*, pp. 44–45; Chekin, *Northern Eurasia*, p. 309.

73 Miller, *Mappaemundi*, vol. 3, pp. 29–37. Patrick McGurk, D. N. Dumville, M. R. Godden, and Ann Knock, ed., *An Eleventh-Century Anglo-Saxon Illustrated Miscellany (British Library Cotton Tiberius B. V. Part 1)* (Copenhagen: Rosenkilde and Bagger, 1983), p. 86; Harvey, *Mappa Mundi*, p. 389.

74 McGurk, *Illustrated Miscellany*, p. 81. These islands also feature on the map prepared by the Arabic scholar Muhammad al-Idrisi at the court of the Norman king Roger II of Sicily in 1154. See S. Maqbul Ahmad, 'Cartography of al-Sharif al-Idrisi', in *The History of Cartography* vol. 2, book 1: *Cartography in the Traditional Islamic and South Asian Societies*, ed. J. B. Harley and David Woodward (London: Chicago University Press, 1992), pp. 156–72.

Icelandic cartography frequently note that the Cotton Map contains this earliest appearance of the name *Iceland*, but seldom observe that it also includes *Tylen*.[75] The map-maker situates Iceland near the Norwegian Peninsula – perhaps recognising some cultural association between the two regions – but locates Thule west of Great Britain and Ireland, in the map's lower left-hand corner.

Iceland and Thule are also both present on the English Hereford Map (*c.* 1300), which shows three islands – 'Ultima Thule', 'Ysland' ('Iceland') and 'Farerei' ('Faroe Islands') – below the Norwegian Peninsula (which the map-maker adorns with an image usually interpreted as a skier) and above the 'Orcades' ('Orkney Islands').[76] The map-maker does not place Iceland and Thule remotely, as on the Cotton Map, but side-by-side. The remaining medieval maps that show Iceland do not show Thule. The English Sawley Map (*c.* 1190) shows a peninsular Iceland connected to the European mainland by a narrow isthmus.[77] A map of Europe preserved in Dublin, National Library of Ireland, MS 700, f. 48r (*c.* 1200), possibly produced at Lincoln, also prominently depicts 'Yslandia', but no Thule, in the waters beyond Ireland.[78] This map is preserved between texts of Gerald of Wales's *Topographia Hibernica* (*Topography of Ireland*) and *Expugnatio Hibernica* (*Conquest of Ireland*), and while the companion map omits Thule, both islands are described as separate entities in the *Topographia*'s survey of the North Atlantic islands.[79] Gerald describes the length of the solsticial day at Thule – information that derives ultimately from Strabo's paraphrase of Pytheas – but does not connect this, as the author of *Landnámabók* had done, with any similar observations on the length of the day in Iceland.

These cartographic parallels complicate our view of the relationship between the names *Iceland* and *Thule* on the larger Viðey Map. While it was in the interests of some medieval authors to present them as synonyms, map-makers tended to leave their relationship unresolved,

75 McGurk, *Illustrated Miscellany*, p. 85.

76 Westrem, *Hereford Map*, pp. 194–95.

77 On the depiction of Iceland and Scandinavia on the Sawley and Cotton Maps, see Hiatt, 'The Sawley Map', p. 122. See also McGurk, *Illustrated Miscellany*, p. 81.

78 See Delano-Smith and Kain, *English Maps*, pp. 15–16; and Chekin, *Northern Eurasia*, p. 139. This map is examined in relation to its accompanying texts in Thomas O'Loughlin, 'An Early Thirteenth-Century Map in Dublin: A Window into the World of Giraldus Cambrensis', *Imago Mundi* 51:1 (1999), 24–39.

79 Gerald takes his information about Thule from Orosius and Solinus, as well as Virgil's *Georgics*. Gerald of Wales, *The History and Topography of Ireland*, trans. John J. O'Meara (Mountrath: The Dolmen Press, 1982) II.13, II.17, pp. 67, 68. Thule and Iceland are also described separately in the Franciscan scholar Bartholomeus Anglicus's encyclopaedia *De proprietatibus rerum* (c. 1240). See Mund-Dopchie, *Ultima Thulé*, p. 120, n. 36.

showing both as separate cartographic entities. It is conceivable that the double placement of these islands on medieval maps has its origins in the uncertainty with which some authors wrote about them. Presented with speculation but no certain information about Thule, a map-maker may have preferred to combine but not reconcile their sources. The proximity of Iceland and Ultima Thule on the Hereford Map may indicate that the map-maker had some awareness of a connection between them, but was, like Theodoricus Monachus, unwilling to pronounce on their precise relationship, or rather, like William of Newburgh, intended to emphasise their distinctiveness.

There is little evidence internal to the map that can help us pronounce on which interpretation is more viable. The map-maker assigns a single landmark two names in just one other instance, locating the Biblical Mount Sinai with the legend 'Syna id est horeth' ('Sinai, which is Horeb'). This double name originates in a discrepancy between the Bible's two books relating the revelation on Mount Sinai. In Exodus 19, the mountain atop which Moses received the Ten Commandments is called Sinai, but in Deuteronomy 1, this mountain is called Horeb. In his *Onomasticon*, Eusebius tells us that Horeb is 'the mountain of God in the land of Midian, located near Mount Sinai beyond Arabia in the wilderness' (*Ono* 946), but whether these were distinct peaks in the same range or separate mountains altogether divided medieval and later exegetes.[80] The Icelandic map assigns both names to the same mountain, though the Hereford Map depicts them separately. It is possible, but not at all required, that if the Icelandic map-maker understood *Iceland* and *Thule* to be synonyms, they may have used a similar formula to the one used here.

Given the Viðey Map's Icelandic origin, and the presumed geographical awareness of its maker, it has proven tempting to wonder whether Thule could represent lands other than Iceland. Indeed, modern identifications of Thule are numerous, and it is conceivable that medieval Icelanders might also have given consideration to its identity. Thule attracted the attention of the map's first commentator, Carl Christian Rafn, who remarks:

> Il est bien remarquable que ce géographe islandais qui adopte presque partout les denominations de lieu créées par les anciens géographes latins, se souvenant de leur 'Ultima Thule' a donné ce nom aux contrées situées dans l'Amérique et découvertes par ses compatriotes. On se rappelle que ces pays qui sont le Groenland, le Helluland, le Markland, le Vinland, ont par les géographes du Nord été rapportés à notre partie du monde. En employant le nom de Tile pour des pays

80 Eusebius, *Eusebii Pamphili Episcopi Caesariensis Onomasticon: Urbum et Locum Sacrae Scripturae. Graece cum Latina Hieronymi Interpretatione*, ed. F Larsow and G. Parthey (London: Williams & Norgate, 1862), p. 375.

situés au-delà de l'Islande, l'auteur du planisphère révêle sa connaissance d'un pays plus éloigné.

(It is indeed remarkable that this Icelandic geographer, who adopts almost all the place-names created by the old Latin geographers, remembered 'Ultima Thule' and gave the name to the lands situated in America and discovered by his countrymen. We remember that these lands, Greenland, Helluland, Markland, and Vínland, had been reported by northern geographers to our part of the world. By using the name of Tile for lands located beyond Iceland, the author of the planisphere reveals his knowledge of a more distant land.)[81]

This is a rather remarkable but indefensible suggestion.[82] If this *Thule* did represent either Greenland or North America, it would be the earliest known cartographic representation of these regions by more than two hundred years: Greenland would not otherwise appear on maps until the publication of Claudius Clavus's *tabula moderna* in a Latin translation of Ptolemy's *Geography* in around 1424–27.[83] Even if we accept that these regions were known to the map-maker, there is no reason to think that they would give the name *Thule* to either of them: as *Landnámabók* and *Breta sögur* indicate, there is no evidence, so far as I am aware, that Icelanders meant anything but Iceland when they wrote *Thule*. It seems unlikely that the map discloses any information about the Norse discovery of North America. The map is generally silent on the Norse explorations of northern waters, and does not show, as the English Hereford Map does, the locations of the Faroe or Orkney islands, which were indisputably known to its maker. As we have seen, the double placement of these islands on maps is traditional and, as their early appearance on the Cotton Map demonstrates, the exemplar from which the Viðey Map was copied may already have incorporated the names *Iceland, Thule*, or both before it was received in Iceland.

81 Rafn, *Antiquités Russes* II, p. 393. My translation.

82 This claim appears to be unknown to Kirsten Seaver, but strengthens her argument that the controversial Vínland Map is an early twentieth-century hoax. See Kirsten Seaver, *Maps, Myths, and Men: The Story of the Vinland Map* (Stanford: Stanford University Press, 2004). Rafn suggests that the earliest map to show America would emerge from the Norse voyages towards the North American continent in the late ninth century. The Vínland Map is accompanied by a confused prose account that conflates the two Vínland voyages of Þorfinnr Þórðarson and Leifr Eíriksson, information that could be derived second-hand from Rafn's volume. It seems to me that this statement in Rafn's account might have inspired the Vínland Map hoax.

83 After Clavus, the north would not be mapped again until the Bavarian-born Jacob Ziegler published his maps of Scandinavia in 1530 and 1536. See Mead, 'Scandinavian Renaissance Cartography', pp. 1781–86. Halldór Hermansson, *Cartography of Iceland*, pp. 14–20; Haraldur Sigurðsson, *Kortasaga Íslands*, p. 11.

Given the complex historiographical and cartographic traditions from which the Icelandic map-maker may have drawn, it is difficult to pronounce on what the relationship between these two names is. It is conceivable that the map is a visual statement of their equivalence, the strength of this interpretation resting in the fact that the motif recurs in the Icelandic *Landnámabók* and *Breta sögur*. The map's cartographic parallels, however, urge caution. Their double placement may be an attempt by the map-maker to preserve the distinctiveness of their sources. The map-maker has taken information from multiple sources – some ancient and some contemporary – and, making no effort to harmonise them, has left them unassembled. *Thule* and *Iceland* were not perfect synonyms. Thule was primarily a *non plus ultra*, an emblem of geographic extremity enlisted by medieval authors, such as Adam of Bremen and Saxo Grammaticus, to aggrandise their territorial claims in the far north, or else, like William of Newburgh, to ridicule those who had used the name *Iceland* to do so. Thule was not a locus of polity or civilisation, but lay beyond their limits. Conversely, Iceland was a contemporary European political entity belonging to western Christendom. Perhaps in writing both Thule and Iceland on the map, the Icelandic map-maker was able to separate these two sets of connotations. Iceland was not Thule: Iceland was civilised, it belonged to western Christendom, and it was European; Thule, which the map-maker plots further from the map's centre, was separate, and its connotations of inaccessibility, disconnectedness, and geographic extremity did not define Iceland. Perhaps instead of a Tilensian precedent the map creates a Tilensian 'other', evoking Thule to excise its connotations from Iceland. The relationship between the map's Iceland and Thule remains uncertain; the map-maker presents these interpretations as alternatives but does not resolve the tensions between them. This is the only map that survives from medieval Iceland that shows Iceland, and we cannot know what forms other examples of Icelandic cartographic self-portraiture might have taken. The combined literary and cartographic evidence urges us, however, to articulate subtler positions on the relations between medieval world maps and their related literatures.

Conclusion

The larger Viðey Map is concise in its description of Scandinavian geography, but presents a version of these regions that centres the Icelanders' European cosmopolitanism. The map is selective in its use of Old Norse, preferring to name places – even those that were a stable part of the Icelandic worldview – in Latin. The map-maker saves Old Norse for areas of Scandinavian settlement, naming Iceland, the provinces of Sweden, and Denmark in the vernacular. Medieval maps frequently used language to distinguish between their regions, enlisting local vernaculars in the depiction of lands known to their audiences, and supraregional Latin to represent places of particular historical or ecclesiastical significance. Old Norse compasses the map's Scandinavian regions, but extends also into Central Eurasia in the place-name 'Tanakvisl fluui*us* maxim*us*'. In *Ynglinga saga,* the name Tanakvísl appears to have been fabricated to strengthen its author's claim that the progenitors of the legendary Swedish Yngling dynasty came to Northern Europe from Troy and the regions surrounding the Tanais Estuary. Tanakvísl or Vanakvísl was the ancestral home of the Norse divinities called the Vanir, whom the saga claims were not true divinities, but people of exceptional ability and cunning who migrated to Scandinavia from Asia, bringing Icelandic vernacular poetics with them. Its appearance on the map evidences the Scandinavians' historic settlements in Central Eurasia, demonstrating the Icelanders' connectedness to other European ruling elites who claimed descent from Troy.

Tanakvísl is not the only mapped place-name that elicits Icelandic prehistory. The name *Thule* appears directly above Iceland, beneath the legend *Europa* that sits outside the *kringla heimsins* on the map's conceptual frame. In the Icelandic *Landnámabók* and *Breta sögur, Thule* was presented as an earlier name for Iceland. To be sure, medieval authors were little concerned with the historical reality of Pytheas's voyage or the factual identity of his Thule. Rather, this elusive name, borrowed from antiquity, enabled medieval historians to extend Scandinavian history into Classical antiquity, and retroactively write Iceland into Roman literature. Through aligning their historical traditions with those of authors who wrote about Thule, Icelanders ensured that they did not write about Icelandic prehistory in the absence of ancient authorities. The map is an important act of self-definition that illuminates how Icelandic thinkers sought to position their writings in relation to wider European literary and historical traditions.

Two of the map's names – Thule and Tanakvísl – were, according to Old Icelandic literature, archaic terms for regions now known by other names. The Icelandic map-maker uses the geographical framework provided by the *kringla heimsins* to write a history of the Scandinavian regions, thematising Iceland's relation to its ancient precursor, Thule, and the archaic Scandinavian presence in Central Eurasia. The map extends Icelandic history into Classical antiquity, drawing the regions depicted at its periphery into those at its centre, and characterises its audiences as members of a cosmopolitan European elite descended from Trojan heroes. The larger Viðey Map is an Icelandic origin myth – a mythopoesis – that uses the cartographic medium to think about the Icelanders' origins, and their contemporary relationships with other European elites.

CHAPTER 5

Forty Icelandic Priests and a Map of the World

The two maps from Viðey, drawn across a single manuscript opening and its following verso, are a combined statement about the order of the world and Iceland's place within it. Both show the lands of the three continents, the *kringla heimsins*, centred on the Mediterranean Sea. The larger map shows in detail the geographic dispersion of the world's people and places, while the smaller map, a simple T-O, bears only the names of the world's three continents. These maps' depictions of the known world are enclosed within cosmological wheels that correlate the four cardinal directions with other natural fours. This conceptual frame conveys a vision of an ordered universe in which both space and time are subject to a fourfold ordering. The correspondences between the four cardinal directions, the four seasons, and the four component stages that measure a human life reveal the order intrinsic to nature.[1] The Viðey Maps are complex visualisations of an ordered and harmoniously proportioned world.

The conceptual frame furnishes the two maps with their main organising principle. On the larger map, the cardinal directions are presented in both Old Norse and Latin at the head of each of the map's quarters.

1 These quaternary phenomena, sometimes called the 'physical and physiological fours' or 'quaternities' in modern scholarship, derived from treatises on natural philosophy steeped in Aristotelian precepts and late antique humourism. See J. A. Burrow, *The Ages of Man: A Study in Medieval Writing and Thought* (Oxford: Clarendon Press, 1988); and Jennifer O'Reilly, 'Patristic and Insular Traditions of the Evangelists: Exegesis and Iconography', in *Le Isole Britanniche e Roma in età Romanobarbarica*, eds. A. M. Luiselli Fadda and É. Ó Carragáin (Rome: Herder Editrice e Libreria, 1998), pp. 49–94. For a detailed overview of these fours and their role in exegesis see Anna C. Esmeijer, *Divina Quaternitas: A Preliminary Study in the Method and Application of Visual Exegesis* (Assen: Gorcum, 1978).

East and west have been placed outside the map's elliptical frame, while north and south (and their associated natural fours) are accommodated by square inserts intruding onto the map's terrestrial circle. Beneath the cardinal directions is a circle of winds, blowing inwards from their points around the map's horizon. The map's other fours are written into the space below the perimeter of winds in each of the map's four quarters.

This quadripartite structure is more prominent still on the smaller map (f. 6v), the central T-O ensconced by three concentric circles divided into four by prominent radials extending from the map's middle. These concentric outlines accommodate groupings of natural fours: the outer circle containing the winds, months, and signs of the zodiac; the middle circle containing the seasons; and the inner circle containing the human microcosm. It would appear that this distinction of the conceptual frame into concentric zones was also intended for the larger map. On the larger Viðey Map, the map's eastern half (f. 5v) bears three circular outlines: the outermost contains the cosmological matters of the conceptual frame, while the inner two enclose two sequences of Asian legends. The outer begins with the Massagetes and ends with Carmania, while the inner begins with Caria and ends with Canaan, the city of Abraham. While three semi-circular outlines have been drawn on the map's eastern half, only the outermost completes its circle into the map's western half. These outlines appear to be an incompletely realised version of the concentric scheme completed on the smaller T-O map overleaf, abandoned, perhaps, because the map-maker felt these circles were restrictive on the inner space available to arrange the map's geographic legends. The cardinal directions and their related fours have been arranged within the map's outermost circle.

Information about the Viðey Maps has been sought only within the narrow horizons of their own geographic outlines. That the Viðey Maps have sometimes been regarded as staidly conventional probably owes much to the cosmological matters disposed around their perimeters. These moribund assemblages of traditional terms have been seen as having least of all to do with map-maker's geographical awareness and the maps' Icelandic reception. The conceptual wheel enclosing the smaller Viðey Map has not, until now, been reproduced in facsimile.[2] When the cardinal points and their associated fours from the map's frame are transcribed, they are usually placed apart from the map's geographical legends.[3] So poorly are the Viðey Maps' forms and contents understood

2 The fours from the larger Viðey Map have been described briefly by Chekin, *Northern Eurasia*, pp. 69–71, and are transcribed, with some small errors, by Kålund, *AÍ* III, p. 71, and Simek, *Altnordische Kosmographie*, pp. 419–20. The inscriptions around the smaller Viðey Map have not been transcribed until now.

3 Rafn, *Antiquités Russes*, pp. 392–94. Kålund transcribed the map's legends

that it has never been observed that they are drawn on the recto and verso of the same folio. Nor have their companion items been seen. The two maps (ff. 5v–6v) are accompanied by a list of Icelandic priests' names (f. 5r), with a coda that dates the list to 1143 and attributes its composition to the preeminent historian of the Icelandic Commonwealth, Ari Þorgilsson the wise (1067–1148). The list comprises a circular survey of forty Icelandic priests' names through the eastern, southern, western, and northern Quarters, Icelandic administrative regions that had been established in the tenth century for representative and juridical purposes. Both map and register convey considerable interest in Iceland's geographical situation, within both a wider world, and more locally in the organisation of its Commonwealth. Both also exhibit a striking formal alignment in their fundamentally quaternary structures, the fourfold order of nature thematised on the two world maps, reflected in the Quartered division of the Icelandic Commonwealth. This chapter will demonstrate that the conceptual frame is not incidental but integral to the map's endeavour.

In bringing these items together, the map-maker juxtaposes a vision of global geography and an image of the Icelandic Commonwealth attributed to its preeminent historian. This chapter alters the lens to focus on what the map might have meant in relation to this companionate depiction of the Icelandic Commonwealth. The conceptual frame thematises the order and symmetry intrinsic to nature, and provided an Icelandic thinker with a means to think about Iceland and its main social institutions in the thirteenth century.

The Great Idea of Nature

The conceptual frame shows that time and space alike are subject to a fourfold ordering. The cardinal directions – south (*suðr, meridies*), west (*occidens, vestr*), north (*norðr, septentrio*), and east (*austr, oriens*) – compass the map, and confer its main organising principle. Beneath the cardinal directions is a circle of winds, comprising the names of the four principal winds – *Subsolanus* (or *Apeliotes*), *Auster* (or *Nothus*), *Zephyrus* (or *Favonius*), and *Septemtrio* – which arose in the four cardinal directions, and their eight satellite winds, which arose either side of them.[4] The

in *Alfræði Íslenzk* III, with a monochrome photographic facsimile on the rear flyleaf. *AÍ* III, pp. 71–72.

4 Treatises about the winds were inherited from classical antiquity, and the most widely available accounts were Pliny's *Historia Naturalis* II.47, Isidore's *De natura rerum* 37 and *Etymologies* 13.11, and Bede's *De natura rerum* 27. Wind diagrams originated with Aristotle's *Meteorologica* 2.6, and are numerous and variable. On the winds, a property of air and a sub-lunary atmospheric phenomenon, see Barbara Obrist, 'Wind Diagrams and Medieval Cosmography', *Speculum* 72 (1997), 33–84, 34.

names of the winds were strongly associated with the directions from which they rose, and combined with the four cardinal directions named on the map to represent the circular horizon.[5] Through the four seasons – spring, summer, autumn and winter – the frame also represents the circular rhythm of the solar year, whose cyclical nature resulted from the succession of changes generated by the annual course of the sun.[6] The smaller iconic map extends this depiction of the solar year with the addition of a *rota mensium* ('circle of months') and the twelve signs of the zodiac.[7] The twelve months, like the twelve winds, belonged to a quaternary scheme, with three months assigned to each of the four seasons: the season begins in one month, matures in a second, and declines in a third.[8] The signs of the zodiac are the constellations through which the sun moves in its annual orbit of the Earth, and accompany the other temporal fours – the seasons and the months – in the map's depiction of time.

The quaternary pattern that structures space and time is perceivable also at human scales. The four *aetates hominum* ('ages of man') extend the analogy between the four cardinal points and other temporal fours in the map's frame, associating east with spring and infancy (*infantia*),

5 The twelve winds and their properties were frequently schematised on circular wind diagrams, an example of which is preserved overleaf from the Icelandic hemispherical world map in AM 732b 4to (f. 2r). On these diagrams, the four principal winds were usually distinguished from their satellites by some means of decoration or ornament. On the English Psalter Map (London, British Library, Add. 28681, f. 9r, c. 1250), for example, the winds are represented by twelve heads blowing in towards its centre: those representing the four principal winds coloured red, while the lesser satellite winds are blue. On the Hereford Map (c. 1300), the four principal winds are shown as small naked figures with grotesque faces, while the eight satellite winds are represented by open-mouthed dragon heads. See Westrem, *Hereford Map*, pp. 12–19; and Harvey, *Mappa Mundi*, p. 3. On the Icelandic wind diagram, see Kedwards, 'Wind Diagrams'.

6 Isidore of Seville explains that 'dicta sunt autem tempora a communionis temperamento, quod inuicem se humore, siccitate, calore et frigore temperent' ('they are called seasons (*tempus*) from the 'balance of qualities' (*temperamentum*) that each shares, because each in turn blends (*temperare*) for itself the qualities of moisture, dryness, heat, and cold') (*Etymologies* 5.35.1, pp. 128–29). The four seasons are among the most recognisable of the map's quaternary schemes. In the Germanic tradition, there were two seasons (ON *misseri*, OE *misēre*) in a year, *vetr* ('winter') and *sumar* ('summer'). The durations of these two seasons are described in the Icelandic computistical treatise *Rímbegla I. AÍ* II, p. 22.

7 This type of diagram and its relationship to the Hereford Map is discussed by Naomi Reed Kline, *Medieval Thought*, p. 14.

8 Isidore, *Etymologies* 5.35.2–3. Isidore describes the conception of the circular year thus: 'It is called a year (*annus*, cf. *anus* "ring") because it wheels back upon itself with the recurring months' (*Etymologies* 5.36.1, p. 129).

south with summer and youth (*juventa*), west with autumn and old age (*senecta*), and north with winter and decrepitude (*decrepita*). The ways in which medieval authors divided the course of the human life into component stages were many and variable. Physiologists tended to discern four component ages, in analogy with the four humours, while astrologers tended to think of seven, each under the patronage of a different planet. Common to all, however, was an attempt to 'integrate the life of man into the larger order of the natural world'.[9] The same four *aetates hominum* that appear in the Viðey Maps' conceptual frame are excerpted for representation on a diagram in the English De Lisle Psalter, drawn in *c.* 1310. This circular diagram shows Christ at its centre, turning a wheel that comprises ten *aetates hominum,* including 'dying', 'dead', and 'entombed'.[10] Around this wheel are depicted four additional figures with banderoles identifying them as personifications of the same four ages of man shown on the Icelandic maps. On the bottom left lies a youthful figure identified as *infantia,* on the top left stands crowned *juventus,* on the top right stands *senectus,* and languishing on the bottom right is *decrepitus.* The ages of man also adorned the fragmentary Duchy of Cornwall Map (*c.* 1286); a gallery of paired roundels at the bottom of the surviving corner showing the remnants of such a scheme.[11] The alignment of the *aetates hominum* with other temporal fours enabled the map-maker to 'relate the ages of man to temporal patterns observable elsewhere – in the cycles of year, month, and day, and the linear time of history'.[12]

Through the four ages of man, the map embodies the human microcosm, attesting that the course of a human life is in balance with the seasonal rhythms that order nature. In the Old Norse *Elucidarius,* a lesson on humanity's physical nature explains how the human body exhibits in miniature the qualities of a 'mi*nn*e hei*m*r', a microcosm or 'small world'. The teacher explains that:

> Af .iiii. hofoþ scepno*m*. oc callasc h*ann* af þui e*nn* mi*nn*e hei*m*r. Þ*ui*at h*ann* hafþe hold af iorþo e*nn* bloþ af vatne blost af lofste e*nn* hita af elde. Hofoþ h*an*s var bollot ígliki*n*g hei*m*ballar. Iþui ero augo tuav se*m* sol *oc* tungl. Ahi*m*ne. Ibrioste es blóstr *oc* hoste se*m* vindar *oc* reiþarþrum-or ílofste. Q*ui*þr tęcr viþ veco se*m* sęr viþ votno*m*. Føtr halda up ollo*m* lica*m* sem iorþ be*rr* allan hofga. Af hi*m*nes-co*m* elde Heuer h*ann* svn. E*nn* af eno øfra lofste he. QavE*nn* hilmni*n*g af eno neþra. Bergni*n*g af vatne. E*nn* ha*nn*-da kenning af iorþo. Harþleik bei-na afsteino*m*.

9 Burrow, *Ages,* p. 2.

10 Burrow, *Ages,* pp. 45–46; Kline, *Medieval Thought,* p. 38.

11 Harvey, *Mappa Mundi,* pp. 33–34; Terkla, 'The Duchy of Cornwall Map', pp. 211–13.

12 Burrow, *Ages,* p. 2.

E*nn* glikeng trea íno-glo*m* hars voxt af g*ra*se. E*nn* h*ann* kenn*er* sin *sem* kvq*ue*nde þ*ui*at þat alt lica*m*lect eþle manz.

(It consists of the four elements and man is therefore called microcosm, the small world. He got his flesh from the earth and his blood from the water, his breath from the air, and his warmth from the fire. His head is ball-shaped just like the globe. In his head are two eyes just like the sun and moon in heaven. In his breast there are breath and cough, like the winds and claps of thunder in the air. His belly takes in liquid as the sea takes in rivers. His feet support his whole body as the earth carries all its weight. He has his sight from the heavenly fire, his hearing from the upper heights and his smell from the lower levels. His taste comes from the water, and his sense of touch from the earth. He gets the hardness of his bones from the stones, and his nails bear the likeness of trees. The growth of his hair he gets from the grass, and he feels like an animal, for all that is the physical nature of man.)[13]

This conceit is elaborately written onto the maps' conceptual frames through the addition of the component parts of the human body – breath (*spiritus*), water (*aqua*), flesh (*corpus*), and blood (*sanguis*) – a bodily scheme reminiscent of the so-called *Annus-Mundus-Homo* ('Year-World-Man') diagrams sometimes encountered in illustrated handbooks and miscellanies based on the works of Isidore and Bede. These diagrams combined the macrocosmic and microcosmic fours to reveal the harmony of the year, world, and body.[14] One such diagram (Baltimore, Walters Art Museum, MS W.73), bearing the central inscription 'ANNUS-MUNDUS-HOMO', shows the four seasons positioned at the four cardinal points – oriented, like its cartographic cousins, with east at the top – with the four opposing contraries (hot and cold; wet and dry) depicted as overlapping circles. This anthropomorphic trope is epitomised by those maps, such as the Ebstorf Map, that transpose the circle of the known world onto the body of Christ – a statement that the entire world was contained within and knowable by his image.[15]

The conceptual frame attests to an ordered universe, making the world understandable by parsing physical nature into its component systems – space, time and the human body – in ways that aver their relatedness. This natural order is revealed to be consistent across multiple cosmic domains, a statement of universal order through quaternary harmony. This notion was widespread in medieval treatises on natural philosophy, and was commonly granted visual expression through tables and diagrams.[16] The

13 *Elucidarius* I.59, pp.14–17.

14 Harry Bober, 'An Illustrated Medieval School-Book of Bede's *De natura rerum*', *The Journal of the Walters Art Gallery* 20 (1957), 64–97, 79; Kline, *Medieval Thought*, p. 14.

15 Akbari, *Idols*, p. 26.

16 The natural books of Isidore and Bede were sometimes illustrated with

fours assembled in the Viðey Maps' conceptual frames were tabulated in Johannes de Sacrobosco's *De anni ratione* (*On the reckoning of years*) (*c.* 1235) in order to show how the four seasons relate to the four cardinal directions, the four winds, the four humours, and the four ages of man.[17] A version of this table survives in an Icelandic miscellany of astronomical and calendrical texts and tables (Stofnun Árna Magnússonar, AM 193 III 8vo, ff. 11r–11v), where it illustrates a short treatise on the world's nations and their languages.[18] A particularly elaborate visualisation of the cosmic fours is the Byrhtferth diagram in the illustrated encyclopaedia Oxford, St John's, MS 17, f. 7v. The diagram schematises, in addition to the four cardinal points (whose names are inscribed in Greek and Latin), the two solstices and two equinoxes, the four seasons, the four elements, the four ages of man (*puericia, adolescentia, iuuentus,* and *senectus*), and the twelve winds. The severely symmetrical diagram in a ninth-century manuscript of Bede's *De natura rerum* (*On the nature of things*) (Munich, Bayerische Staatsbibliothek, Clm. 210, f. 132v) resembles our Icelandic maps in that it comprises a schematic T-O map, with south at the top, embedded in a diagram that reconciles the cardinal directions and other spatial and temporal fours.[19]

The most compelling analogue for the conceptual frame, however, is not an image *per se*, but a description of one in the form of a treatise known as *The Mystic Ark* (*De arca Noe mystica*), written *c.* 1125–30 by the Parisian theologian and scholastic philosopher Hugh of Saint Victor (*c.* 1096–1141). *The Mystic Ark* is an enormously detailed description of a monumental painting of Noah's Ark, described in Genesis 6, which served as a basis for Hugh's lectures at the school of St Victor on the Christian understandings of time, space, and humanity's salvation. The described image, and its written description, are remarkably complex. Like Augustine, Hugh recognised that an understanding of natural philosophy could deepen one's understanding of divine revelation, and his treatises on the Biblical Ark – which also include *On the Moral Interpretation of Noah's Ark* and *On the World's Vanity* – compel an understanding of Scripture that

circular diagrams of cosmic conformity. See Bober, 'Illustrated Medieval School-Book', 64–97; and Kline, *Medieval Thought.*

17 Sacrobosco, *De anni ratione,* p. 39. This treatise was known to some degree in Iceland, having been excerpted and translated in 1812 I on folio 10v.

18 Simek, *Altnordische Kosmographie,* p. 501.

19 Like the Viðey Maps, this diagram has south at the top, and contains a world map marked with the positions of the three continents The cardinal directions are placed at the points of the inner square. The four elements (earth, water, air, and fire) are disposed in roundels at the spaces between the corners of the two squares, with the four contraries (hot, cold; wet, dry) between them. On the sides of the inner square are written the names and qualities of the seasons as they appear on the two Viðey Maps. See Woodward, 'Medieval *Mappaemundi*', p. 335.

encompasses all areas of human knowledge. Although Hugh's complex image is its focus, no manuscripts of *The Mystic Ark* contain so much as a partial attempt to draw it; rather its aim may have been to compose the image mentally and thus reveal the lesson it teaches, not to describe an image that is already made.[20] Like the lectures on which it was based, these scrupulous instructions guide the pupil to spiritual betterment through focussed mental imaging of the Ark's layered components.

Hugh's treatise culminates in a description of 'world map' (*mappa-mundi*) contained within the form of the Ark.[21] This map comprises a 'circle of the earth' (*orbis terrae*) set amid a hierarchy of nested ovals (*circulus oblongus*) representing the zone of air (*aer*) and the region of ether (*ethera plaga*).[22] Hugh writes the history of humanity's salvation onto the world map at the Ark's centre.[23] The map is aligned in such a way that the bow of the Ark points eastwards and its stern westwards, so that the map's geographical contents proceed, as Hugh explains, 'in sequence with the events of time'.[24] The commencement of human history is depicted at the Ark's bow, in the earthly paradise, while its end is depicted at the Ark's stern, in scenes from the Last Judgement and resurrection.

Hugh is little concerned with the geographic features of the central map, but rather focuses on what he calls the *machina universitas* ('machine of the universe'), arranged into two concentric circles around the *orbis terrarum*: the inner zone of air (*aer*), and the outer region of ether (*ethera plaga*). These two circles accommodate cosmological matters in much the same way as the Viðey Maps' conceptual frames. The peripheral circle, the region of ether, accommodates the twelve months and the twelve signs of the zodiac, like the outer circle on the smaller Viðey Map. The

20 Mary Carruthers and Jessica Weiss, 'A Little Book about Constructing Noah's Ark', in *The Medieval Craft of Memory: An Anthology of Texts and Pictures*, ed. Mary Carruthers and Jan M. Ziolkowski (Philadelphia: University of Pennsylvania Press, 2002), pp. 41–70, pp. 41–42.

21 On Hugh's theological cartography and the importance of his *Libellus de formatione arche* see Dan Terkla, 'Introduction: Where to Fix Cadiz?', in *A Critical Companion to English Mappae Mundi of the Twelfth and Thirteenth Centuries*, ed. Dan Terkla and Nick Millea (Woodbridge: Boydell, 2019), pp. 1–20. On the Ark treatises see pp. 10–15.

22 Latin citations from the Ark treatise are taken from Hugh of Saint Victor, *Hugonis de Sancto Victore De archa Noe, Libellus de formatione arche*, ed. Patrice Sicard (Turnhout: Brepols, 2001), p.157; translations follow Conrad Rudolph, *The Mystic Ark: Hugh of Saint Victor, Art, and Thought in the Twelfth Century* (New York: Cambridge University Press, 2014), pp. 397–502, pp. 482–83. As Rudolph observes, the oval shape enables the 'visual contiguity of Ark and earth', *The Mystic Ark*, p. 483, fn. 377.

23 Hugh of Saint Victor, *Libellus*, pp. 157–60; Rudolph, *The Mystic Ark*, pp. 482–93.

24 Rudolph, *The Mystic Ark*, pp. 483–85.

circle of air accommodates the twelve winds, [25] but also shows how the cardinal directions relate to other spatial and temporal fours, a vision of the cosmos 'in quo spatio secundum quatuor partes mundi quatuor anni tempora desponuntur' ('in which the four seasons are arranged in conformity with the four quarters of the world').[26] The spatial and temporal fours coalesce in these depictions of the four seasons, each of which is personified as one of the four ages of man. Spring is depicted as a 'boy' (*puer*); summer, as a 'young man' (*juvenis*); autumn 'portrayed in a manly, mature period of life' (*virili etate expressus*); and winter, 'in the form of one grown old' (*formam senis gestans*). These are described figurally:

> Ver pingitur a lumbis sursum, puer fistulam manu tenens et cantat. Estas iuuenis flores aspicit. Autumpnus in uirili etate expressus, fructus naribus admouet et olfacit. Hiemps formam senis gestans, fructus comedit. Et omnes a lumbis sursum piguntur, singuli in principiis suis. Ver habet delectationem aurium, estas oculorum, autumpnus olfaciendi, hiemps gustandi, quia oblectamentum pueritie est in auditu doctrine, oblectamentum iuuentutis in exemplo operis, oblectamentum uirilis etatis in appetite probitatis, oblectamentum senectutis in experientia uirtutis.
>
> (Spring is painted from the hips up as a boy who plays a shepherd's pipe being held in his hand. Summer, a young man, gazes at some flowers. Autumn, portrayed as a man of middle age, lifts some fruit to his nose and smells it. Winter, having the form of an old man, is eating fruit. All are painted from the hips up, each one at that point at which his particular season begins. To Spring belongs delight of the ears, to Summer of the eyes, to Autumn of smelling, and to Winter of tasting. For the pleasure of childhood is in hearing doctrine, the pleasure of adolescence in the imitation of [good] works, the pleasure of middle age in the desire for uprightness, the pleasure of old age in the experience of virtue.)[27]

The image epitomises the divine order of nature through combining the seasons, ages of man, the senses (whose number has been altered to align with the quaternary scheme),[28] and the course of a person's spiritual life, from their early encounters with Scripture to their experience of virtuous deeds. Hugh's explanation of this complex image illuminates the cosmic fours concentrated in the Viðey Maps' conceptual frames. On the smaller

25 The orientation of the perimeter of winds does not quite accord with the other fours disposed within this circle. Rudolph has suggested that the author derives his information from various sources, perhaps a wind diagram. Rudolph, *The Mystic Ark*, p. 490, fn. 395.

26 Hugh of Saint Victor, *Libellus*, p. 157; Rudolph, *The Mystic Ark*, p. 486.

27 Hugh of Saint Victor, *Libellus*, p. 158; Rudolph, *The Mystic Ark*, p. 487.

28 Rudolph, *The Mystic Ark*, p. 487, fn. 390.

Viðey Map, especially, the natural fours are accommodated by two concentric zones that conform with Hugh's description of the region of ether and zone of air. The outer circle contains the winds, months, and signs of the zodiac; the middle circle contains the seasons; and the inner circle contains the components of the human microcosm.

The Viðey Maps' conceptual frames epitomise what Hugh called the *magna nature ratio* ('great idea of nature').[29] The frame synthesises nature's component systems and combines them into a 'series of categories that could be used to make sense of the world, facilitate memorization of its properties, and analyse the relationship of the parts to the whole'.[30] In bringing these systems together, the conceptual frame is at once a representation of the observable horizon, the seasonal rhythm of a solar year, the measure of a human life, and the human body. That the map's frame could be made to stand for all these things demonstrated the divine symmetry of creation, and of humanity's place within it. The fours are not a structure observed from above, but one that is intrinsic to the world the map shows. Through the combination of the four elements that make up all matter in the sublunary universe, the map's geographical contents are in a literal sense built out of them.[31] The conceptual frame is a statement of quaternary harmony. The map shows that this order was not imposed but rather discerned; these quaternary patterns were naturally, not humanly, constructed.

The cardinal directions are the maps' main organising principle, with their names written in both Old Norse and Latin at the head of each of the world's quarters. The naming of the cardinal directions in two languages, Latin and Greek, was an occasional conceit on medieval maps; the initial letters of their Greek names spell, when the sign of the cross is made over them, the name 'Adam', as we see on the Byrhtferth diagram and on the world map in the same manuscript (Oxford, St John's College, MS 17 on f. 6r and on f. 7v).[32] This conceit was known in medieval Iceland. In the Old

29 Hugh of Saint Victor, *Libellus*, p. 159.

30 Akbari, *Idols*, p. 20.

31 These fours all belong to the sublunary world, and are phenomena inherent to the world depicted inside the terrestrial circle. The nature of the sublunary world is impermanence and change. The four contraries that meet in the seasons combine also to make the four elements (earth, which is cold and dry; water, cold and moist; air, hot and moist; fire, hot and dry), which in turn combine to make infinite, unstable variety. The diagram of planetary orbits, preserved alongside the hemispherical world map in both of its manuscripts, exemplifies this. The representation of the Earth at the centre of the seven planetary spheres is inscribed with the names of the four classical elements that compose all matter up to the lunar sphere, from which point all is composed of quintessence, or ether. Lewis, *Discarded Image*, p. 95.

32 Faith Wallis, '2. Computus Related Materials: 20. Byrhtferth's Diagram', *The Calendar and the Cloister: Oxford, St John's College MS17*. 2007 (McGill

Norse *Elucidarius*, the student asks the teacher how Adam got his name. The teacher responds:

> At. iiii. ott*om* hei*ms* þ*at* es austr *oc* uestr norþr oc suþr. Enn at g*ri*ksco male a d a m callasc anatole disis artos mes*em*bria þat es se*m* g*ri*plor hende til nafns ada*ms*. E*nn* at þui toc h*ann* nafn at fioro*m* otto*m* hei*ms* at fvn h*ans* atte at co*m*a íallar hei*ms*.
>
> (From the four compass points of the world, that is from East and West, North and South. In Greek these are called *Anatole, Dysis, Arktos, Mesembria,* and they form an acrostic of Adam's name. He got his name from the four corners of the world, since his kin was expected to spread out in all directions over the world.)[33]

Adam's name enshrined Christianity's globalising mandate, the expectation that Adam's kin would extend to the four corners of the world, and that the Apostles would take Christianity to them (Matthew 28:19 and Mark 16:15). Elsewhere in the *Elucidarius*, we are told that Christ was dead for forty hours in order to revive the four parts of the world, which had been befouled ten times over through of the violation of the Ten Commandments (I.156–157). On the larger Viðey Map, the four cardinal directions are written in both Old Norse and Latin. These four terms are some of the map's few vernacular contributions, and may have been written in both languages to emphasise the map's quaternary scheme.

The conceptual frame has been assumed to be a relic of a traditional scheme that has nothing to do with the map-maker's geographical awareness, or the map's historical moment in the thirteenth century. The importance of this structuring principle, however, emerges through an examination of the other item preserved alongside the two Viðey Maps in this thirteenth-century fragment. The register of forty Icelandic priests' names that comes before the paired maps, on folio 5r, comprises a circular survey of ten priests from each of the four Quarters of Iceland, attributed to Ari Þorgilsson inn fróði ('the wise'), the preeminent historian of the Icelandic Commonwealth. The conceptual frame is an authoritative statement of order, whose spatial argument resonates with the accompanying register. It remains in this chapter to consider their relationship.

Forty Icelandic Priests

The two Viðey Maps are conjoined, through their placement on the recto and verso of the same manuscript folio, into a single cartographic statement. The larger map teems with geographic detail and variety,

University Library, Digital Collections Program, http://digital.library.mcgill.ca/ms-17) (accessed 27th July 2019).

33 *Elucidarius* I.64, p. 17.

which the smaller map parses into its essential structures, revealing its T-O framework and the quaternary structure of the conceptual frame. These companion maps are preserved with a third item, written in the same hand: a register of forty highborn Icelandic priests allegedly compiled in 1143.[34] The making of world maps was not a distinct or separate activity in the medieval period. Most maps appear in manuscript books and are written in the same hand as the texts alongside them.[35] The thinker responsible for the Icelandic Viðey Maps was also interested in the Icelandic Commonwealth, and it is appropriate these combined items are inspected in combination.[36] The register survives only in this one manuscript witness, and probably owes its survival in the Viðey book compilation to the visually interesting world map that begins on its verso. These two folios (ff. 5r–6v) appear to have been excerpted from an earlier composition and inserted into this highly illustrated compilation for their inclusion of two world maps, and to complement the zonal map and planetary diagrams preserved in the fragment 1812 I, into which these two folios have been bound.

The register (figure 16) contains the names of forty Icelandic priests, with ten drawn from each Quarter of Iceland. The survey advances in a clockwise motion, beginning in the east (*austr*) and proceeding through south (*suðr*), west (*vestr*) and north (*norðr*).[37] Before becoming part of

34 This register is sometimes called 'the Prestaskrá of 1143', and is printed in the *Diplomatarium Islandicum*, volume 1 [no. 29], pp. 183–190, edited by Jón Sigurðsson.

35 Woodward, 'Medieval *Mappaemundi*', p. 286. The hand responsible for the lettering on the monumental Hereford map has also been detected in texts written in manuscript books. M. B. Parkes, 'The Hereford Map: The Handwriting and Copying of the Text', in *The Hereford World Map: Medieval World Maps and their Context*, ed. P. D. A. Harvey (London: British Library, 2006), pp. 107–117, p. 107.

36 It might also be appropriate to examine the Viðey Maps in relation to the texts contained in the second bifolium written in this hand. However, these two fragments, bound together into GkS 1812 4to, can only be related speculatively. It has been assumed that since these two fragments are written in the same hand they must derive from the same manuscript – an assertion possible, but by no means necessary: the fragments may just as well have been written by the same scribe but to form parts of two different manuscripts. The second bifolium contains the first two months from a calendar, in addition to a short time-reckoning treatise titled *Bócarbót*. The treatise describes the division of time into days, months, and years, in addition to the solar and lunar cycles and the intervals over which they achieve parity. Since all other maps from medieval Iceland are preserved in illustrated encyclopaedias, we might reasonably suppose that the original context of the 1812 III maps was similar.

37 A register of the names of the fjords in Iceland headed 'Þessi eru fjarðanöfn á Íslandi' ('these are the fjords' names in Iceland') is preserved in AM 415 4to, a learned miscellany that dates to the beginning of the fourteenth century. The

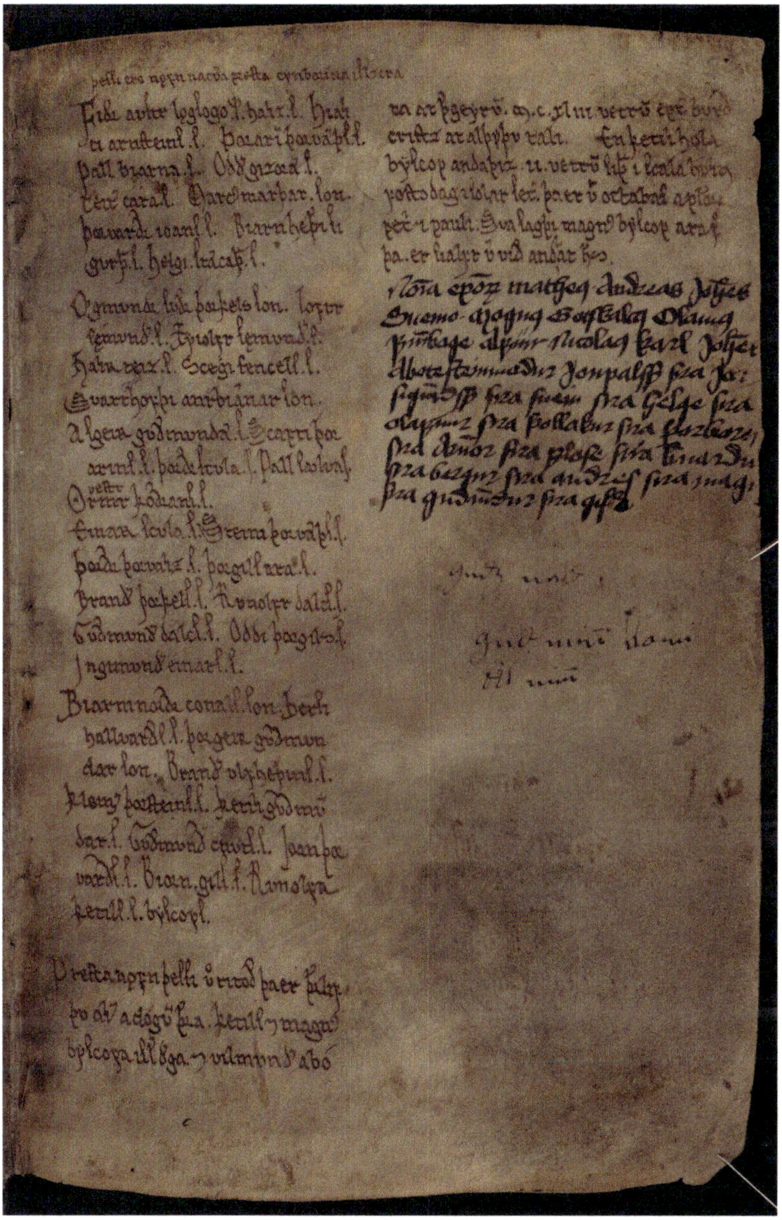

FIGURE 16: The sole manuscript witness to the register of forty highborn Icelandic priests. The *titulus* is at the top of the left column. The register then comprises a circular survey of forty Icelandic priests' names through the eastern, southern, western, and northern Quarters. The coda begins at the bottom of the left column and concludes in the right column. Below, in a hand dated to the 1480s, is a list of Icelandic bishops that includes one abbot of Viðey. It is partly on this basis that 1812 III has been associated with the Augustinians at Viðey. Reykjavík, Stofnun Árna Magnússonar, GkS 1812 III 4to, f. 5r, *c.* 1225–50.

the Norwegian kingdom in 1262/64, the Icelandic Commonwealth was characterised by its decentralised distribution of power.[38] In place of a sovereign or executive power, Iceland was governed by a variable number of *goðar* (chieftains) who presided over the regional Quarter Courts and the annual national assembly.

The register of forty highborn Icelandic priests

(Reykjavík, Stofnun Árna Magnússonar, GkS 1812 III 4to, f. 5r) *c.* 1225–50

Þessi ero nǫfn nacv*era* presta cynbori*nn*a islenzcra
(These are the names of some highborn Icelandic priests)

Fiðr avstr lǫgsogo *maðr* hallz *son*
Hiallti arnsteins *son*
Þorari*n* þorvarþs *son*
Pall biarna *son*
Odd*r* gizora*r* *son*
Teit*r* cara *son*
Marc*us* marþar son
Þorvarðr ioans *son*
Biarnheþi*n* sigvrþ*ar* *son*
Helgi sta*r*caþ*ar* *son*
Ögmundr svðr þorkels son
Loftr sæmvnd*ar* son
Eyiolfr sæmvnd*ar* *son*
Hallr teit *son**
Sceg*g*i fencels *son*
Svarthofþi anrbia*r*nar son
Asgeir*r* gvðmvnda*r* *son*
Scafti þorarins *son*
Þorðr scvla *son*

list follows the same order as the register, beginning with the eastern fjords, then naming those in the south, west and finally north. *AÍ* III, pp. 4–5. This is also the geographical sequence followed by the list of ten chieftain-priests in *Kristni saga. Biskupa sögur*, vol. 1, ed. Sigurgeir Steingrímsson, Ólafur Halldórsson and Peter Foote, Íslenzk Fornrit 15.2 (Reykjavík: Hið Íslenska Fornrit Félag, 2003), pp. 42–43.

38 Stephen Pax Leonard, 'Social Structures and Identity in Early Iceland', *Viking and Medieval Scandinavia* 6 (2010), 147–59, 148.

Pall sælvas*on*
Ormr vestr koðrans s*on*
Einar*r* scvla s*on*
Steini þorvar*þ*s s*on*
Þorðr þorvallz s*on*
Þorgils ara s*on*
Brand*r* þorkels s*on*
Rvnolfr dalcs s*on*
Gvmvnd*r* dalcs s*on*
Oddi þorgils s*on*
Jngimvnd*r* einars s*on**
Biarni norðr conals son
Bersi hallvarðs s*on*
Þorgeir*r* gvðmvndarson
Brand*r* vlfheþins s*on*
Klæng*r* þorsteins s*on*
Ketill gvðmv*n*dar s*on**
Gvmvnd*r* cnvts s*on*
Joan þorvarðs s*on**
Biorn gils s*on*
Rvnolfr ketils s*on* byscops

Presta nǫfn þessi *voro* ritoð þa er þ*eir* lifþv all*er* á dǫgvm þ*eir*ra ketils *oc* magn*us* byscopa isl*e*ndi*n*ga *oc* vilmvnd*ar* abóta at þ*in*geyrv*m* m c xliii vetru*m* ept*ir* burð cristz at alþyðu tali. en Ketill hola byscop andaþiz . ii . vetru*m* siþ*ar* i scalaholltj fostodag i solar set*r* þa er v*ar* octabas apostolorvm pet*ri et* pauli. Sva sagþi magn*vs* byscop ara fr*o*þa er sialfr v*ar* við andlat h*ans*.

(These priests' names were written down while they all lived, in the days of Magnús and Ketill, bishops of the Icelanders, and Abbot Vilmundr of Þingeyrar, 1143 winters after the birth of Christ by the common reckoning. Bishop Ketill of Hólar died two winters later at Skálhólt, at sunset on a Friday when it was eight days after the feast of the Apostles Peter and Paul. So said Bishop Magnús to Ari the wise when he himself [Bishop Magnús] was near the end of his life.) [39]

* mentioned as a prominent chieftain-priest in *Kristni saga* (Ch. 17).

The register is arranged into two columns. At the top of the folio, a

[39] *DI* I, p. 181. My translation.

line introduces the register, 'þessi ero nǫfn nacv*er*a presta cynbori*nna* islenzcra' ('these are the names of some highborn Icelandic priests'). The register lists the names of forty priests, ten drawn from each administrative Quarter of Iceland. The first entry for each Quarter comprises the priest's forename (e.g. 'Fiðr'), followed by the cardinal direction from which that Quarter takes its name (e.g. 'austr'), and the priest's patronymic (e.g. 'hallz son'). The other nine priests in each quarter are named without the addition of the cardinal direction. Only the first and last entries of the register contain any additional biographical information about the priests named. The first named priest is identified as a *lögsögumaðr* (lawspeaker): 'Fiðr avstr logsogo maðr hallz son' ('Finnr, Eastern Quarter, son of Hallr, lawspeaker'); and the last is identified as the son of a bishop: 'Rvnolfr ketils son byscops' ('Rúnólfr son of Bishop Ketill').

The register is concluded by a coda of nine half-lines, which describes the register's authorship and the circumstances of its composition with considerable care. It dates the register to 1143, a date that it corroborates with reference to the incumbent bishops of Iceland's two episcopal sees, Ketill Þorsteinsson who was bishop at Hólar (1122–45) and Magnús Einarsson at Skálhólt (1133–48). The third person mentioned in the coda, Vilmundr Þórólfsson, was abbot at the Benedictine monastery at Þingeyrar in the diocese of Hólar from its consecration in 1133 to his death in 1148. It would seem that the register was compiled from information provided by these men, or at least Bishop Magnús, to Ari Þorgilsson (b. 1067– d. 9th November 1148). Ari was the preeminent historian of the Icelandic Commonwealth, whose history of the Icelanders, the *Libellus Islandorum*, or more commonly *Íslendingabók* (1122–33), is the earliest known example of narrative prose in a Scandinavian language.[40] Beneath the coda is a list of seven Icelandic bishops from Skálholt and Hólar written in a hand dated to the 1480s. The bishops held their offices between 1440 and 1480.[41]

Jón Sigurðsson was confident in his view that Ari authored the register, and suggested that the coda's 'Ara fróða' had been adapted from the first person 'mér' at some stage in its earlier transmission.[42] The register resembles Ari's other known work on a number of counts. In the prologue to *Íslendingabók*, Ketill Þorsteinsson, the bishop of Hólar mentioned in the coda, stands opposite Bishop Þorlákr Runólfsson, who

40 In the prologue to *Íslendingabók*, Ketill Þorsteinsson, mentioned in the coda, stands opposite Bishop Þorlákr Runólfsson, who was bishop of Skálhólt between 1118 and 1133. From the reigns of these, bishops it can be determined that *Íslendingabók* was composed by Ari between 1122 and 1133. *Íslendingabók, The Book of the Icelanders; Kristni Saga, The Story of the Conversion*, trans. Siân Grønlie (London: Viking Society for Northern Research, 2006), p. 15.

41 *DI* I, p. 184; *AÍ* II, p. ccxii.

42 *DÍ* I, p. 188.

was bishop of Skálhólt between 1118 and 1133. The committee of learned persons who appear to have collaborated in the register's composition resembles the team responsible for *Íslendingabók*. [43] The line that introduces the register, 'Þessi ero nǫfn nacv*er*a presta cynbor*inn*a islenzcra' ('these are the names of certain highborn Icelandic priests'), resembles the line 'Þessi eru nǫfn langfeðga Ynglinga ok Breiðfirðinga' ('these are the names of the ancestors of the Ynglings and the people of Breiðafjörðr') that heads the appendices to *Íslendingabók* in both of the seventeenth-century manuscripts that preserve it.[44]

A further characteristic of Ari's historical method that can be seen in the coda is its use of dating 'at alþyðu tali' ('by the common reckoning', i.e. CE). The use of CE dating in *Íslendingabók* may, according to Grønlie, be its 'main debt to European learning'.[45] Over the course of his history Ari supplies three CE dates: the settlement of Iceland in 870, the conversion to Christianity in 1000, and the death of Bishop Gizurr in 1120.[46] Since CE reckoning was not the only method of dating available (most events were dated relatively) Pernille Hermann has argued that the selective use of dates after the birth of Christ was a conscious attempt to synchronise Icelandic historical events with 'an international and worldwide perspective'.[47] Hermann argues that CE dates should be understood as references to the shared temporal perspective of Iceland and wider Europe, and the date supplied by the coda should perhaps be seen in similar terms.

The similarities between the coda and the prologue to *Íslendingabók* may endorse the coda's claim that Ari was involved in the register's composition.[48] The coda is careful to establish that the names were written down in 1143 ('presta nǫfn þessi voro ritoð … m c liii vetru*m* ept*ir* burð cristz'), but must itself have been written sometime after 1148, since

43 On the role of the learned committee in the composition of *Íslendingabók* see Pernille Hermann, 'Spatial and Temporal Perspectives in *Íslendingabók*: Historiography and Social Structures', *Viking and Medieval Scandinavia* 1 (2005), 73–89, 73–74.

44 *Íslendingabók Ara Fróða, AM. 113a and 113b, fol.*, ed. Jón Jóhannesson (Reykjavík: Háskóli Íslands, 1956), p. xx.

45 Grønlie, *Íslengingabók*, xx.

46 Hermann, 'Perspectives', 76. The death of Pope Gregory I in 604 is mentioned in the chronology at the end of *Íslendingabók*, but this is outside its main historical purview and usually not mentioned.

47 Hermann, 'Perspectives', 76; see also Grønlie, *Íslengingabók*, p. xx.

48 Richard Sharpe observes that information about the composition and authorship of a work was often incorporated into the text's prologue or coda, 'where it would be less susceptible to omission or change' than it would in a standalone incipit or colophon, which may become unattached when the text is copied. The coda is certainly an important feature of the overall composition of the register and contains valuable information about its origins and composition. Sharpe, *Titulus*, p. 31.

it mentions the deaths of Bishop Ketill Þorsteinsson (b. 1075–77 – 7th July 1145) and Bishop Magnús Einarsson (b. 1092 – 30th September 1148).[49] If Ari authored the coda, as well as compiled the preceding register, he would have to have done so before he died on 9th November 1148. Patricia Pires Boulhosa considers the value of author attributions made to Ari elsewhere. A rubric in the Fríssbók manuscript of *Heimskringla* attributes the work to Ari, even though it describes the reign of Magnús Erlingsson, who ascended the throne in 1162, fourteen years after Ari's death.[50] Whether or not such author attributions are genuine, Boulhosa argues that 'an attribution, as much as a prologue, adds meaning to the text to which it is attached'.[51] The register's attribution to Ari situates it among the wider historiographical output of medieval Iceland, drawing its reader's attention to a formative period in the history of the Icelandic Church in the previous century.

The Forty Highborn Icelandic Priests

The register provides a portrait of the Icelandic priesthood in the first half of the twelfth century. In the first two centuries of the Icelandic Church, there was a close association between ecclesiastical and secular power.[52] Orri Vésteinsson has argued that in the twelfth century, the Icelandic chieftains (*goðar*) assumed roles in the Church to strengthen their grasp on the new institutionalised power. The ecclesiastical identities secured by Iceland's chieftains enabled them to lead their people spiritually, as well as politically, and was a statement of their social and political pre-eminence.[53] In *Kristni saga*, a history of the Christianisation of Iceland written in the early thirteenth century (and so broadly contemporary with

49 The prologue to Ari's *Íslendingabók* likewise identifies the work as a second edition, produced under the guidance of a committee of learned persons. There is a fragment resembling *Íslendingabók* Ch. 5, the account of Þorsteinn surtr's reform of the Icelandic civil calendar, in the section of the Viðey book dated *c.* 1200 (1812 IV, f. 25v). See Larsson, *Äldsta delen, 2 I*, pp. 7–8; *AÍ* II, pp. 65–66.

50 Patricia Pires Boulhosa, *Icelanders and the Kings of Norway: Medieval Sagas and Legal Texts* (Leiden: Brill, 2005), pp. 10–11.

51 Boulhosa, *Icelanders and the Kings*, p. 12; see also Sharpe, *Titulus*, p. 23.

52 Jón Viðar Sigurðsson, *Chieftains and Power in the Icelandic Commonwealth*, trans. Jean Lundskær-Nielsen (Odense: Odense University Press, 1999), p. 185; Jón Viðar Sigurðsson, 'Kings, Earls, and Chieftains. Rulers in Norway, Orkney and Iceland c.900–1300', in *Ideology and Power in the Viking and Middle Ages: Scandinavia, Iceland, Ireland, Orkney, and the Faeroes*, ed. Gro Steinsland, Jon Vidar Sigurdsson, Jan Erik Rekdal and Ian B. Beuermann (Leiden: Brill, 2011), pp. 69–108, p. 91; Orri Vésteinsson, *The Christianisation of Iceland: Priests, Power, and Social Change 1000–1300* (Oxford: Oxford University Press, 2000), p. 193.

53 Orri Vésteinsson, *Christianisation*, p. 193.

the Viðey Maps), it is written that around a century earlier, in the days of Bishop Gizurr (d. 1118), it was the custom that 'flestir virðingamenn lærðir ok vígðir ok lærðir til presta þó at hǫfðingjar væri' ('most men of high rank were educated and ordained priests, even though they were chieftains'). The saga then proceeds to name ten prominent examples of chieftain-priest, five of whom are named also in the register. Chieftain-priests named in both *Kristni saga* and the register are: Hallr Teitsson from the Southern Quarter; Ingimundr Einarsson from the Western Quarter; and Ketill Guðmundarson (d. 1158) and Jón Þorvarðarsson (d. 1150) from the Northern Quarter.[54] The register seems to have been composed a couple of decades after *Kristni saga*. Ketill Þorsteinsson is named as a chieftain-priest in *Kristni saga*, but from 1122 was bishop at Hólar (his son, Rúnólfr Ketilsson, however, is one of the highborn priests named in the register). The renowned priest and scholar Sæmundr Sigfússon inn fróði is named as a prominent chieftain-priest in *Kristni saga*, but by the time the register is compiled, Sæmundr (d. 1133) has been succeeded by his sons Loptr and Eyjólfr Sæmundarson.[55]

The close association between secular and ecclesiastical power was also observed by the author of the thirteenth-century history of the bishopric at Skálhólt, *Hungrvaka* ('hunger-waker', 'appetiser'), in which it is said of Bishop Gizurr (d. 1118) that 'var rétt at segja at hann var bæði konungr ok byskup yfir landinu meðan hann lifði' ('it was right to say that he was both a king and a bishop over the land while he lived').[56] We see from Adam of Bremen's *Gesta* that this characterisation of the Icelandic episcopate had become known already in the eleventh century, Adam noting that 'episcopum suum habent pro rege' ('they hold their bishop as king').[57] A century later, Gerald of Wales notes in the *Topography of Ireland* (*Topographia Hibernica*) (*c.* 1187) that in Iceland 'their priest is their king, and their king is their priest. The bishop has the powers of both kingship and priesthood'.[58] These statements are consistent with the characterisation of the Icelandic priesthood in the line at the register's head, which states that the priests listed are 'cynborinna' ('highborn').

54 *Biskupa sögur* vol. 1, pp. 42–43.

55 Grønlie describes this overlap, but omits Ketill Þorsteinsson, whose name appears in the coda. *Íslendingabók*, pp. 71–72, fn. 106. The register and similar list in *Kristni saga* may derive from a common ancestor or tradition. In the final line of the paragraph naming these chieftain-priests, its author remarks: 'ok margir aðrir þó at eigi sé ritaðir' ('and many others, though their names are not written down [here]'). *Biskupa sögur*, vol. 1, p. 43.

56 *Biskupa sögur* vol. 2, ed. Ásdís Egilsdóttir, Íslenzk Fornrit 16 (Reykjavík: Hið Íslenska Fornrit Félag, 2002), p. 16.

57 Adam of Bremen, *Gesta* IV. 36, p. 273; Adam of Bremen, *History*, p. 217.

58 Gerald of Wales, *Topographica Hibernica* II.46.

The forty priests named in the register probably constituted no small proportion of the total number of clergymen of Iceland in 1143. Of course, the apportioning of ten priests to each Quarter of Iceland is not random, and has little to do with the actual number of priests who ministered in each Quarter.[59] Orri Vésteinsson estimates that the named priests probably amounted to more than ten percent of Iceland's clerical population, and as such is a valuable source for the prosopography of the Icelandic priesthood in this period.[60] Of the sixteen priests in the register that are known from other sources, thirteen of them were chieftains, at a time when there would have been around twenty-seven *goðorð* (chieftaincies) in the country as a whole.[61] Whether or not these men held *goðorð*, their careers and lineages, where known, indicate that all would have been politically influential figures.[62]

The highborn priests known from other sources – such as the Icelandic annals, the *Biskupa sögur*, and the *Sturlunga saga* compilation, written about Icelandic domestic affairs in the thirteenth century – have impressive lineages, and could claim descent from Iceland's primary colonists and other prominent people in the history of the Icelandic Commonwealth.[63] From the Southern Quarter hail both Loptr Sæmundarson and his brother Eyjólfr, sons of the renowned priest and scholar Sæmundr the Wise (1056–1133). One of these sons, Loptr, married Þóra, daughter of the Norwegian king Magnús III (r. 1093–1103). The son he had with Þóra was the famous chieftain Jón Loptsson (d. 1197), who was in turn the foster father of the statesman and literary impresario Snorri Sturluson (d. 1241). [64] Another priest named in the Southern Quarter who would have an illustrious career in the Church was Hallr Teitsson, the nephew of the famous Bishop Gizurr. Hallr was elected bishop of Skálhólt after Magnús Einarsson's death in 1149, but died in Utrecht on his way to Rome the year after in 1150.[65] Klængr Þorsteinsson, named among the priests in the Northern Quarter, was consecrated bishop in his place. Also connected with the Icelandic bishops are Runólfr and Guðmundr Dálksson from the Western Quarter, nephews of Bishop Ketill Þorsteinsson of Hólar (named in the coda). A priest of particular note from the Southern Quarter is Þorgils Arason, son of Ari Þorgilsson the Wise, who is named as the register's compiler. Ari Þorgilsson would have been at least seventy-five years

59 The householders in the four Quarters of Iceland were counted by Bishop Gizurr after the establishment of Iceland's second Episcopal see, the northern diocese of Hólar in 1106 (*Íslendingabók* Ch. 10). Iceland's southern quarter was the most populous.

60 Orri Vésteinsson, *Christianisation*, p. 188.

61 Jón Viðar Sigurðsson, 'Chieftains', p. 91.

62 Orri Vésteinsson, *Christianisation*, p. 188.

63 *DI* I, pp. 188–194.

64 *DI* I, p. 188.

65 *DI* I, p. 190.

old in 1143 and, by this time, would likely have passed on his *goðorð* to his son.[66] The exclusion of Ari from the survey has been taken, *ex negativo*, as evidence of his authorship, and might also demonstrate that the list was intended to show the current political climate in 1143, by which time his role may have been diminished.

The number of elite Icelandic landowners and statesmen who held both *goðorð* and positions within the Church declined towards the end of the twelfth century, until the dual role of chieftain-priest was prohibited by Bishop Eiríkr Ívarsson in 1191, from which time chieftains were only allowed to assume roles in the Church lesser than deacon.[67] In the thirteenth century, when the Viðey Maps and their companion register were produced, aristocratic families less frequently assumed ecclesiastical offices. When the version of the register paired with the two Viðey Maps was written, a survey of highborn Icelandic priests could no longer have been produced.

In addition to this quaternary structure, the register is framed by references to the two most important institutions of the Icelandic Commonwealth. As noted above, the only two names in the list furnished with any additional biographical information are the first and last: the lawspeaker 'Fiðr avstr logsogo maðr hallz son' ('Finnr, Eastern Quarter, son of Hall, lawspeaker'); and the bishop's son 'Rvnolfr ketils son byscops' ('Rúnólfr son of Bishop Ketill'). These two names enclose the register with references to the most important institutions of the Icelandic Commonwealth: the office of lawspeaker (*lögsögumaðr*) and the Icelandic episcopate.[68]

Like other men named in the register, the lawspeaker Finnr Hallsson was of aristocratic descent. Finnr could trace his lineage through his maternal family to Iceland's very first colonist, Ingólfr of the settlement, through Eyvindr, the foster father of Steinuðr the Old, who was Ingólfr's cousin.[69] Steinuðr's settlement is described in *Landnámabók*, where it is stated that she was given a considerable tract of land by Ingólfr himself.[70] The lawspeaker was the sole government office of the Icelandic

66 Orri Vésteinsson, *Christianisation*, p. 188. Chieftaincies were not hereditary, but naturally associated with influential families.

67 Orri Vésteinsson, *Christianisation*, p. 190; Gúðrun Nordal, *Tools of Literacy: The Role of Skaldic Verse in Icelandic Textual Culture of the Twelfth and Thirteenth Centuries* (London: University of Toronto Press, 2001), p. 20.

68 The Icelandic Commonwealth is the period after the establishment of the *alþingi* in 930 up until Iceland's loss of independence to the Norwegian Crown in 1262–64. New constitutions were drafted in 1271 and 1281, and the chieftaincies were abolished. Jón Viðar Sigurðsson, *Chieftains*, p. 9.

69 *DI* I, p. 187.

70 On the importance of Steinuðr's endowment by Ingólfr see Christopher Callow, 'Putting Women in their Place? Gender, Landscape, and the Construction of *Landnámabók*', *Viking and Medieval Scandinavia* 7 (2011), 7–28, 21–22.

Commonwealth, and presumably chosen by consensus of the holders of wealth and lineage. The role of the lawspeaker was to preside over the national assembly, deliver official pronouncements, recite the law from memory and offer legal advice.[71] Pernille Hermann describes the term of the lawspeaker as a 'national point of reference', sometimes used to date events relatively in Old Norse historical writings.[72] Finnr was lawspeaker for two terms between 1139 and 1145, and his inclusion at the beginning of this list (since all the priests are said in the coda to be living at the time it was composed) would likely have been understood as a reference to the date of its composition.

The last priest named in the list, Runólfr Ketilsson, was also of noble descent. His father was Bishop Ketill Þorsteinsson, who is named in the register's coda as the incumbent bishop of Hólar, and who, a decade or two earlier, had been one of the bishops to receive a draft of Ari's *Íslendingabók*. Ketill could claim descent from Síðu-Hallr and the Síðumenn, prominent early advocates of Christianity in Iceland whose fortunes are described in a number of the *Íslendingasögur*. His son Runólfr is also known for composing a skaldic stanza in 1154 about the construction of the church at Skálholt comissioned by Bishop Klængr (also named in the register, in 1143 still a priest in the Northern Quarter), the only stanza featured in *Hungrvaka*.[73]

The function of this register, both at the time of its composition after 1143 and at the time of its reproduction in the Viðey book (*c.* 1225–50), is difficult to discern. Jón Sigurðsson notes that the priests' names were probably recorded in some informative order, but examines the structure of the register no further. He suggests that the list was compiled to demonstrate that in the 1140s it had been customary for men of aristocratic birth to be ordained to the priesthood.[74] The register may have been a precursor to other thirteenth-century histories, such as *Hungrvaka*, which show the close association between ecclesiastical and secular power in the preceding century. Jón Jóhannesson observes that the register's uses appear to be 'very limited', but comments on it no further.[75] Orri Vésteinsson agrees with Jón Sigurðsson that the aim of the list was to document the lineages of prominent families and their connections to the Church, but suggests further the list is an assertion of ecclesiastical power in the four administrative Quarters of Iceland.[76] The Quarter

71 Jón Jóhannesson, *A History of the Old Icelandic Commonwealth*, trans. Haraldur Bessason (Winnipeg: University of Manitoba Press, 1974), pp. 47–48; Grønlie, *Íslendingabók*, pp. 19–20, fn. 34.

72 Hermann, 'Perspectives', p. 76.

73 *Biskupa sögur*, vol. 2, p. 36.

74 *DI* I, p. 184.

75 Jón Jóhannesson, *Íslendingabók*, p. xxii.

76 Orri Vésteinsson, *Christianisation*, p. 188.

division of the list has been perceived as incidental rather than central to the register's design, but is the basis on which it relates to the other items preserved with it in the Viðey book – two maps that thematise the fourfold ordering of nature.

Myth and the Four Quarters of Iceland

The register of forty priests' names maps onto the Icelandic administrative landscape through its use of the four Quarter scheme. Of course, the division of the list along these lines reflects the real geographical boundaries drawn between the administrative Quarters, and is reified by the need to be associated with, and attend regularly, one of the Quarter Courts. These geographical and political boundaries would have been well-known to Icelanders both when the list was allegedly composed in 1143 and at the time it was copied *c.* 1225–50. The Eastern Quarter extended from Helkunduheiði and Skoravíkurbjarg at Langanes to the Jökulsá River at Sólheimasandur; the Southern Quarter extended from Jökulsá to the Hvítá River; the Western Quarter extended from Hvítá to the Hrútafjarðará River; and the Northern Quarter completes the circle between Hrútafjarðará and Helkunduheiði.[77] Through its survey of forty highborn priests in these four Quarters of Iceland, the register becomes a projection of both ecclesiastical and secular authority onto a geographical framework.

The *landnám* ('settlement') episodes in Old Norse historical writings, and most of the *Íslendingasögur* ('sagas of Icelanders'), show that the Icelandic settlement required the socialisation of the new geographical area, as 'immigrant society was obliged to 'produce' its own social space in an entirely new environment'.[78] Ari's *Íslendingabók* documents the Icelanders' cultivation of their land into a new social space through the establishment of the *alþingi* (chapter three) the determination of the calendar (chapter four), the division of Iceland into administrative Quarters (chapter five), and the conversion to Christianity and acceptance of a common law (chapter seven).[79] The division of Iceland into Quarters, and the establishment of the Quarter Courts for juridical and representative purposes in *c.* 965, conventionally marks the end of the Icelandic settlement period and the beginning of the Icelandic Commonwealth. However, while Ari describes the establishment of the Quarters in chapter five of *Íslendingabók*, each Quarter had already been assigned its own primary colonist in chapter two. Hrollaugr, son of Earl Rögnvaldr in Mærr and half-brother of Rollo, settles in the East at Síða; Ketillbjörn the Old settles in the South at upper Mosfell; Auðr the Deep-minded, daughter of Ketill Flat-nose, settles in the West at Breiðafjörðr; and Helgi

77 Jón Jóhannesson, *History*, p. 50.

78 Clunies Ross, 'Land-taking', p. 159.

79 Following Hermann, 'Perspectives', p. 74.

the Lean settles in the North at Eyjafjörðr. The appearance of these primary colonists in *Íslendingabók* before the formal establishment of the Quarter Courts implies that their emergence, and with it the establishment of an Icelandic Commonwealth, was inevitable.

The four Quarters acquire a mythic status that resonates beyond their putative role in the administration and jurisdiction of the Icelandic Commonwealth. In the *Heimskringla* version of *Óláfs saga Tryggvasonar* ('the saga of Óláfr Tryggvason') (*c.* 1230), each Quarter is assigned a mythic protector in the form of one of the four *landvættir* ('land-wights'). When King Haraldr Gormsson ('blue tooth') of Denmark sends a wizard on a reconnaissance mission to Iceland in the form of a whale, his magical scout finds that each Quarter of Iceland is under the protection of a guardian-wight (*landvættr*). In Vopnafjörðr in the Eastern Quarter, the land is defended by a dragon; in Eyjafjörðr in the Northern Quarter an eagle; in Breiðafjörðr in the Western Quarter a great bull; and in Reykjanes in the South a giant.

> Haraldr konungr bauð kungum manni at fara í hamfǫrum til Íslands ok freista, hvat hann kynni segja honum. Sá fór í hvals líki. En er hann kom til landsins, þá fór hann vestr fyrir norðan landit. Hann sá, at fjǫll ǫll ok hólar váru full af landvættum, sumt stórt en sumt smátt. En er hann kom fyrir Vápnafjǫrð, þá fór hann inn á fjǫrðinn ok ætlaði á land at ganga. Þá fór ofan eptir dalnum dreki mikill, ok fylgðu honum margir ormar, pǫddur ok eðlur ok blésu eitri á hann. En hann lagðist í brott ok vestr fyrir land, alt fyrir Eyjafjǫrð. Fór hann inn eptir þeim firði. Þar fór móti honum fugl svá mikill, at vængirnir tóku út fjǫllin tveggja vegna, ok fjǫldi annarra fugla, bæði stórir ok smáir. Braut fór hann þaðan ok vestr um landit ok svá suðr á Breiðafjǫrð ok stefndi þar inn á fjǫrð. Þar fór á móti honum griðungr mikill, ok óð á sæinn út ok tók at gella ógurliga. Fjǫlði landvétta fylgði honum. Brott fór hann þaðan ok suðr um Reykjanes ok vildi ganga upp á Víkarsskeiði. Þar kom í móti honum bergrisi ok hafði járnstaf í hendi, ok bar hǫfuðit hæra en fjǫllin, ok margir aðrir jǫtnar með honum. Þaðan fór hann austr með endilǫngu landi – 'var þá ekki nema sandar ok ørœfi ok brim mikit fyrir utan, en haf svá mikit millum landanna', segir hann, 'at ekki er þar fœrt langskipum'.

> (King Haraldr told a wizard to go to Iceland in an assumed shape, and to see what he could learn there to tell him. He went in the shape of a whale. And when he came to the land he went by way of the north to the west side of Iceland, where he saw that all the mountains and hills were full of land-wights, some big and some small. And when he came to Vapnafjörðr he went down the fjord and intended to make land, a huge dragon, followed by snakes, frogs, and toads came down the valley towards him and blew poison at him. Then he turned to go westward around the land as far as Eyjafjörðr. He went into the fjord. Then a bird came against him, which was so large that its wings

stretched over the mountains on either side of the fjord, and many birds, big and small, followed it. Then he swam farther west, and then south into Breiðafjörðr. When he came into the fjord a large grey bull came against him, wading into the sea, and bellowing fearfully, and he was followed by a crowd of land-spirits. From thence he went round by Reykjanes, and wanted to land at Vikarsskeið, but a mountain-giant came down against him with an iron staff in hand. He was a head higher than the mountains, and many other giants followed him. He then swam eastward along the land: 'there was nothing', he said, 'but sand and skerries, and surf beyond them, and the ocean between the countries was so great', he said 'that a longship could not cross it'.)[80]

The wizard reports back on Iceland's supernatural defences, and the natural oceanic expanses between Iceland and Denmark that make the island invulnerable to invasion. The suggestion that the sea is so wide that no boat can cross it is not so much a statement of Icelandic isolationism or remoteness as it is a statement of Icelandic invulnerability to foreign political encroachment and the Icelanders' right to their own sovereignty.

This episode and its depiction of the Icelandic *landvættir* shows familiarity with Patristic traditions on the four Gospels and their authors. The four Evangelists were linked in exegesis to the so-called four living creatures, heavenly beasts described in Ezekiel 1:10 and in John's vision in Revelation 4:7.

Et animal primum simile leoni et secundum animal simile vitulo et tertium animal habens faciem quasi hominis et quartum animal simile aquilae volanti.

(And the first living creature like a lion; and the second living creature like a calf; and the third living creature having a face like a man; and the fourth living creature like an eagle flying.)

The four Evangelists are not explicitly associated with the four living creatures in the Biblical revelations, but were connected in exegeses in both word and image. The most influential patristic source for the equivalence of the four living creatures and the four Evangelists was Jerome's *Plures fuisse*, and the introduction to his commentary on Matthew's gospel (written 398).[81] Jerome uses the number four to defend the four canonical gospels against heretics who would alter their number with the addition of more spurious books, linking them explicitly to what Jennifer O'Reilly describes as 'those other sets of four whose diverse components also form a unity: the four elements, the four seasons, the four cardinal virtues, the four directions or parts of the earth'.[82] The four Evangelists

80 *Heimskringla*, p. 271.
81 O'Reilly, 'Evangelists', p. 53.
82 O'Reilly, 'Evangelists', p. 57.

and the four heavenly beasts of Revelation were also evidence of the divine order intrinsic to nature, which had been 'assimilated to existing cosmological concepts in which space, time and matter were seen as part of a fourfold ordering'.[83]

On medieval world maps, the four corners of the world are often associated with the four Evangelists and their animal avatars: Mark, the lion, in the East;[84] Matthew, the man, in the North; John, the eagle, in the West; and Luke, the ox, in the South.[85] In a section of the Icelandic Teiknibók (AM 673a III 4to, f. 16r) dated 1450–75, a simple representation of the world, a T-O map, features amid a complex composition showing Christ in Majesty within a mandorla (figure 17). This image completes a series of roundels that show the six days of Creation.[86] On the preceding folio (f. 15v), four roundels show the separation of light and darkness; the placing of the sun, moon, and stars in the firmament; the separation of land and water and the creation of the plants; and the creation of fish and birds. On folio 16r, the scheme is completed *bas-de-page* with depictions of God's creation of the land animals, and lastly humanity, in Adam and Eve. The simple T-O map, representing all miniaturised creation, lies at the foot of Christ's throne. Surrounding the Christ and the T-O map are representations of the four living creatures, symbols of the Evangelists, whose likenesses, in Iceland, paralleled the four *landvættir*. This composition is a concise statement of the world and time that begins with Creation, and looks forward, through the four living creatures of Revelation, to the end of time and humanity's redemption. The four Evangelists were also incorporated into Henry III of England's Painted Chamber at Westminster in 1243, to complement a world map he had commissioned for those walls in 1236.[87] According to Isidore, the Evangelists' number corresponded with their mission.

> Hi sunt quattuor Euangelistae, quos per Ezechielem (1. 10) Spiritus sanctus significauit in quattuor animalibus. Propterea autem quattuor animalia, quia per quattuor mundi partes fides Christianae religionis eorum praedicatione disseminata est.
>
> (These are the four Evangelists, whom the Holy Spirit symbolized through Ezechiel (1:10) as four animals. The animals are four because, by their preaching, the faith of the Christian religion has been disseminated through the four corners of the earth.)[88]

83 O'Reilly, 'Evangelists', p. 54.

84 The Icelandic tradition has a dragon as the northern living creature, where we would expect the lion of St. Mark. I have not been able to find any other examples, in text or image, of the dragon.

85 Woodward, 'Medieval *Mappaemundi*', p. 336.

86 Guðbjörg Kristjánsdóttir, *Íslenska Teiknibókin*, pp. 138–43.

87 Birkholz, *Two Maps*, p. 17.

88 Isidore, *Etymologies* 6.2.40, p. 138.

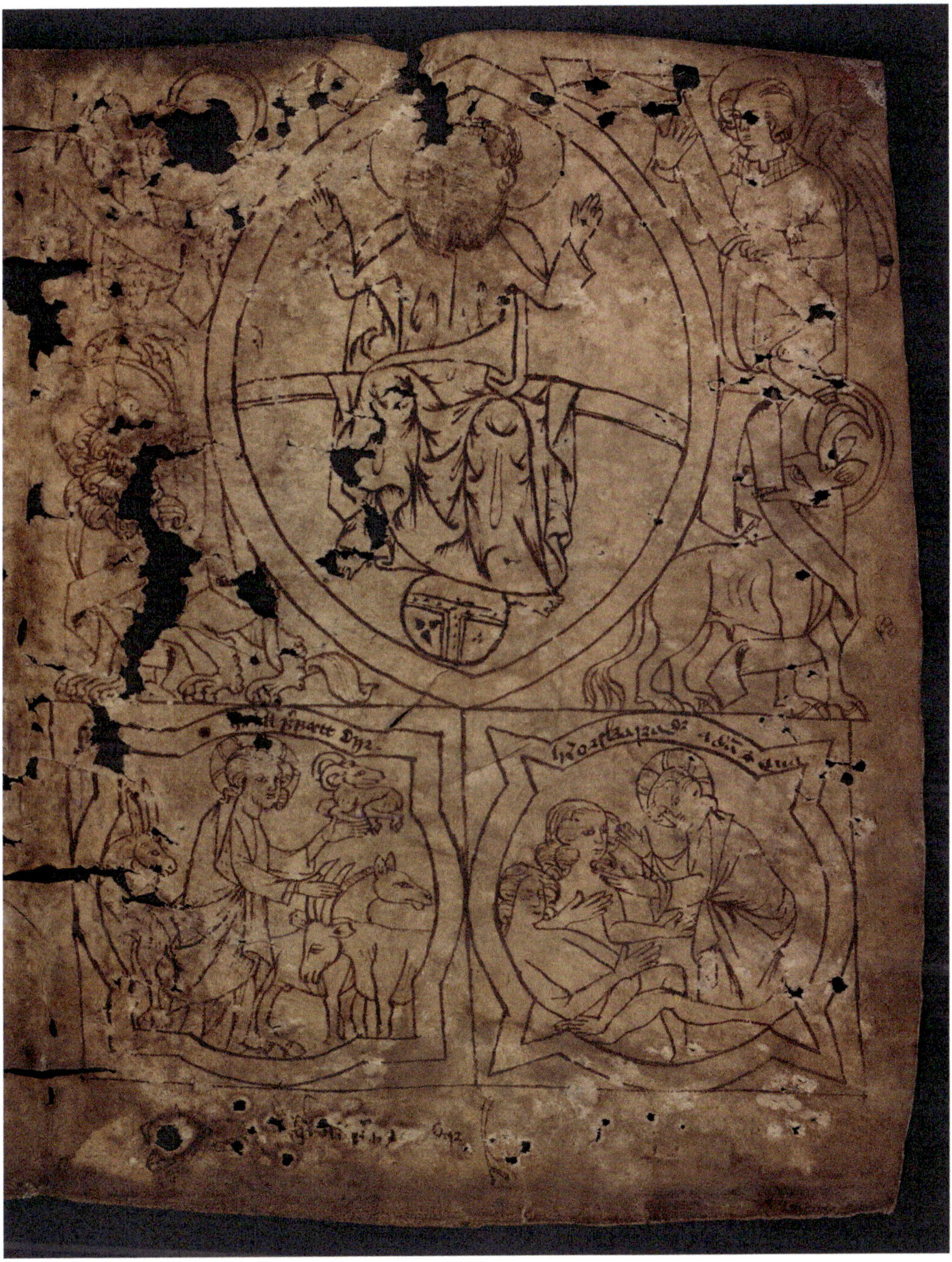

Figure 17: Icelandic depiction of Christ in Majesty, with the T-O map at his feet. The four living creatures, symbols of the Evangelists, are stationed around him. This composition is the culmination of a series of drawings showing the six days of Creation. Shown here are God's creation of the land animals, and Adam and Eve. Reykjavík, Stofnun Árna Magnússonar, AM 673a III 4to, f. 16r, 1450–75.

This knot of fours is once again explained in Hugh's *Mystic Ark*, which assigns an Evangelist to each of the Ark's four corners.

> Sunt etiam in quatuor angulis arche quatuor Euangeliste depicti: in frigore orientis leo ut terreat elatos, in frigore occidentis aquila ut illuminet cecos, in calore occidentis uitulus ut carnem mactet, in calore orientis homo ut hominem ad originem suam reuocet.
>
> (Also, the Four Evangelists are depicted at the four corners of the Ark: in the cold of the east the Lion, so that it may inhibit the exalted with fear; in the cold of the west, the Eagle, so that it might illuminate the blind; in the heat of the west, the Calf, so that it might slay the flesh; in the heat of the east, the Man, so that he might summon the human race back to the place from where it originally came.)[89]

The Evangelists were four because the directions in which the word of God was to be taken was also four. Their depiction on or by world maps averred their globalising mission.

The correspondence between the four living creatures and the Icelandic *landvættir* reveals that the political institutions of the Icelandic Commonwealth were divinely sanctioned and protected. As there are four Evangelists, there are four Quarters of Iceland. Because of their manifold associations in natural philosophy, Scripture, and exegesis, the cardinal points became more than a simple means of orientation but, as David Woodward writes, 'mythical entities in their own right'.[90] The same may be said of the four Quarters of Iceland. In addition to the juridical and representative functions that are the officially cited reasons for their establishment, the Quarters became a more fundamental cornerstone of the distinctiveness of Icelandic social institutions. In *Heimskringla*, the division of the four Quarters is sanctioned by their divine protectors modelled on the Evangelists; in *Íslendingabók*, these divisions are sanctioned by the etiological narratives associated with the Quarters' outstanding primary colonists: Hrollaugr in the east, Ketillbjörn in the south, Auðr the Deep-minded in the west, and Helgi the Lean in the north. The Quarters emerge in these works as natural and inevitable.

When we examine the register and the two Viðey Maps in sequence, we see that all three have the four cardinal points as their main structuring principle, calling attention to the fourfold ordering intrinsic to both Iceland and the world. This common structure legitimises the main social and legal institutions of the Icelandic Commonwealth by showing that the map's authoritative spatial argument is replicated in miniature within them. The Icelandic Quarters are elevated to the distinction of being a

89 Hugh of Saint Victor, *Libellus*, p. 147; Rudolph, *The Mystic Ark*, p. 457.

90 Woodward, 'Medieval *Mappaemundi*', p. 337.

natural four on a par with the cosmological matters assembled in the maps' frames. On the larger Viðey Map, the cardinal directions appear to be named twice, in both Latin and Old Norse. However, given the strong formal affinity between register and map, we might suggest that the map's *suðr, vestr, norðr,* and *austr* do not denote the four cardinal points, but the Icelandic Quarters, which had been depicted on the previous folio. The forms in which these words appear on the register and on the larger Viðey Map are identical; the register names the Quarters, as *austr, suðr, vestr,* and *norðr,* without elaborating upon them with the Old Norse term *fjórðungar* ('Quarters', English *farthings*), as they are commonly called. The register and the two Viðey Maps show that the Icelandic Quarters have universal significance because they appear to derive from, or even anticipate, the natural order of the world.

One tentative condition on which the Viðey Maps and the register of forty Icelandic priests' names can be associated further is presented in the coda's statement concerning Bishop Magnús Einarsson's role in its composition. In *Hungrvaka* it is written that Bishop Magnús Einarsson, the register's primary informant, was particularly accomplished in the areas of 'búnað ok farar' ('household management and travel').[91] When Magnús is elected bishop, we are told that he travels via Norway south to Denmark, where he is consecrated as bishop by Archbishop Özurr, on the feast day of St Simon (28th October). He returns to Iceland via Sarpsborg in Norway, and is particularly well received on his homecoming for the tidings he brings from Norway and other places he journeyed. That year he had the church at Skálhólt expanded and renovated, and 'lét tjalda kirkju borða þeim er hann hafði út haft, ok váru þat inar mestu gersemar' ('had the church hung with those boards which he had brought out [to Iceland], and they were the greatest of treasures').[92] What these precious 'borð' ('boards') depicted is not elaborated upon. When this term appears in compounds it frequently denotes *table* (Latin *tabula*), which is occasionally used in medieval Latin to describe a map (e.g. the *Tabula Peutingeriana* and the *Tabula Rogeriana*). Wall-mounted maps, such as those at Hereford and Ebstorf, are known to have decorated the walls of medieval churches.[93] *Hungrvaka* supplies no indication of what these *borð* depicted, or whether they were indeed *tabulae* or wall-mounted maps. Bishop Magnús was celebrated as a traveller, and the conjecture that his 'greatest treasures' were cartographic accessories to this reputation is tempting but not demonstrable.

91 *Biskupa sögur,* vol. 2, p. 29.

92 *Biskupa sögur,* vol. 2, p. 30.

93 Edson, *Time and Space,* p. 7.

Conclusion

A wide-angle view of both the Viðey Maps and their companion register of forty Icelandic priests reveals formal and functional affinities between them that cannot be seen when they are inspected in isolation. The register comprises a topographical survey of the Icelandic Commonwealth, whose Quartered structure reproduces in miniature the authoritative image of universal conformity in the two world maps on the following folios. Its medieval compiler may well have referred to the register itself as a *mappa mundi,* in much the same way that Gervase of Canterbury used the term to describe his topographical survey of bishoprics and ecclesiastical foundations in England, Wales, and parts of Scotland.[94] The register of forty Icelandic priests presents a counterpoint and a local perspective to the two world maps, and derives its authority from the fourfold ordering that repeats across the natural world.

The coda invests the register with an epochal tone that looks back to a formative period in the history of the Icelandic Church in the previous century. The register is conspicuously dated to 1143, and is therefore timely with other twelfth-century writings on Icelandic ecclesiastical history, among them *Íslendingabók* and the Christian Law Section of the Icelandic lawcode *Grágás,* written by the Bishops Ketill Þorsteinsson and Þorlákr Runólfsson in the same decade. These writings have been interpreted as attempts to entrench the national policies of Bishop Gizurr (d. 1118) and other Icelanders who had contributed to the strengthening of the Church in Iceland.[95] The register may have been an attempt to entrench these national policies into a documentary tradition, extending the history of the Icelandic Church from the beginning into the middle of the twelfth century. This version of the register, copied into sequence with the Viðey Maps in *c.* 1225–50, is contemporary with the Icelandic ecclesiastical histories *Kristni saga* and *Hungrvaka,* and partakes in their preoccupation with the close association between secular and ecclesiastical power in Iceland in the preceding century.

The maps and register were copied amid the intense social transformations of the thirteenth century, the so-called *Sturlungaöld* (Age of the Sturlungs). From the 1220s, the stability of the Norwegian Crown empowered King Hákon IV Hákonarson (r. 1217–63) to extend his presence in the North Atlantic. In Iceland, political power had already begun to concentrate in the hands of the most prominent Icelandic families, among them the Sturlungar to whom Snorri Sturluson belonged, who exploited their relationship with the Norwegian king to secure their positions in Iceland. Hákon tasked his vassals with bringing Iceland under his sovereignty, the result being a period of increasing violence

94 Knowles, 'Mappa Mundi', 237–47.

95 Jón Jóhannesson, *Íslendingabók*, p. xvi.

and instability that would culminate in Iceland's annexation to Norway in 1262/64.[96] *Hákonar saga Hákonarsonar*, the saga of King Hákon IV, states that Hákon's territorial aspirations had the support of the Church. When Cardinal William of Sabina came to Norway in 1247, he suggested that it was desirable for Icelanders to submit to a king.[97] Iceland was already dependent on Norway in a number of areas before its eventual submission to Norwegian rule: Iceland's lay and clerical elites were closely tied to those in Norway, many prominent Icelanders had joined the king's retinue in order to secure their status at home, and from 1238 onwards, Norwegians occupied the two bishoprics of Iceland. By the middle of the thirteenth century, King Hákon, through his vassals, controlled most of the *goðorð* in the Northern, Western, and Southern Quarters.[98] Boulhosa posits that the Norwegian king exercised 'administrative and punitive power over Icelanders' earlier, and to a greater degree, than is generally recognised.[99] In *Íslendingasögur* composed *c.* 1220–60, Theodore Andersson observes 'a will to identify what is peculiar to Icelandic institutions, Icelandic law, and Icelandic character'.[100] The register and two Viðey Maps, *c.* 1225–50, are a timely statement about the distinctiveness of Icelandic governance under highborn Icelanders in the preceding century. The Icelandic Quarters, depicted in the register and sanctioned by the two Viðey Maps, were the very institutions whose roles in Icelandic society were imperilled by civil war and the threat of Norwegian rule.

The larger Viðey Map is an image of Iceland and the world. Recent scholarship has brought to light the ways in which maps can be invested in secular interests and enshrine national identities. Lavezzo has shown how maps produced in medieval England cultivated a trope of English geographical marginalism in order to transform England's relative geographical isolation into a marker of exceptionalism.[101] Birkholz similarly calls attention to the importance of world maps in English

96 Bagge, 'Scandinavian Kingdoms', p. 723.

97 *Hákonar saga Hákonarsonar*, ed. Þorleifur Hauksson, Sverrir Jakobsson, Tor Ulset, Íslenzk Fornrit 31 (Reykjavík: Hið Íslenzka Fornritfélag, 2013), p. 144. On William of Sabina's alleged support, see David Ashurst, 'The Ironies in Cardinal William of Sabina's Supposed Pronouncement on Icelandic Independence', *Saga-Book* 31 (2007), 39–45; and David Brégaint, 'Conquering Minds: *Konungs Skuggsjá* and the Annexation of Iceland in the Thirteenth Century', *Scandinavian Studies* 84 (2012), 439–66, 441.

98 Erika Sigurdson, 'The Church in Fourteenth-Century Iceland: Ecclesiastical Administration, Literacy, and the Formation of an Elite Clerical Identity' (PhD diss., University of Leeds, 2011), pp. 16–18.

99 Boulhosa, *Icelanders*, p. 1.

100 Theodore M. Andersson, 'The King of Iceland', *Speculum* 74 (1999), 923–34, 933.

101 Lavezzo, *Angels*.

political culture, showing how maps commissioned for display at seats of royal power were used in thirteenth-century England to support the Crown's geopolitical ambitions.[102] Icelanders likewise used cartographic images to make statements about their nation and that which threatened it. The maps and their companion register present an image of a decentralised, balanced and harmoniously proportioned national polity, at a time when Iceland was threatened by the concentration of political power and violent internecine strife.

On the larger Viðey Map, Iceland and Thule are accorded more cartographic space than any other of the map's legends. This space might imply a desired emphasis on the place where the map was produced. However, the maps appear at precisely the time when Norwegian encroachment threatened Icelandic sovereignty, and the mapped distance between Iceland and the Scandinavian Peninsula may signify a map-maker's desire to maintain their political distinctiveness. As the wizard reports to the Danish King Haraldr Gormsson in *Óláfs saga Tryggvassonar*, 'haf svá mikit millum landanna, segir hann, at ekki er þar fœrt langskipum' ('the sea between the lands is so large', he said, 'that no ship can cross it'). In combining the Viðey Maps and their companion register, an Icelandic thinker held a mirror up to nature – and Iceland to the world – to reveal that the Icelandic Commonwealth derived from, or even anticipated, the 'self-evident quadripartite formation of the whole universe'.[103]

[102] Birkholz, *Two Maps*.
[103] O'Reilly, 'Evangelists,' p. 55.

Conclusion

It is conventional in the opening pages of books on medieval maps for their authors to define the central object of their study. *Mappa mundi* is a term that comprises the Latin words *mappa* ('cloth') and *mundus* ('world'). Map historians duly scrutinise the word *mappa*, from which we derive the modern English word 'map', but seldom do *mundus* the same service. If the intellectual work of maps is to make the world, or an aspect of it, understandable, what 'world' was it that preoccupied medieval Icelanders?

This book has been about the Icelandic conception of *mundus*. The small corpus of Icelandic world maps give full expression to the range of conventions for visualising the known and theorised world available to medieval Europeans. The Icelandic hemispherical and zonal maps show how Icelanders theorised global space, using the cartographic image to think about the Earth's cosmic position, its sphericity, and the latitudinal distinction of its climatic zones. The paired Viðey Maps focalise the tricontinental land area known to medieval Europeans, their complex spatial arguments creating a space for Icelanders to consider their place in human history and civilization. These maps engage Icelandic understandings of *mundus* that resonate variously with *world, Earth,* and *globe*.[1]

I have not been overly concerned with what a map is and is not, but rather have tried to show that an overreliance on 'map' as an analytical category, and the tendency to study maps only alongside other maps, has concealed the scale of their interactions with other literatures and histories. This book's analyses of the Icelandic maps in their restored manuscript contexts enable us to revise prevailing views about maps and the cultural and historical movements they intersect. They do not relate only to each another, as variant and complementary attempts to visualise the world and its places, but to other contemporary literatures about the Icelanders' ancestry and social origins, their Commonwealth, and their place in Europe.

It has been assumed that the motivation for drawing these maps must have been to hold a mirror to the world, as Icelanders documented

1 Cosgrove, *Apollo's Eye*, pp. 7–8.

their own geographic awareness in map form. They were scrutinised, especially, as historical sources for areas of Scandinavian settlement in North America and the eastern Baltic, areas in which Icelanders were presumed to possess intelligence that surpassed their Latin exemplars. Thule, according to the map's earliest commentator, represented places associated with the Norse explorations in North America; while other commentators examined Russia and Kiev for intelligence about the Slavic and Scandinavian regions explored by the Vikings. However, while thirteenth- and fourteenth-century Icelanders possessed a capacious geographic imaginary, their world did not always centre on their own northern waters, or the regions coextensive with the Viking diaspora.

Icelanders used the cartographic medium to think about their place in the wider world, and to locate their geographic experiences within European frameworks for conceptualising global space. On the hemispherical and zonal maps, Icelanders saw the cold northern regions they had recently discovered and settled as an integral and stable part of the world described by ancient authoritative authors. That their northern regions had been theorised in antiquity enabled Icelandic thinkers to mitigate anxieties about Iceland as a *terra nova* with no human prehistory. These maps, oriented with south at the top, place the northern hemisphere at the bottom of the folio, and position their Icelandic viewers to look bodily from their own regions southwards towards the wider world. The legend *norðr bygilig hálfa* ('northern inhabitable region') on the Icelandic zonal map may have conveyed to Icelanders both the global concept of the northern hemisphere, and their own northern regions more narrowly.

The larger Viðey Map conveys a minimalist depiction of the islands of the North Atlantic, the map-maker giving fuller expression to Central Eurasia and the lands venerated by medieval Christianity than to the world's northern waters. The map's focal regions are positioned elsewhere and far away, as Icelanders, in common with other European elites, located the origins of their cosmopolitan and religious cultures eastwards. In Icelandic literature, the legendary progenitors of the Scandinavian royal families are shown to have had their ancestral homelands in Central Eurasia, coming to European shores as refugees from the Trojan war. The Icelanders were also distanced from the sacred places of western Christianity, which were situated far away in the regions surrounding Palestine and Jerusalem. The Icelanders did not fixate on their own regions, but turned their attentions outwards; the centres and peripheries they mapped onto their world are different from the ones we imagine for them.

This book's main work has been to revise some previously under-informed assumptions about maps and their geographical interests, recognising that they did not function only, or perhaps even primarily, as vehicles for geographical information. Geography, to be sure, is a powerful and evocative modern category that makes some aspects of these maps visible, but conceals others. This book revaluates the objects of their historical inquiry by looking at maps not only as sources for past geographical understandings, but as registers of Icelandic national self-perception and imagining. The larger Viðey Map is not only a visualisation of world geography, but of history, granting visual expression to the medieval legend of the Trojan diaspora, which Icelanders' enlisted to conceptualise their origins in common with other European elites. World maps provided Icelanders with tools with which to think, and Icelandic authors used the cartographic medium to think about Iceland, its people, and their history. The larger Viðey Map is a work of mythopoesis, an Icelandic origin myth, through which its maker writes the history of Iceland – from Tanavísl to Thule – within a T-O framework.

The Icelandic maps propose a version of Icelandic identity that centres their Europeanness. The hemispherical and zonal maps, appearing among treatises on the cosmos and *computus*, seek to synchronise Icelandic and wider European perspectives on the world and time. It was the aim of the encyclopaedias in which maps were drawn to enter Iceland into a clocked community whose observance of the liturgical calendar was consistent across European Christendom. The Icelanders' European self-fashioning is also emphasised on the Viðey Maps. These maps show the division of the *kringla heimsins* into its continental thirds, with the legends Asia, Africa and Europe written outside the map's terrestrial circle on the conceptual frame. In the map's European quarter, the names Thule and Iceland are positioned prominently beneath the legend Europe. Through this sequence of names – Europe, Thule, Iceland – the Icelandic map-maker creates a European Iceland through *Thule*, the name through which medieval authors established an Icelandic presence in the Roman literature on which the European curriculum was based. Rome, the centre of western medieval Christianity, is positioned in the middle of the map's representation of Europe. The map's European geometry emphasises the Icelanders' belonging within a European space centred on the Church; while Iceland was at the head of Europe, Rome was at its centre.

The drawing of these maps is timely with the rise of Iceland's vernacular literary culture, and the maps are themselves texts in varying degrees of translation. The lessons these maps teach about the world's shape and the dispersion of its lands are productive, as we have seen, across literary genres. The Icelandic sagas may have been one of the most significant channels through which geographical information was disseminated in Icelandic culture. Icelanders may have achieved their knowledge of distant places from the northern *fornaldarsögur*, which locate their action in a legendary Scandinavia before the discovery of Iceland, and the southern *riddarasögur*, whose narrative actions sometimes span continents, from the conventionalised courtly settings of southern Europe, to exoticised eastern kingdoms in Constantinople and India. These romance sagas were the fictional vehicles for a mass of matter about world geography, differing from their Continental and English relations, as Geraldine Barnes has shown, in their openness to interpolations from bookish and scholarly sources. Icelandic romance authors drew on their learned sources for information on the world's places to create 'opportunities for the narrative players themselves to step directly into the world as constructed in learned texts', or indeed, in the *mappae mundi*.[2] This creative engagement with bookish sources shows that Icelanders were able mentally to sort these received ideas – descriptions of the three continents, the latitudinal distinction of the world's climates, and maps themselves – into new and mobile forms. Maps may have played a role in this opening up of the Icelanders' geographical imaginary, but Icelanders may also have approached maps with an awareness of their conventions drawn from literature.

The Icelandic maps relate also to literatures and histories whose concerns lay outside world geography. The Icelanders' preoccupation with their ancestry and social origins, which we see in the mapped names Thule and Tanakvísl, motivates much of the Icelanders' vernacular literary culture, including the saga narratives that constitute its canonical works.[3] The larger Viðey Map, as we have seen, is a pioneering work of Icelandic historical writing that bears a close relationship to contemporary histories of the Icelandic Church, namely *Hungrvaka* and *Kristnisaga*, as well as those historical undertakings whose aim was to canonise the formation of Icelandic institutions in the settlement period, among them *Landnámabók*. The Viðey Maps are contemporary with the fractious political developments of the thirteenth century, and may have been the expressive vehicles of the Icelanders' sovereign fantasies in this period. The Icelandic maps were not culturally inert copies of ancient articles of learning, but relate to the contemporary cultures that produced them.

This book has shown that the Icelandic maps neither engage passively with inherited ideas, nor move through the culture that produced them without changing it. There has been a tendency in scholarship theorising Icelandic engagement with Latin and foreign literatures to assume a relationship in which Icelanders were inspired by or borrowed from European models, while maintaining a distinctiveness between their indigenous and Latin literary cultures.[4] Icelanders may have borrowed from, rewritten, or made occasional forays into European culture, but seldom are they recognised as Europeans writing European literature. What we see in the maps, however, is not so much a comingling of

2 Barnes, *Bookish Riddarasögur*, p. 28.

3 Margaret Clunies Ross, *The Cambridge Introduction to the Old Norse-Icelandic Saga* (Cambridge: Cambridge University Press, 2010), p. 11.

4 Annette Lassen, 'Indigenous and Latin Literature', pp. 82–83.

Icelandic and European cultures as Icelanders making the case for their own innate Europeanness. The appearance of these maps in Icelandic manuscript books is no belated response to an outside culture, but is timely with the apogee of map production in medieval Europe. The thinkers who copied these maps used them uniquely to think about Icelandic culture, but at all times emphasised its European cosmopolitanism. Icelandic map-makers did not change or elaborate upon their exemplars so much as tilt them towards a new light.

The Icelandic environment in which these cartographic images were copied assigned its own meaning to them, their conventional characteristics – like Thule, the Tanais, or the northern polar circle – becoming particularly charged sites of Icelandic self-imagining. The maps illustrate global concepts like the midnight sun, the latitudinal distinction of the Earth's climates, and the ocean's tides. These phenomena were hypothetical to ancient authorities, but here existed in the Icelanders' lived experience. Iceland was a place of unusual natural extremes in its cool climate and the length of its day, and the Icelanders' mundane experience of these phenomena may uniquely have provided the cue for, and enhanced the relevance of, contemplation of their world's global condition. This, however, did not simply produce natural philosophical enquiry for its own sake. Of all the literatures Icelanders inherited with the advent of Christianity and bookish culture in *c.* 1000, it was writings on natural philosophy in which they were able to see themselves and study their experiences. The *mappae mundi* provided for Icelanders an opportunity to rationalise their geographical position – on a northern island, where the sun only rose for half a year – and the historical circumstances of their settlement there, without breaking with established European geographical and historical narratives. They wrote their history through, not against, the bookish authorities who theorised the world's northern regions, and tentatively described the conditions experienced there. Medieval Icelanders had the world on their minds, and the maps they drew worked to deprovincialise Iceland by drawing it into conventional ways of thinking about global space. Above all, world maps contributed to the Icelanders' understanding of themselves.

MAP TEXTS AND TRANSLATIONS

I. The Icelandic Hemispherical World Maps

II. The Icelandic Zonal Map

III. The Larger Viðey Map

IV. The Smaller Viðey Map

The Icelandic Hemispherical World Maps

Copenhagen, Arnamagnæan Collection, AM 736 I 4to, f. 1v (*c.* 1300), and Copenhagen, Arnamagnæan Collection, AM 732b 4to, f. 3v (*c.* 1300–25)

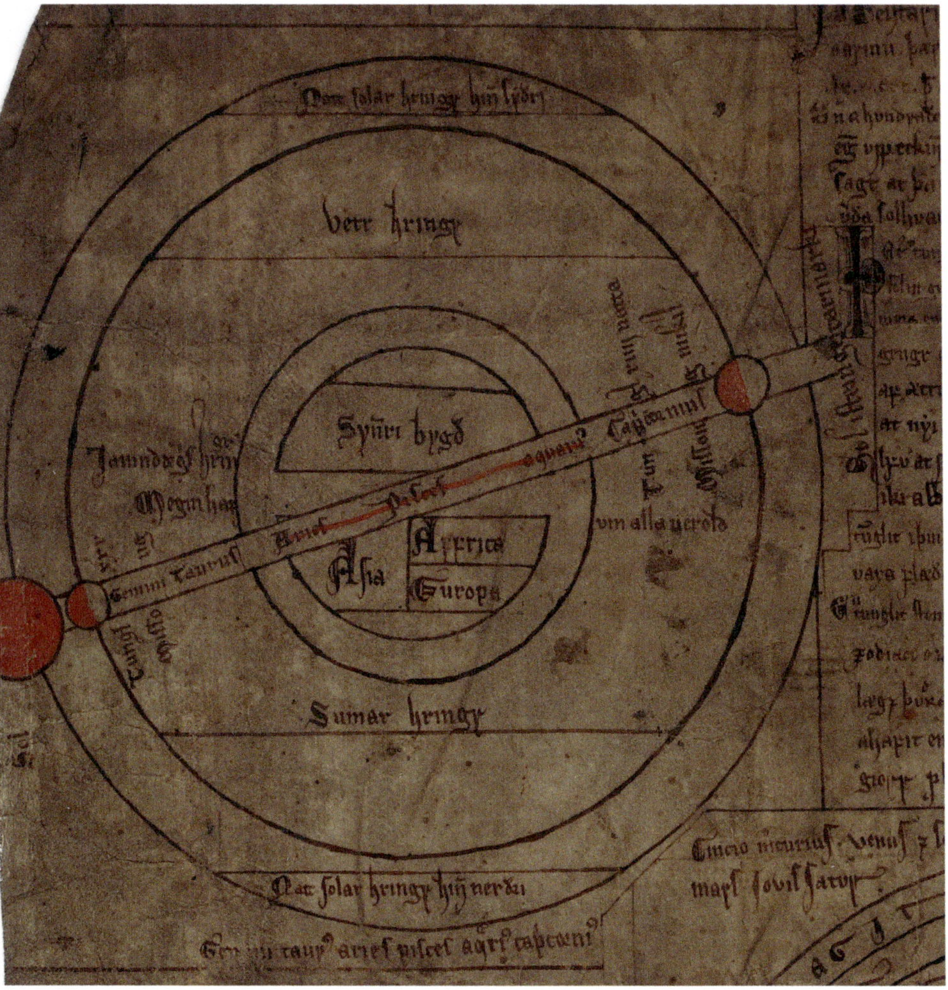

AM 736 I 4to, f. 1v

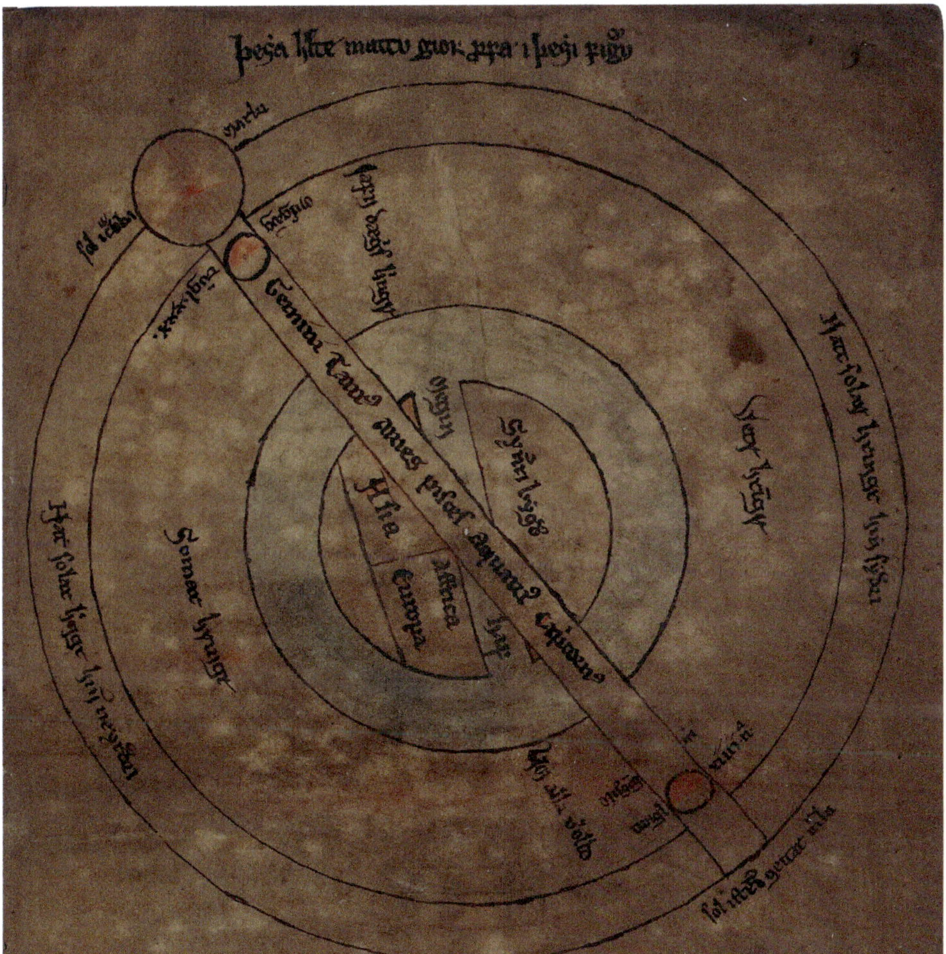

AM 732b 4to, f. 3v

Previous editions and commentaries: Rafn, *Antiquités Russes*, vol. 2, p. 391; Miller, *Mappaemundi*, vol. 3, p. 125; *AÍ* II, pp. 118–19; Destombes, *Mappemondes*, p. 175; Simek, *Altnordische Kosmographie*, p. 410.

The Icelandic hemispherical world map, with south at the top, shows the globe divided into climates along five lines of equal latitude, or *parallels*: the two polar circles, the two tropics, and the Equator. In the northern hemisphere, the temperate region between the Earth's cold polar regions and hot equator is divided into the three continents of the known world, Asia, Africa, and Europe. The Earth's southern hemisphere bears an Old Norse legend that denotes the theorised temperate region south of the Equator. The sun and moon are shown in two positions in their orbits around the Earth, in configurations that cause seasonal variation in the tides. These two depictions of the sun are connected by a narrow band inscribed with the signs of the zodiac, representing the ecliptic. The map is inscribed with twenty-one legends, written in a combination of Latin and Old Norse. The map is extant in two manuscript copies. The transcription is based on the version of the map in AM 736 I 4to, with one addition from AM 732b 4to. On both versions of this map, see Chapter 1.

Natt solar hringr hin*n* syðrj	Antarctic Circle
Vetr hringr	Winter Tropic / Tropic of Capricorn
Sol isteingeitarmarki	Sun in Capricorn
tungl xiiij natta	Moon at 14 nights
missong mikil	High springs
Sol [i krabbi marki][1]	Sun [in Cancer]
tungl xxx	Moon [at] 30 [nights]
missong	Springs
Gemini Taurus Aries Pisces	Gemini, Taurus, Aries, Pisces
Aquari*us* Capricornius	Aquarius, Capricorn
Syn*n*ri bygð	Southern inhabitable land

1 The bracketed reading is absent from the AM 736 I 4to map, but present in AM 732b 4to. It does not necessarily, however, indicate that the more complete 732b version does not derive from 736 I, since the explanation of the map, in the tidal note, would have provided the copyist with the means to expand or complete it. The tropical signs (Cancer and Capricorn) are those in which the sun appears to stand still in its course around the Earth and reverse direction, and contain the sun's course within their bounds. The note tells us that, in this configuration, 'solin gengr iþvi marki er vær kaullum krabba mark 7 norðaz er isolar hring' ('the sun goes into the sign we call Cancer and is northernmost in its cycle') *AÍ* II, pp. 118-19.

Iamndægris hringr	Equator
um alla uerold	around the whole world
Megin haf	Ocean
Asia	Asia
Affrica	Africa
Europa	Europe
Sumar hringr	Summer Tropic / Tropic of Cancer
Natt solar hringr hin*n* nerðri	Arctic Circle
Gemini Taurus Aries Pisces	Gemini, Taurus, Aries, Pisces,
Aq*arius* Cap*ri*corni*us*	Aquarius, Capricorn

The Icelandic Zonal Map

Reykjavík, Stofnun Árna Magnússonar, GkS 1812 I 4to, f. 11v (1315–*c.* 1400)

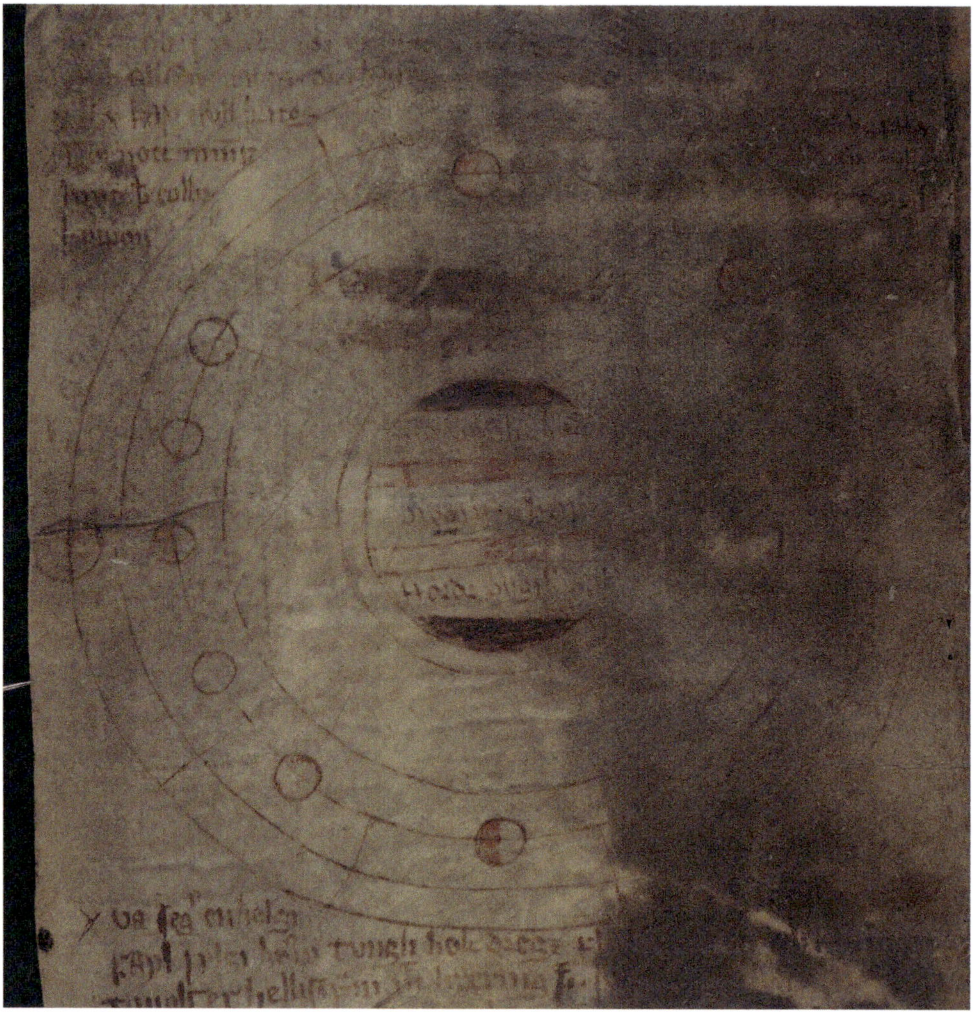

Previous editions and commentaries: Rafn, *Antiquités Russes*, vol. 2, p. 390–91; *AÍ* II, pp. ccxiv–ccxv; Simek, *Altnordische Kosmographie*, p. 590–91.

The Icelandic zonal map, with south at the top, shows the latitudinal distinction of the Earth's climatic zones. The map uses colour to distinguish the cold polar regions and the hot equatorial regions. Human civilization was presumed to be limited to the northern habitable region. The map shows the theorised subequatorial continent south of the equatorial ocean, which may (medieval authorities are divided) have been inhabited. The map is accompanied by three other diagrams between ff. 10r and 12v that focalise the Earth and its cosmic position. On this map, and its diagrammatic companions, see Chapter 2.

sudr bygilig halfa	Southern inhabitable region
megin haf	Ocean
nordr bygilig halfa	Northern inhabitable region

The Larger Viðey Map

Reykjavík, Stofnun Árna Magnússonar, GkS 1812 III 4to,

ff. 5v–6r (*c.* 1225–50)

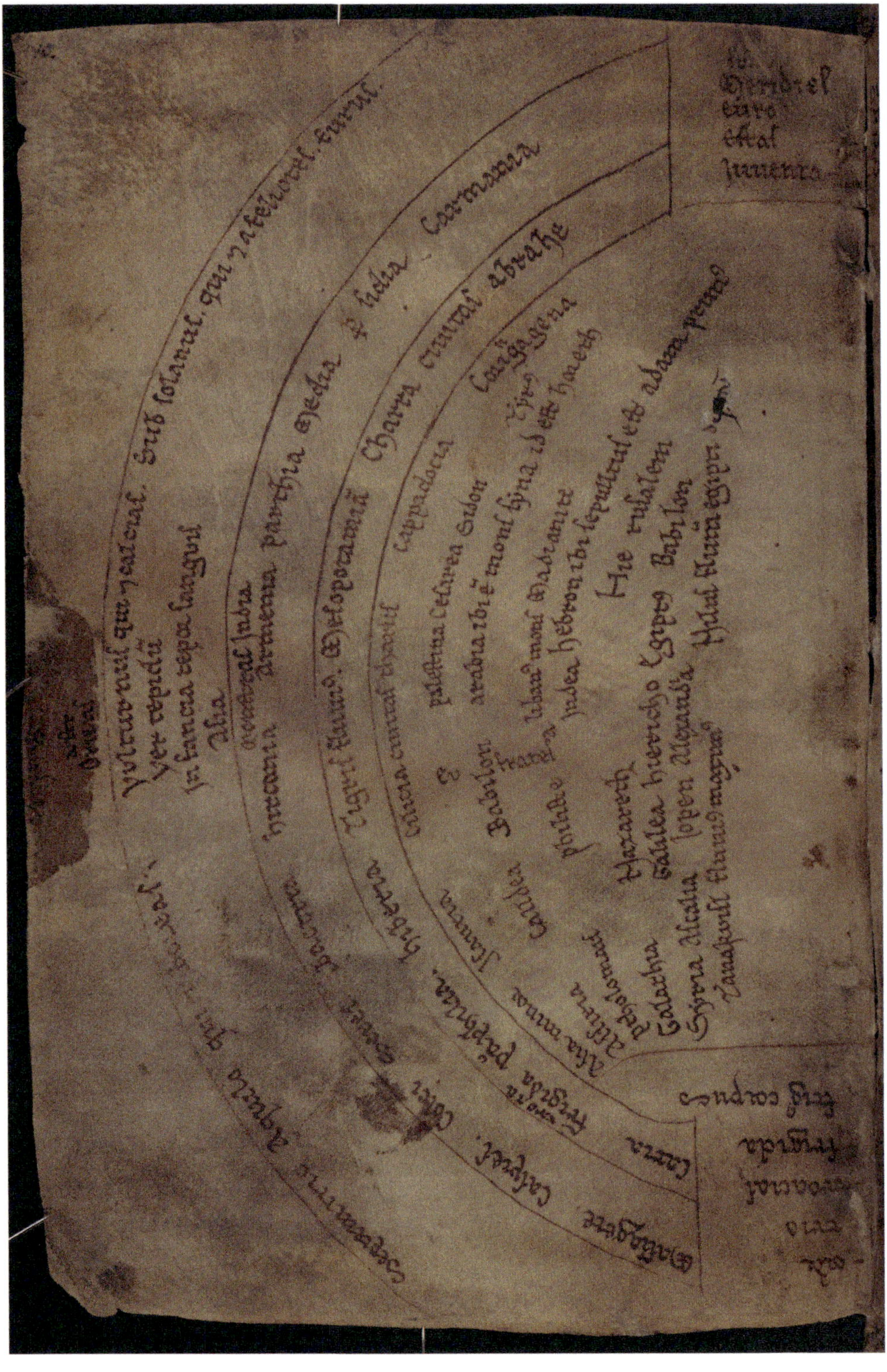

Carmania
Parthia
Armenia
Cappadocia
Babilon
Iudea
Nazareth
Babilon

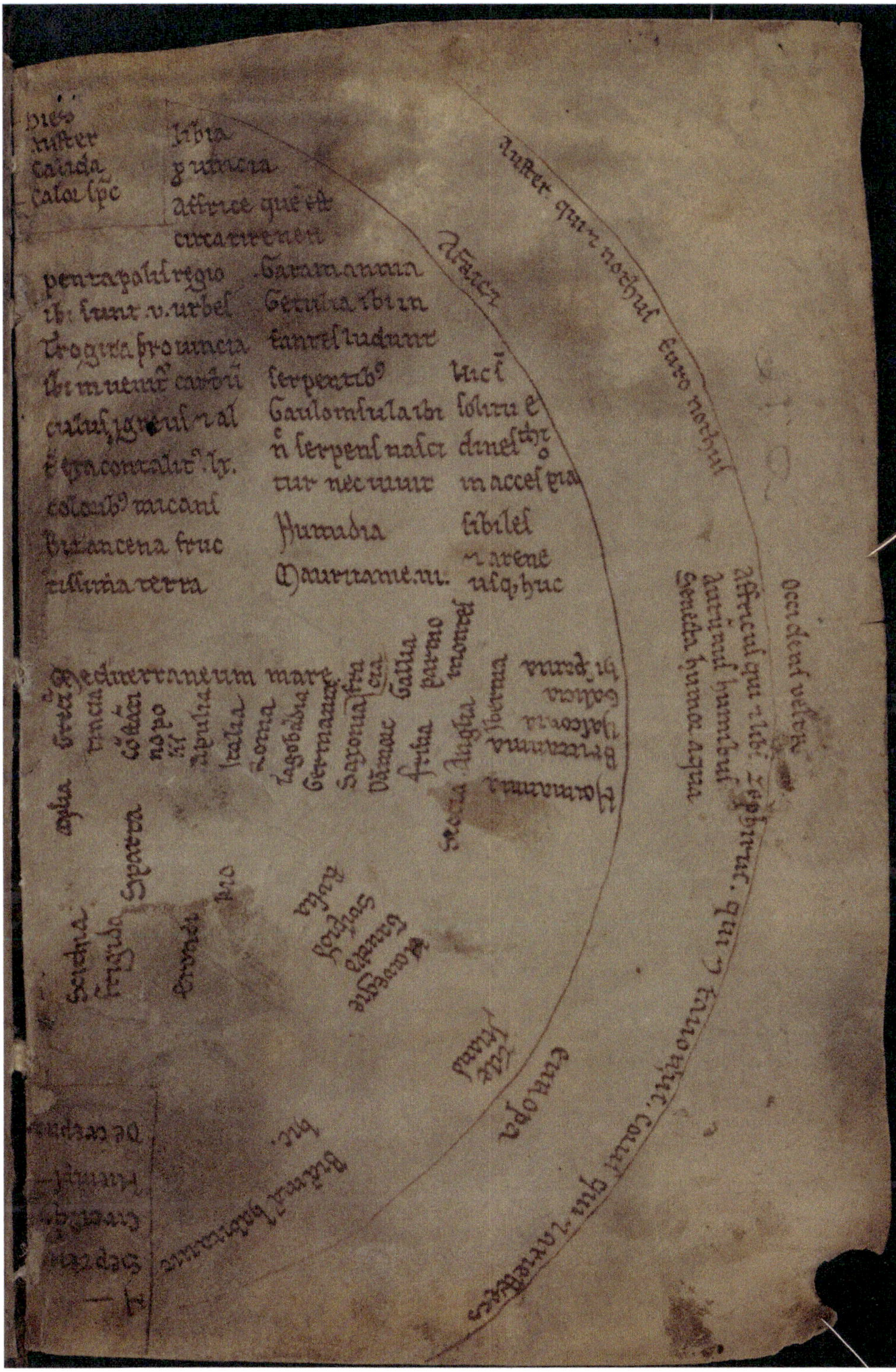

Auster
Calida
libia
pentapolis regio
ibi sunt v. urbes
Garamantania
Mauritanie
Mediterraneum mare
Grecia
Italia
Germania
Gallia
Anglia
Hibernia
Britannia
Hispania
Scithia
Frigida
Europa
Auster qui et nothus
Euro nothus
Occidens

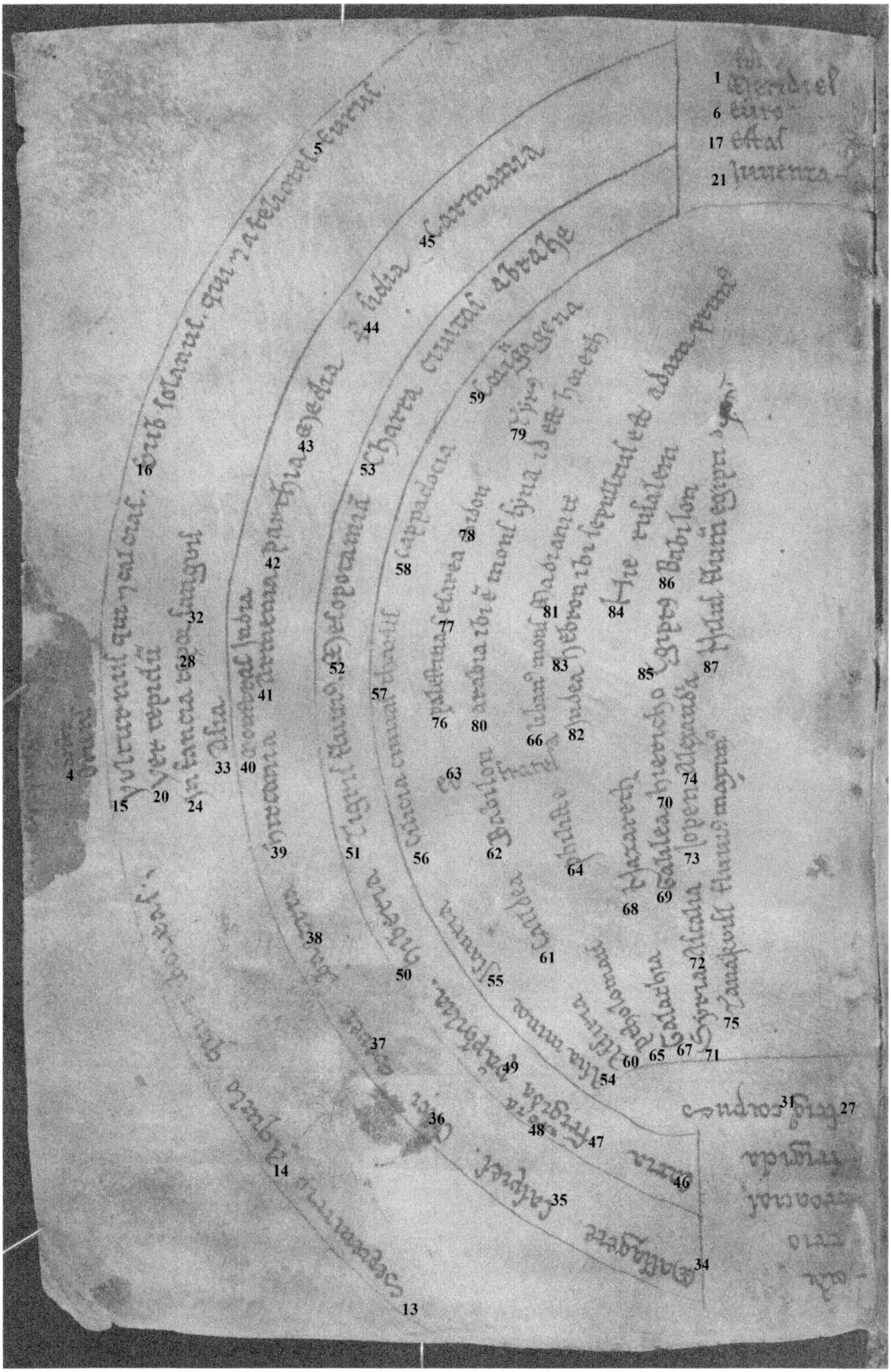
1 6 17 21 5 45 44 59 79 43 53 16 78 42 58 32 81 84 86 77 28 52 83 85 87 41 57 76 80 66 82 4 33 40 63 74 15 20 24 70 39 51 56 62 73 64 69 68 38 61 72 50 55 75 37 49 60 65 67 71 54 31 27 36 48 47 14 46 35 34 13

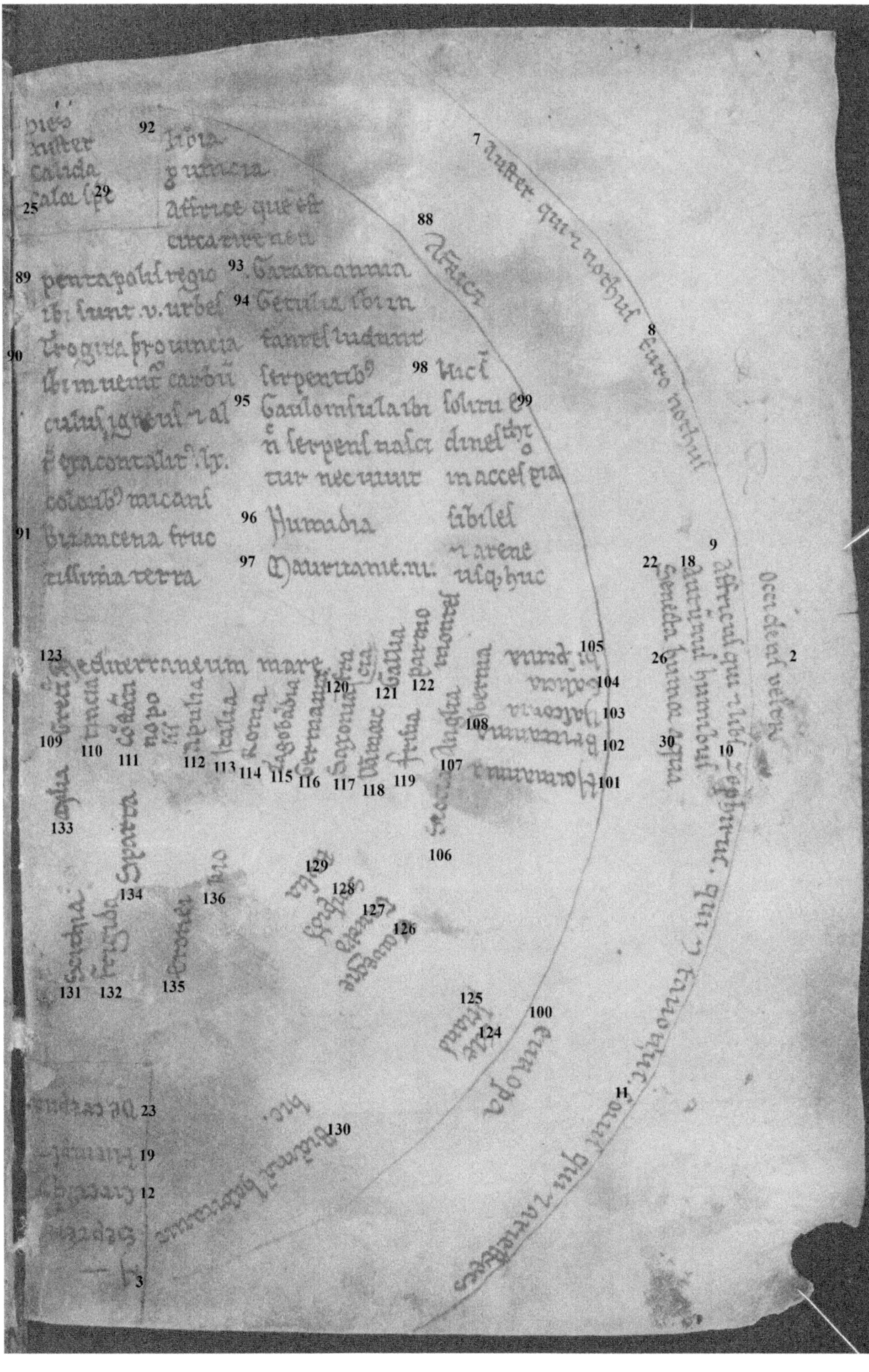

Auster qui et nothus
Affrica
Euro nothus
pentapolis regio
ibi sunt .v. urbes
Garamantia
Getulia
Humadia
Mauritanie
Mediterraneum mare
Apulia
Roma
Gallia
Anglia
Scotia
Europa

Previous editions and commentaries: Rafn, *Antiquités Russes* vol. 2, pp. 392–94; Kålund, *Alfræði Íslensk* III, pp. 71–72; Destombes, *Mappemondes*, p. 175; Haraldur Sigurðsson, *Kortasaga Íslands*, p. 54; Pritsak, *The Origin of Rus'* , vol. 1, pp. 514–16; Simek, *Altnordische Kosmographie*, pp. 419–23; von den Brincken, *Fines Terrae*, pp. 128–29; Chekin, *Northern Eurasia*, p. 69–71.

Headwords, in bold, are diplomatic transcriptions of legends as they appear on the map. These transcriptions provide the most certain indication of what the map's readers saw, but do not answer the question of what Icelanders may have understood by them. The normalised names below the headwords are the forms by which they are conventionally indexed in the editions of the ancient and medieval texts in which they originate. What follows a headword is more an interpretation than a translation. Some of the map's legends are corrupted, so that it is only possible to know what was likely intended through comparison with earlier maps and treatises.

For some names I have supplied cross-references to the Hereford Map (these references begin with H, and are numbered following Westrem, *Hereford Map*), and two major medieval descriptions of world maps that no longer survive: Roger of Howden's *Expositio mappa mundi* (these references begin E, and follow Gautier Dalché's numbering), and Hugh of Saint Victor's *Descriptio mappa mundi* (these references begin D, and follow Gautier Dalché's numbering). The aim of these cross-references is not to facilitate a full comparison of the larger Viðey Map with these works, whose lengths and complexities far exceed it, but to corroborate the interpretations that follow the headword. The relevance of other of the map's legends have more salient parallels with other texts and traditions.[1] On this map's depiction of world geography, see Chapter 3. On its depiction of Scandinavia and Iceland, see Chapter 4. On the conceptual frame and the circumstances of the map's thirteenth-century conception, see Chapter 5.

Transcription and Commentary

The Conceptual Frame	1–32
Asia	33–87

[1] The editions and translations referenced are those cited earlier in this book and listed in the bibliography. Additionally, reference is made to Aethicus Ister, *The Cosmography of Aethicus Ister: Text, Translation, and Commentary*, ed. Michael W. Herren (Turnhout: Brepols, 2011); and Martianus Capella, *Martianus Capella and the Seven Liberal Arts*, vol. 2: *The Marriage of Philology and Mercury*, trans. William Harris Stahl (New York. Columbia University Press, 1977). Book and chapter numberings follow these editions.

Africa	88–99
Europe	100–136

Key

a	animal (zoological or fabulous)
c	city, town, or settlement
cm	cosmological matter or concept
e	ethnic group or race (historical or fabulous)
fl	river
h	human individual (including Biblical figures)
i	island
m	mountain
mr	marine toponym (ocean or sea)
o	other item
r	region, realm, or territory

The Conceptual Frame

The Cardinal Directions

1 **Suðr Meridies dies**
South (cm)

The map opens with south at the top, though there is no dominant textual orientation or way up from which to read its inscribed names. The cardinal directions are written in both Old Norse and Latin, emphasising the conceptual frame's quaternary structure. The Old Norse names may echo their appearance on the other item preserved alongside this map, a register of forty Icelandic priests preserved on f. 5r, where they refer to the Quarters of Iceland (see Chapter 5). The repetition of 'dies' seems to be a homeoteleuton.

2 **Occidens vestr**
West (cm)

3 **Norðr Septentrio**
North (cm)

4 **austr Oriens**
East (cm)

The Twelve Winds

5 **Eurus**

Eurus (cm)

6 **euroauster**

Euroauster (cm)

7 **Auster qui *et* nothus**

Auster, which is Nothus (cm)

One of the four principal winds, which rises due south. Like the other winds on the map, it is one place clockwise out of alignment. Its satellite winds are Euroauster (*sinister*) and Euronothus (*dexter*).

8 **Euro nothus**

Euronothus (cm)

9 **Affricus qui *et* libs**

Africus, which is Libs (cm)

10 **Zephirus. qui *et* fauon*i*us**

Zephyrus, which is Favonius (cm)

One of the four principal winds, which rises due west. Like the other winds on the map, it is one place clockwise out of alignment. Its satellite winds are Africus or Libs (*sinister*) and Corus or Argestes (*dexter*).

11 **Corus qui *et* ariestes**

Corus, which is Argestes (cm)

12 **Circi*us* q*ui* *et* troacias**

Circius, which is Thracias (cm)

13 **Septemtrio**

Septentrio (cm)

One of the four principal winds, which rises due north. Like the other winds on the map, it is one place clockwise out of alignment. Its satellite winds are Circius or Thracias (*sinister*) and Aquilo or Boreas (*dexter*).

14 **Aquilo qui *et* boreas**

Aquilo, which is Boreas (cm)

15 **Vulturnus qui *et* calcias**

Vulturnus, which is Calcias (cm)

16 **Sub solanus. qui *et* afeliotes**
Subsolanus, which is Apeliotes (cm)

One of the four principal winds, which rises due east. Like the other winds on the map, it is one place clockwise out of alignment. Its satellite winds are Vulturnus or Calcias (*sinister*) and Eurus (*dexter*).

The Four Seasons and their Qualities

17 **Estas Calida**
Hot Summer (cm)

These and other fours depicted on the map are presented in a table in an Icelandic astronomical and calendrical miscellany (Stofnun Árna Magnússonar, AM 193 III 8vo, ff. 11r–11v), which illustrates a short treatise on the world's nations and their languages (Simek, *Altnordische Kosmographie*, p. 501). Summer is associated with the south in accordance with theories about the Earth's climatic zones.

18 **Autu*m*nus humibus**
Humid Autumn (cm)

Autumn is associated with the west. In the northern hemisphere, the sun rises due east and sets due west on the spring and autumn equinoxes.

19 **Hiemps frigida**
Cold Winter (cm)

Winter is associated with the north in accordance with theories about the Earth's climatic zones.

20 **Ver tepidu*m***
Warm Spring (cm)

Spring is associated with the east. In the northern hemisphere, the sun rises due east and sets due west on the spring and autumn equinoxes.

The Four Ages of Man

21 **Iuuenta**
Youth (cm)

The four ages of man contribute to the map's depiction of the human microcosm. South is associated with summer and youth.

22 Senecta
Old Age (cm)

West is associated with autumn and old age.

23 Decrepita
Decrepitude (cm)

North is associated with winter and decrepitude.

24 Infancia
Infancy (cm)

11 East is associated with spring and infancy.

The Four Physical Qualities

25 Calor
Heat (cm)

This quaternary scheme resembles the four contraries (hot and cold; wet and dry), whose interactions produce change and variety in the sublunary world. The four qualities that appear on this Icelandic map are visualised on a strongly symmetrical diagram of universal concord in a ninth-century manuscript of Bede's *De natura rerum* (*On the nature of things*) in Munich, Bayerische Staatsbibliothek, Clm. 210, f. 132v.

26 humor
Moisture (cm)

27 frig*us*
Cold (cm)

28 tepor
Warmth (cm)

The Components of a Human Body

29 sp*iritus*
Breath (cm)

A continuation of the map's representation of the human microcosm.

30 **aqua**
Water (cm)

31 **corpus**
Flesh (cm)

32 **sanguis**
Blood (cm)

Asia

33 **Asia**
Asia (r)

One of the world's three continents (from Latin *terra continens*, 'continuous land'), whose notional boundary with Europe was marked by the River Tanais, and with Africa by the River Nile. The names of the three continents are placed outside the terrestrial circle on the map's conceptual frame. The map's Asian legends occupy its transverse half on folio 5v.

34 **Massagete**
The Massagetes (e)

A nation described by ancient geographers, numbering among those peoples descended from the Scythians (Herodotus, *Histories* 1.215–6; Isidore, *Etymologies* 9.2.63; cf. H216). The Massagetes and the Parthians were said by Orosius to inhabit the region between the source of the Tigris and Carrhae, an ancient city in Upper Mesopotamia (*Histories* 1.2.41). The historical Massagetes were an Iranian nomadic confederation who inhabited the steppes of Central Asia.

35 **Caspies .**
The Caspian Gates (?) (m)

This legend may be a corruption of *porte Caspiae*, a staple name in antique descriptions of the Caucasus range, the mountain massif that reaches across Central Asia from the Black Sea (Orosius, *Histories* 1.40; Isidore, *Etymologies* 14.3.10; cf. H218, EI.15, D301, D303). Simek suggests that the inscription may derive from *regio caspia*, *caspius mons*, or *caspium mare* (*Altnordische Kosmographie*, p. 203), but the Gates feature in a similar position, next to the Massagetes, on the Hereford Map, as well as on the Sawley and Ebstorf Maps, and in the *Expositio* and *Descriptio mappa mundi*. The historical Caspian Gates may correspond to the Dariel Pass, a gorge between the Black and Caspian Seas in modern Georgia.

36 **Colci**
Colchis (r)

Orosius situates Colchis, and its native Colchians, in the region where the Caucasus range begins to rise (*Histories* 1.2.36; cf. H219, EI.7). The Colchians also appear near the Caspian Gates and Massagetes on the Hereford Map. The historical Colchis was an ancient region in the southern Caucasus on the eastern coast of the Black Sea.

37 **seres**
The Seres (e or c)

The Seres (or Serians) were, according to Isidore, a nation in the Far East, named for their home city and known to 'weave a kind of wool that comes from trees' (Isidore, *Etymologies*, 9.2.40, p. 194; 14.3.29; cf. H146, H92, EI.39, [EI.32]). The Seres are usually understood to be the Chinese, or the people of East Asia generally. The Seres appear in two legends on the Hereford Map, which may have obtained its information from Pliny's *Historia Naturalis* (IV.54), Solinus's *Collectanea* (15.4, 50.2–4), or Isidore's *Etymologies* (19.22.14). The longer legend reads: 'Seres primi homines post deserta occurrunt – a quibus serica uestimenta mittuntur' ('the Seres, the first men who come after the desert, export silk garments'); while the shorter reads 'Seres civitas' ('City of the Seres'). The Seres appear near Bactria on both the Hereford and larger Viðey Maps.

38 **bactria**
Bactria (r)

Bactria adjoins the lands of the Seres, and was notable, antique sources agree, for its camels (e.g. Isidore, *Etymologies* 14.3.30; Aethicus Ister, *Cosmographia* 32.28; cf. H160, D213). The historical Bactria encompassed modern-day Afghanistan and Tajikistan.

39 **hircania**
Hyrcania (r)

A forested land to the south of the Caspian Sea on the River Tigris that borders Armenia, Albania, and Hiberia (Isidore, *Etymologies* 12.1.7; 14.3.33; cf. H151, H153, EI.34, D176, D297). Aethicus Isther tells us that it lies between Scythia and Asia and extends to the Caspian Gates (*Cosmographia* 69.1). Hiberia, Armenia and the River Tigris are all nearby on the map.

40 monstras India

The Mountains of India (?) (m)

India reaches as far north as the Caucasus range and borders Hyrcania in the north (Isidore, *Etymologies* 14.3.5; *Stjórn* 146–49; cf. H50, H51, H53, H61). It is probable that the legend is a corruption of *montes Indiae* ('mountains of India'), which appears on the Hereford Map ('montes Yndie'). Isidore situates India between the Caucasus and Parthia (14.3.5), which corresponds with its placement here. India is the easternmost name on the larger Viðey Map, immediately beneath the continental headword *Asia* and the cosmological matters grouped around *Austr Oriens* ('east'). This placement resonates with Orosius's account of the subjugation of India by Alexander, 'in order to make the Ocean and the furthermost east the borders of his empire' (*Histories* 3.19.1). It is conceivable that the legend is a corruption of 'monstra Indiae' ('monsters of India'), since Isidore populated the country with 'dragons, griffins, and human monsters of immense size' (14.3.7).

41 armenia

Armenia (r)

Armenia is a region in the south Caucasus, known in the Bible as the place where Noah's Ark settled (Isaiah 37:38; Eusebius, *Onomasticon* 182). Armenia is described by Isidore in a sequence of places also depicted on the map (Bactria, Hyrcania, Armenia, Hiberia, and Cappadocia) that lie between the lands of the Seres and the River Tigris (*Etymologies* 14.3.29–35; cf. H166, D305, D307, D172, D194, D169, D191, D297).

42 parthia

Parthia (r)

Isidore tells us that Parthia was a large and powerful nation between India and Mesopotamia, whose neighbours Assyria, Media, and Persia sometimes came under its suzerainty (*Etymologies* 14.3.8; *Stjórn* pp. 261–63; cf. H114, H167, D241, D245). Orosius placed Parthia, with Assyria, Persia, Media, and Arachosia (not on the map) between the River Indus in the east and the River Tigris in the west, and claims that this whole region is often referred to as Media in Scripture (*Histories* 1.2.17–18). The historical Parthia, also known as the Arsacid Empire, was a major political and cultural power in ancient Iran and Iraq that extended, at its height, from modern-day central-eastern Turkey to eastern Iran.

43 Media

Media (r)

Media shared borders with Parthia, Persia, and Armenia, all situated nearby on the map (*Etymologies* 14.3.8, 14.2.11; *Stjórn* 286–99; cf. H114, H167, D221, D242, D246). Orosius claims that in Scripture the name Media encompassed the wider area between the Rivers Indus and Tigris (*Histories* 1.2.17–18). The historical Media, home of the Medes, was situated in the mountainous north-west of modern-day Iran.

44 p(er) sidia

Persia (r)

Persia and Media are close neighbours that adjoin the Parthian kingdoms, according to Isidore (*Etymologies* 14.3.8; *Stjórn* 299–310; cf. H113–14, H119, H167, D244, D249). This part of the map follows Isidore in placing Persia next to Carmania (14.3.12). Orosis has Persia as a region within Media between the Indus and Tigris (*Histories* 1.2.17)

45 Carmania

Carmania (r)

Carmania was a desert region that neighboured Persia (Isidore, *Etymologies* 16.10.6; cf. H128), situated in the Kerman province of modern Iran.

46 Caria

Caria (r)

A Roman administrative district, and one of the nine provinces of Asia Minor (Isidore, *Etymologies* 14.3.38; cf. H337).

47 Frigida

Phrygia (r)

A Roman administrative district, and one the nine provinces of Asia Minor (Isidore, *Etymologies* 15.1.38; cf. H235, H316). The map contains two inscriptions that read *frigida,* one in Europe and one in Asia.

48 troia

Troy (c)

An ancient city in the province of Phrygia, famed for its destruction and described on the Hereford Map as 'civitas bellicosa' ('the warring city'). The Troy legend has been added in a cramped hand above Phrygia, where Isidore said it was located (*Etymologies* 14.3.41). That its omission was noted and corrected likely avers its importance to the map's historical argument. European elites, Icelanders among them, frequently traced their dynastic lines to Trojan heroes. In his *Edda*, Snorri Sturluson offers *Enea*, a name that appears nowhere else, as an alternative name for Europe, perhaps to associate it with its eponymous founder Aeneas, the storied Trojan exile and Brutus's ancestor (see Chapter 4) (cf. H345, EI.102).

49 pa*m*philia
Pamphylia (r)

A Roman administrative district, and one of the nine provinces of Asia Minor (Isidore, *Etymologies* 14.3.44; cf. EI.112).

50 hiberia
Iberia (r)

An ancient kingdom in the Caucasus, which Isidore situates near Armenia (*Etymologies* 14.3.36; cf. H164).

51 Tigris fluui*us*
River Tigris (fl)

A staple on medieval maps, being one of the four rivers believed to flow out of Paradise (Eusebius, *Onomasticon* 901). The Tigris, Euphrates, and Nile are also named on the map, but the Phishon (or Ganges) is absent. The Tigris and the Euphrates mark the eastern and western boundaries of ancient Mesopotamia (cf. H67, H114, H117, H169, [EI.61]).

52 Mesopotamia*m*
Mesopotamia (r)

A large historical region situated between the Rivers Tigris and Euphrates. The map places Mesopotamia next to the River Tigris (Isidore, *Etymologies* 37.2.6; *Stjórn* 310; cf. H167, H173).

53 **Charra civitas Abrahe**

Carrhae (?) city of Abraham (c, h)

The Biblical Abraham left Ur of the Chaldeans, his homeland, to go to Canaan, but settled in the Mesopotamian city of Charrae (Genesis 11.31; Eusebius, *Onomasticon* 751, 941). This legend has a confused parallel on the Hereford Map, which reads 'Hur habet et patria et Caldea' ('Ur has… and the homeland and Chaldea') (H183), next to what appears to be a depiction of Abraham. Westrem notes that the Jerome Map has 'Ur, home of Abraham' ('Hur patria Abrae') and suggests that the Hereford map is a corruption of 'Hur Caldee patria Abrahe' ('Ur of Chaldea, home of Abraham') (*Hereford Map*, pp. 88–89). Unlike the Hereford and Jerome Maps, the Viðey Map situates Abraham in the city in which he settled; his 'patria', Chaldea, appears elsewhere on the map.

54 **Asia minor**

Asia Minor (r)

A peninsular supraregional entity made up of nine Roman administrative districts. Orosius is the earliest attested author to use this term (Orosius, *Histories*, p. 39, n. 55). It is made of nine regions: Caria, Phrygia, Galatia, Cicilia, Pamphylia, and Isauria are present on the map, but Lycia, Bithynia, Lydia are not (Isidore, *Etymologies* 14.3.38; Aethicus Isther, *Cosmographia* 71). Geographically, it occupied the region in south-western Asia between Black Sea and the Mediterranean, coextensive with modern-day Turkey.

55 **Isauria**

Isauria (r)

A Roman administrative district, and one of the nine regions of Asia Minor (Isidore, *Etymologies* 14.3.44; cf. H335, EI.115).

56 **Cilicia**

Cicilia (r)

A Roman administrative district, and one of the nine regions of Asia Minor (Isidore, *Etymologies* 14.3.45; cf. H334, EI.118, [EI.123]).

57 **civitas tharsis**

City of Tarsus (c)

The capital city of the Roman province of Cilicia, which is adjacent on the map (Orosius, *Histories* 3.17.1–4; cf. H327, EI.120).

58 cappadocia
Cappadocia (r)

A region in the southern Caucasus, and neighbour to Asia Minor, Syria, Cicilia, and Isauria (Isidore, *Etymologies* 14.3.37–38; cf. H312).

59 Cori*n*gagena
Commagene (r)

The kingdom of Commagene was an ancient kingdom located between Asia Minor and the Partian Empire, south of the Black Sea (cf. H240). The garbled form of this name that appears on the map may show that the map-maker attempted to copy the word *ad literalis*, but was unfamiliar with it.

60 assiria
Assyria (r)

Assyria was a major Mesopotamian kingdom of the ancient Near East and the Levant (cf. H167, H174). It appears to have been misplaced, and should be nearer the Parthian kingdoms named in the south-east of the map, as per Isidore's description (*Etymologies* 14.3.8).

61 Calldea
Chaldea (r)

Isidore notes that in the course of its long history Chaldea had sometimes been a part of, and sometimes ruled, neighbouring Babylon (*Etymologies* 14.3.13; cf. H183). The historical Chaldea was in south-eastern Mesopotamia, on the Euphrates.

62 Babilon
Babylon (c)

Babylon was a major city of ancient Mesopotamia, built on the banks of the Euphrates (cf. H181, H294, H182). On the map, the name Babylon bisects the river name Euphrates, showing its maker's awareness of the city's geographical position. There are two Babylons on the map, Babylon the Old and Babylon the Babylon the New (Cairo). The Icelandic geographical treatise, preserved in one manuscript with the hemispherical world map, is similarly aware of the close proximity of Babylon to the Euphrates ('Eufrates fellur igegnum babílon inafornu') (*AÍ* II, p. 235).

63 eu frates

Euphrates (fl)

One of the four rivers reported to have their source in Paradise (Genesis 2:14). The legend is bisected by the name Babylon, the major city on its banks. The Euphrates is named next to Babylon on the Hereford Map (H181, cf. H66, H184).

64 philiste

Philistis (c)

This legend may be a corruption of Philistis, the capital city of the province of Palestine, which Isidore notes was called in his day Ascalon (*Etymologies* 14.3.19; cf. H236).

65 ptholomais

Ptolomais (c)

Ptolomais is an alternative name for Acre, on the coastal Levant, a fortress for the crusader armies named after the Egyptian ruler who built it, according to Isidore (*Etymologies* 15.1.36). It was the Christian stronghold in the Levant until it fell to Muslim forces in 1291. The Hereford Map depicts Ptolomais and Acre as separate cities (H375, cf. E1.136).

66 liban*us* mons

Mount Lebanon (m)

According to Isidore, Mount Lebanon was the principal peak of a mountain range in Phoenicia (*Etymologies* 14.3.20; cf. H243, [E1.129, E1.134]).

67 Galathia

Galatia (r)

A Roman administrative district, and one the nine provinces of Asia Minor (Isidore, *Etymologies* 14.3.40). This legend has become separated from the other six provinces of Asia Minor mentioned on the map.

68 Nazareth

Nazareth (c)

The map contains a detailed depiction of those lands described in the New Testament. The Biblical Nazareth was located in the Galilee, Matthew 2:23, and is shown next to it on the map (cf. H365).

69 **Galilea**
Galilee (r)

The Biblical Galilee was a region in Palestine (Isidore, *Etymologies* 14.3.23; cf. H368).

70 **hiericho**
Jericho (c)

The Biblical Jericho was an ancient city raised by Joshua, the leader of the Israelite tribes, after the death of his father Moses, Joshua 12:9 (cf. H381).

71 **Syria**
Syria (r)

Syria was a regional term sometimes used broadly in pilgrimage accounts for territories between Egypt and Asia Minor, including the lands venerated by Christianity (Westrem, *Hereford map*, p. 108; cf. H233).

72 **Ascalia**
Ascalon (c)

Ascalon was a city near Gaza in Palestine. Isidore tells us that it was formerly known as Philistis (*Etymologies* 14.3.19; cf. H396, E1.142).

73 **Iopen**
Jaffa (c)

Iope, or Jaffa, was an ancient port city in Palestine (Eusebius, *Onomasticon* 774). The name appears in the form 'Tope' on the Hereford Map (cf. H377, E1.138).

74 **Alexand*ria***
Alexandria (c)

An ancient city in Egypt named for Alexander the Great (cf. H424, H875, E1.142).

75 Tanakuisl fluui*us* maxim*us*

Tanakvísl, the greatest river (fl)

The Tanais is one of the four rivers that flows out of Paradise, and also figured in the traditional T-O schema as the boundary between Europe and Asia. This macaronic legend combines the Latinate *Tanais*, which has been adapted to the inflectional system of Old Norse, with the noun *kvísl* ('river branch', 'tributary'). In *Ynglinga saga*, the forms Tanakvísl and Vanakvísl were fabricated to strengthen the author's thesis of the Icelanders' Trojan origins, creating a link through the Tanais to the subgroup of Norse divinities called the *Vanir*, who, the saga maintains, had their ancestral home on its banks. Tanakvísl forms the boundary between Asia and Europe, and its vernacularised form evokes the Vanir's transcontinental migrations from Central Eurasia into Northern Europe. This legend was translated by Chekin as 'the great Tanakvisl river' (*Northern Eurasia*, p. 71) (cf. H67, H114, H117, H169).

76 palestina

Palestine (r)

The Biblical Palestine incorporated many of the lands venerated by Christianity, and occupies a central position within the map's depiction of Asia (H410 E1.149).

77 Cesarea

Cesarea (c)

Cesarea was the chief city of ancient Cappadocia, which is placed above this legend (cf. H311).

78 Sidon

Sidon (c)

A famous city in the province of Phoenicia at the northern boundary of Canaan and later Judea (Eusebius, *Onomasticon* 803). Sidon is one of few Asian inscriptions that appears in the *Expositio* but not on the Hereford Map (E1.134).

79 Tyr*us*

Tyre (c)

Tyre was a metropolis of Phoenicia (Eusebius, *Onomasticon* 894). It appears here in the vicinity of Sidon, another Phoneician city (cf. H374, E1.135).

80 arabia ibi e*st* mons Syna id est horeth

Arabia: in that place is Mount Sinai, which is Mount Horeb (r, m)

In Exodus 19, Moses received the Ten Commandments atop a mountain called Sinai, but in Deuteronomy 1, he receives them on Mount Horeb. In his *Onomasticon*, Eusebius tells us that Horeb is 'the mountain of God in the land of Midian, located near Mount Sinai beyond Arabia in the wilderness' (*Onomasticon* 946). The Icelandic map identifies Sinai and Horeb as alternate names for the same mountain, but the Hereford Map shows both as separate mountains (Arabia, H185, H189; Mount Sinai, H267; Mount Horeb, H399).

81 Madianite

The Midianites (e)

The Biblical Midianites inhabited the region on the Sinai Peninsula east of Sinai in Genesis 25:2 (Eusebius, *Onomasticon* 650; cf. H265).

82 Iudea

Judah (r)

The Biblical Judah was situated in the southern Levant (cf. H393).

83 hebron ibi sepulltus est adam prim*us*

Hebron, where Adam the first man buried (c, h)

According to Isidore, Hebron was the place where Adam was buried (*Etymologies* 15.1.24). In the Old Norse *Elucidarius*, we are told that Adam was born 'istaþ þei*m* es ebron hei*ter*. þar es h*ann* do siþan *oc* vas graven' ('in that place which is called Hebron where he died later and was buried') (*Elucidarius*, trans. Firchow, pp. 18–19). Hebron was also where the Biblical Patriarchs – Abraham and his wife Sarah, Isaac and Rebekah, Jacob and Leah – were buried (Westrem, *Hereford Map*, p. 164) (Hebron, H368; Adam, H64, H71).

84 Hie rusalem

Jerusalem (c)

Jerusalem is among the most prominent of the map's legends, situated amid the map's depiction of the Biblical lands (cf. H389).

85 Egipt*us*
Egypt (r)

In medieval tradition, Egypt comprised part of Asia. Here it is depicted with Alexandria, the new Babylon (Cairo), and the River Nile (cf. H287–88, H410, H423, E1.150, E1.149, [E1.162]).

86 Babilon
Cairo (c)

One of the map's two Babylons. In an Icelandic geographical treatise, Babylon the New (Cairo) was a city in lower Egypt located on the Nile: 'Nilus heitir eþr geon oðro nafni. hín fiorða a su er fellr or paradiso. hon skilr asiam oc africam. hon fellr umhueruis egipta land. A egipta landí er babilon in nyia. oc hofu - borg su er alexandría heitir' ('The fourth river that flows out of Paradise is called the Nile, or by its other name Geon. It divides Asia and Africa, and flows through the whole of Egypt. In Egypt is Cairo [Babylon the New], and the capital city which is called Alexandria') (cf. *AÍ* I, p. 9–10). Their arrangement on the map conforms to this tradition (cf. H181, H294, H182).

87 Nilus flum*en* egipti
The Nile, the river of Egypt (fl)

The River Nile is one of the four rivers that flow out of Paradise, and the river that separates Asia from Africa. On the map, it features prominently with the other two waters that divide the *kringla heimsins* into its continental thirds: the Tanais, and the Mediterranean Sea. The Tanais and Nile are written along the manuscript seam, between the map's Asian half and its depiction of western Europe and Africa. The confluence of these three waters forms the 'T' in the map's T-O framework (cf. H194, H427, H441).

Africa

88 AFRICA
Africa (r)

One of the world's three continents, mostly comprising the Mediterranean regions subdued by the Roman Empire. The legend sits outside the terrestrial circle on the map's conceptual frame; its south-easterly position at the head of the continent mirrors the north-westerly placement of the legend Europe.

89 Pentapolis regio ibi sunt .v. urbes.

Pentapolis region, in that place are five cities (r)

A grouping of five cities – one of several in the ancient world – probably formed for political, commercial, and military reasons. On the Hereford Map, this is situated below Libya Cyrenensis. Cyrenaica was also called Pentapolis (Martianus Capella, *De Nuptiis*, 672) (cf. H873, E II.5).

90 Trogita prouincia ibi in uenit*ur* carbu*n*culus igneus *et* alt*er* exacontalit*us* .lx. colorib*us* micans

Province of the Troglodytes, in that place is discovered fiery carbuncles and also 'sixtystone', which sparkles in sixty colours (e, r, o)

Carbuncles were frequently said to abound in the territories of the Troglodytes (Isidore, *Etymologies* 16.14.1). *Hexacontalithos* ('sixtystone') is also found in Africa (16.12.5), in Libya among the Troglodytes. The Troglodytes are a people well attested in late antique geographies, said to live in caves (cf. H890–91, H914).

91 Bizancena fructissim*m*a terra

Byzacium, most fruitful land (r)

Byzacium was 'a region… whose sowings are repaid by hundredfold harvests' (Martianus Capella, *De Nuptiis*, 670), famed for the richness of its soil. The *Expositio* likewise tells us that Byzacium is a region 'fert fructum centesimum' ('bearing fruit hundredfold'), but the Hereford Map says this of Zeugis, the northern part of Tunis and a territory Isidore places between Byzacium and Numidia (*Etymologies* 14.6.7). Some authorities, including Orosius (*Histories*, 1.2.91–92), Hugh (*Descriptio*, 459–60) and the *Expositio* (2.19), consider Zeugis and Byzancium to be the same region. The Hereford Map instead calls Byzacium the 'nobilissimis oppidis' ('noblest of towns') (H907, cf. H905, E2.30, E2.19).

92 Libia p*r*ouincia Affrice que est circa cirenen

Libya, province of Africa, which is around Cyrene (r, c)

Cyrene is the chief city of Libya (Isidore, *Etymologies* 14.6.4; cf. H870, H873, E2.2).

93 **Garamannia**

Garamannia (r)

The Garamantes are a nation described by Herodotus, and situated in the south-west of ancient Libya. Their region was called Garamannia.

94 **Getulia ibi infantes ludunt serpentib*us***

Getulia, where children play with snakes (r, a)

The Getuli appear on the Hereford Map (cf. H888) among other *barbari* ('barbarians') – the Natabres and Garamantes – but the observation that they play with snakes may be, as Simek postulates, an Icelandic innovation (*Altnordische Kosmographie*, p. 71). There is no shortage of snakes in the Hereford Map's depiction of Africa: the men among the Philli (H969–70) expose their new-borns to snakes to test their wives' fidelity, while Mount Ardens abounds with snakes.

95 **Gaulo insula ibi n*ec* serpens nascitur nec uiuit**

The island of Gaulo, where snakes are neither born nor live (i, a)

This legend originates in Isidore's *Etymologies* (9.2.124), and appears almost identically on the Hereford Map as 'Gauloena, ubi serpents nec vivunt nec nascuntur' ('Gaulo, where serpents neither live nor are born') (H977).

96 **Numidia**

Numidia (r)

A coastal North African kingdom that warred with Rome, becoming a Roman province in 46 BCE. The war is recounted in Sallust's *The Jugurthine War* (17.1–18.12) which forms part of the Icelandic saga of the Romans, *Rómverja saga* (on African geography, Ch. 7) (cf. H930, H939, E2.40, E2.73).

97 **Mauritanie .III.**

Mauritania III (r)

Orosius divides Mauritania into three parts (*Histories* 1.2.93–94), and these are marked separately on the Hereford Map: Nauritania Cesariensis, Mauritania Tingitana, and Mauritania Sifiensis (H936, H931, H934; cf. E2.52, E2.63, E2.42).

98 **Hic *sunt* solitu dines inaces sibiles *et* arene usq*ue* huc**

Here are inaccessible wildernesses and deserts all the way up to this place (r)

This legend echoes the description, from Orosius's *Histories* (1.2.92) and Isidore's *Etymologies* (14.5.11), of the region around Mount Astrixis and the deserts that extend to the Ethiopic Ocean. This is on the Hereford Map as: 'Mons Astrixis dividit vivam terram et arenas iacentes usque ad occeanum Ethiopiam, in quibus oberrant Gangines Ethiopes' ('Mount Astrixis divides arable land and the desert stretching all the way to the Ethiopic Ocean, where the Gangines Ethiopians wander about') (H958). This legend enjoins Ethiopia at Africa's edge.

99 **E thi o pia**

Ethiopia (r)

A region in Africa that was a staple feature in the works of ancient geographers and ethnographers, but was sometimes vague or fabulous in the works of medieval authors (Friedman, *Monstrous Races*, pp. 64–65) (cf. H440, H962).

Europe

100 **EUROPA**

Europe (r)

One of world's three continents, separated from Africa by the Mediterranean Sea, and from Asia by the River Tanais, both waters prominently depicted on the map. Europe has been written outside the terrestrial circle on the map's conceptual frame. The legends Thule and Iceland are written, on the same plane of orientation, immediately below.

I. Europe's Western Seaboard

101 **Normannia**

Normandy (r)

The places assembled here on Europe's western seaboard – Normandy, Brittany and Gascony – were those lands that had, until the early years of the thirteenth century, comprised the Continental reaches of the Angevin Empire. These three regions are separated from Francia and Gallia on the map by England, Scotland and Ireland. Their depiction on the map may be a remnant of a tradition that emphasised the distinctiveness of the French and English realms some decades earlier (cf. H686).

102 Brittannia

Brittany (r)

The north-western promontory of France, south of Normandy. Brittany had formerly been among the Plantagenet holdings, but from *c.* 1213 was held in homage by the dukes of Burgundy for the King of France.

103 Vasconia

Gascony (r)

A region in south-western France, south of the Pyrenees, depicted in the vicinity of Galicia. The map's names proceed in geographical sequence down the Iberian Peninsula (cf. H700).

104 Galicia

Galicia (r)

One of the map's few Hispanic territories, in the north-west Iberian Peninsula.

105 Hispania

Spain (r)

The southernmost of the regions named on Europe's western seaboard, and nearest to Africa. For Martianus Capella, 'the starting point and commencement of Europe is assigned to Spain' (*De Nuptiis*, 627) (cf. H842, H855, EII.63, EIII.136).

II. Britain and Ireland

106 Scocia

Scotland (r)

Britain and Ireland are represented by three names – Scotland, Ireland, and England. Their arrangement is conventional, and conveys no particular interest in, or updated information about, regions that were certainly known to the Icelandic map-maker (cf. H760).

107 Anglia

England (r)

A conventional name that couples with Scotland and Ireland in the map's depiction of Britain. An Icelandic geographical treatise describes England and Scotland as 'ei eín' ('one island') (*AÍ* II, pp. 231–40) (cf. H790).

108 Ibernia

Ireland (i)

Orosius places Ireland between Britain and Iberia, corresponding with its position on the map (*Histories* 1.2.80–81). Simek omits this legend (*Altnordische Kosmographie*, pp. 419–23) (cf. H832).

III. Mediterranean and Continental Europe

109 Grecia

Greece (r)

The map's depiction of the Byzantine world comprises three conventional names – Constantinople, Thrace, and Greece – proceeding in sequence towards the map's centre. These names are also in the near vicinity of Moesia and Sparta. Greece is one of the few place-names named here, but not on the Hereford Map.

110 tracia

Thrace (c)

A historical region in south-eastern Europe, which the map-maker places in sequence with Greece and Constantinople (cf. H478, EIII.170).

111 Co*n*sta*n*ti nopo lis

Constantinople (c / r)

Constantinople was a stable part of the Icelandic world image, with Icelanders reputed to have held esteemed positions in service of the Byzantine emperors in a number of the Icelandic sagas (e.g. *Laxdæla saga*, Ch. 77 and *Njáls saga*, Ch. 80). In Old Norse literature, Constantinople was frequently known as Miklagarðr ('great city'), which an Icelandic geographical treatise describes as a 'kingdom': 'ríkis þes er constantinopel heitir, er ver kaullum míklagarðr' ('that kingdom called Constantinople, which we call Miklagarðr') (*AÍ* I, p. 10). The map was likely drawn after the western or Latin conquest of Byzantium that had taken place in 1204, but before the collapse of the Latin empire in 1261 (Jacoby, 'Latin Empire of Constantinople', p. 528) (cf. H479, EI.98, 100, EIII.7).

112 Apulia

Apulia (r)

The southern region of the Italian Peninsula and one of the map's four Italian legends (cf. H645).

113 Italia

Italy (r)

Italy has been placed, with Rome, near the map's European centre (cf. H695).

114 Roma

Rome (c)

Rome was the centre of the western Church, and its special status is written onto the map through its position at the middle of the map's representation of Europe, occupying a central position on the half of the map on folio 6r. On the Hereford Map, Rome is described as 'Roma, caput mundi, tenet orbis frena rotundi' ('Rome, head of the world, holds the bridle of the spherical earth') (H680).

115 Lagoba*r*dia

Lombardy (r)

The Lombards ruled the Italian Peninsula between the sixth and the eighth centuries, and were unknown to ancient geographers (*Expositio*, 252). On the Hereford Map, a legend associates Lombardy and the Italian north-western coastal region Liguria ('Langobardia, hec et Ligria') (H627). The older name *Ligria* also appears in the *Exposito* (EIII.121).

116 Germania

Germania (r)

The map bears the names of two Roman provinces on the Rhine. For Isidore, Germania was divided into two parts: *Germania superior* and *Germania inferior*, the latter also called *Saxonia*. Germania superior extended from the Danube to the Rhine, and reached north as far as Scythia (*Etymologies* 14.4.14) (cf. H538, EIII.179).

117 Saxonia

Saxony (r)

A region also called Germania inferior ('Germania inferior, hec et Saxonia') (*Etymologies* 14.4.14). It borders Denmark on the map (cf. H544, EIII.186).

118 Da*n*morc

Denmark (r)

Denmark is depicted away from the map's other Scandinavian lands, adjoining Saxony. Denmark is one of the few places to be named here, but not on the Hereford Map. The Hereford Map does, however, mention the Danes in a legend on the River Elbe identifying 'Terminus Danorum et Sanonum' ('the frontier between the Danes and the Saxons') (H543, see also EIII.182). The Danes are mentioned in the *Expositio* also in relation to their city of Schleswig ('Slessic ciuitas Danorum') (EIII.180). This is one of few names on the map written in the vernacular Old Norse. Denmark was the most populous Scandinavian region for most of the Middle Ages.

119 Frisia

Frisia (r)

The Frisians were unknown to ancient geographers, but appear on the Hereford Map as 'Frisones, qui inter Saxones deputantur' ('the Frisians, who count among the Saxons') (H555). They are also mentioned in relation to the Saxons in the *Expositio* (EIII.226, cf. EIII.197).

120 Fracia

Francia (r)

Francia in this period was limited to the Île-de-France and sometimes its nearest environs, but was certainly to be explicitly contrasted with areas such as Brittany and Normandy. Simek omits this legend (*Altnordische Kosmographie,* pp. 419–23). Chekin has 'Francia' but the legend reads 'Fracia' (*Northern Eurasia*, p. 70).

121 Gallia

Gaul (r)

A traditional name representing a historical region of Western Europe, situated here in the vicinity of France and a mountain range that may be the Pyrenees.

122 Parmo montes

Pyrenees (?) (m)

These mountains are positioned between Gaul and Spain and probably represent the Pyrenees. Chekin suggests this legend may be 'Parma mountains', the Apennines, noting that the Apennines are named on the Ebstorf Map as 'Par… montes' (*Northern Eurasia*, p. 70, fn. 8). If 'Parmo' is a rendering of 'Parma', it has become separated from the map's other Italian place-names.

123 Mediterraneum mare

Mediterranean Sea (mr)

The Mediterranean Sea is one of three waters that, with the Tanais and Nile, divides the three continents, and marks the conventional boundary between Europe and Africa. It has been written on a plane of orientation perpendicular to the map's Continental European legends.

IV. Northern Europe

124 Tile

Thule (i)

Thule originates in Classical European literature relating the discovery of an island in the North Atlantic (H459, EIII.188). In Roman literature, it was an emblem of geographic extremity and remoteness, but in the medieval literatures of Scandinavia and England was sometimes associated with Iceland. The Icelandic *Landnámabók* and *Breta sögur* claim *Thule* was an earlier name for Iceland (see Chapter 4).

125 Island

Iceland (i)

Iceland, with its double Thule, are depicted at the head of Europe, immediately beneath the legend *Europa* on the map's conceptual frame. According to the Icelandic *Landnámabók*, Iceland was named by one of its earliest discoverers and settlers, Hrafna-Flóki ('Raven-Flóki'), in the ninth century. The name's earliest written attestation is on the English Cotton Map, *c.* 1050 (see Chapter 4) (cf. H460).

126 Norvegie
Norway (r)

Norway is depicted at the head of the Scandinavian Peninsula, nearest Iceland (H457). It is the only Scandinavian land named in Latin instead of the vernacular *Nóregr*. The opening chapters of the *Íslendingasögur* frequently relate the migrations of the early Icelanders from Norway from the second half of the ninth century. Norway was closely connected to Iceland linguistically and culturally throughout the Commonwealth period. From around 1220, the Norwegian Crown exerted increasing power over Icelanders until they acceded to Norwegian rule in 1262/64 (see Chapter 5).

127 Gautl*and*
Götaland (r)

The historical southern region of Sweden. Adam of Bremen recognised two separate entities in Sweden, *Gothia* ('Götaland') and *Sueonia* ('Svealand'), entities that were not unified until the thirteenth century. This may be the earliest map to show Sweden unambiguously. Westrem suggests that the Hereford Map's *Ganzmir*, positioned east of the Scandinavian Peninsula, is also from Old Norse *Gautar* (*Hereford Map*, pp. 192–93) (cf. H455 (?), EIII.176 (?)).

128 Suiþioð
Svealand (r)

The historical middle region of Sweden (Brink, 'Naming the Land', p. 60). Simek omits this legend (*Altnordische Kosmographie*, pp. 419–23).

129 Rusia
Russia (r)

Russia was a stable part of the medieval Icelandic worldview, sometimes called Garðaríki or Garðaveldi ('kingdom of towns') in Old Norse literature. The map names Rus', the Scandinavian- and Slavic-inhabited region on the eastern Baltic littorals, and positions it next to the provinces of Sweden – Götaland and Svealand – on the Scandinavian Peninsula. The capital of Kievan Rus', Kiev, is also displayed on the map. On the

Hereford Map, Russia is associated with the central European Roman province Dacia ('Dacia, hec et Rusia') (H471).

130 Bia*r*ma*r* habitauit hic

Bjarms live here (e)

This macaronic legend relates to the inhabitations of the *Bjarmar* (Old English *Beormas*), an apparently Finnic speaking people whose presence in the White Sea region was reported to King Alfred the Great by the Norwegian traveller Ohthere, his account incorporated into the Old English translation of Orosius's *Histories.* Their inclusion on the map is very likely an Icelandic addition, the Bjarms a regular presence in Scandinavian historical writings as well as the Icelandic *fornaldarsögur*. The Bjarms do not feature on any other medieval world map mentioned in Chekin's *Northern Eurasia,* but do feature on the *Carta Marina* (1539) of the Swedish cartographer Olaus Magnus, where they are placed in northern Finland. The formula *hic habitant* ('live here') appears a number of times on the Hereford world map (Westrem, *Hereford Map*, pp. 75; 347).

V. The Baltic and Eastern Europe

131 Scithia

Scythia (r)

In ancient geographies, Scythia was a region north and west of the Black Sea, but by the Middle Ages it had become a nebulous term for Northern Eurasia. The map's depiction of the Baltic Sea region conforms with Adam's *Gesta,* in which a Baltic gulf is 'mare Barbarum seu pelagus Scithicum vocatur a gentibus quas alluit barbaris' ('named the Barbarian Sea or Scythian Lake, from the barbarian peoples whose lands it laps') (IV.10). This gulf extends through the Scythian regions and as far as Greece (H40, H48, H89, H167, H208–09, H212, H217, H308–09, H450, EIII.4).

132 frigida

Phrygia (r)

One of two Phrygias on the map, the other located in Asia Minor. Kålund and Simek both couple the European *frigida* with the neighbouring legend *Scythia* to form *Scythia frigida* ('Scythia the Cold'), while Chekin sees the two *frigida* legends as a 'rudiment of zonal structure' inherited from an earlier map, perhaps of the Macrobian type, that influenced it (*Northern Eurasia,* p. 70). The Asian *frigida* is certainly a corruption of *Phrygia,* however, and there tends to be multiple Phrygias also in ancient descriptions of Central Eurasia (cf. H253, H316, H340). Orosius, for instance, has a Greater and Lesser Phrygia (*Histories* 3.23.8–9).

133 Misia

Moesia (r)

A Roman province situated on the Danube that extended as far as Macedonia (Aethicus Isther, *Cosmography* 74; cf. H482).

134 Sparta

Sparta (c)

An uncommon name on medieval maps, here located in the vicinity of the map's Mediterranean legends, near Moesia, Greece and Thrace (cf. EIII.36).

135 eronei

nomads (?) (e)

This legend, from Latin *erroneus* ('wandering', 'vagrant') is placed near Kiev, the capital of Rus', and may indicate the nomadic peoples of the Eurasian Steppe, the Mongols who lay siege to Kiev in 1240 (Chekin, *Northern Eurasia*, p. 224). There are, alternatively, several Scythian tribes that appear on maps of whose names 'eronei' may be a misreading, e.g. the 'essedones' (H212 and EI.7, from Solinus, *Collection* 15.13) and the 'eunochi' (H217).

136 Kio

Kiev (c)

The map positions Rus', the Scandinavian- and Slavic-inhabited region on the eastern Baltic littorals, next to the provinces of Sweden – Götaland and Svealand – on the Scandinavian Peninsula. Kiev was its capital.

The Smaller Viðey Map

Reykjavík, Stofnun Árna Magnússonar, GkS 1812 III 4to,
f. 6v (*c.* 1225–50)

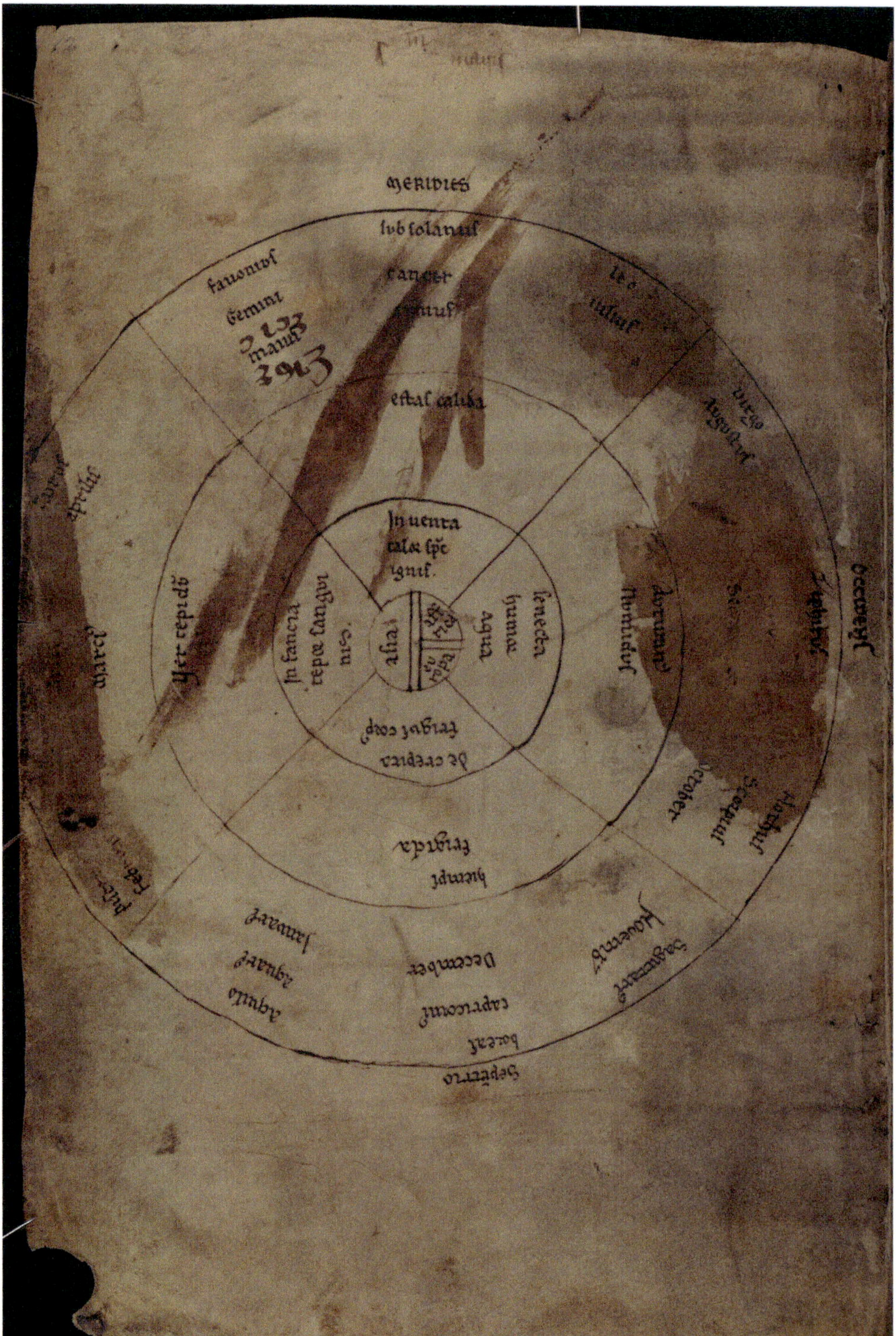

Meridies
Sub solanus
cancer
leo
iulius
estas calida
Juuenta
senecta
humida
aqua
autumnus

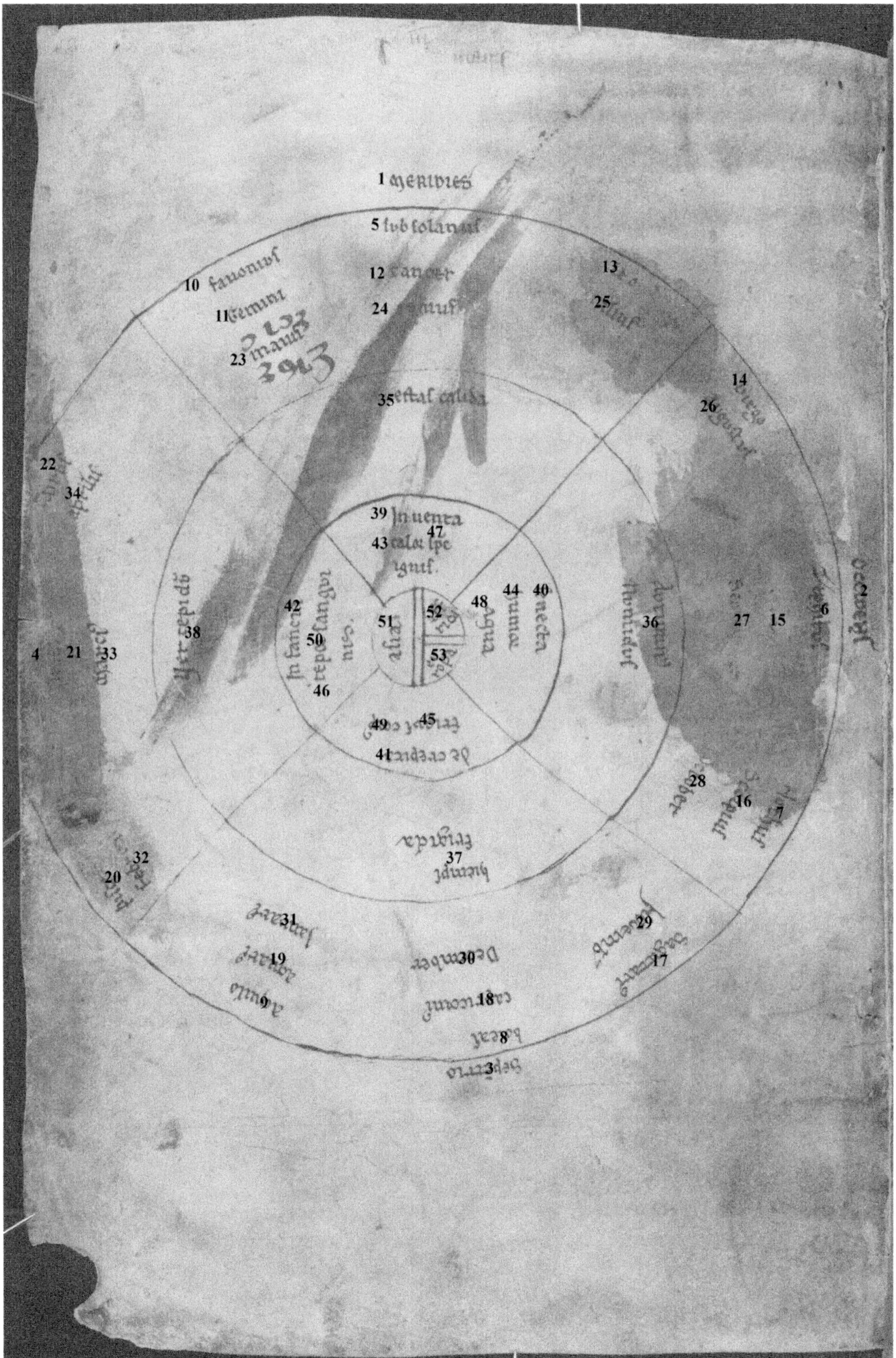

Meridies
sub solanus
cancer
favonius
estas calida
aqua

Previous editions and commentaries: Rafn, *Antiquités Russes*, vol. 2, p. 391; Simek, *Altnordische Kosmographie*, pp. 508–09.

This map is preserved overleaf from its larger companion. Its simple and iconic form may have done an exegetical service to the larger map, enabling its T-O structure, and the quadripartite formation of its conceptual frame, to be seen at a glance.

The Cardinal Directions

1 **MERIDIES**
South (cm)

The map is oriented, like its larger companion, with south at the top.

2 **OCCIDENS**
West (cm)

3 **SEPTE*M*TRIO**
North (cm)

4 **[destroyed by trimming]**
East (cm)

The Winds

5 **subsolanus**
Subsolanus (cm)

One of the four principal winds, which rises due east. The winds on the smaller map are arranged with no obvious order, so that this easterly wind appears in the south. Its placement at the top of the map may indicate that the winds were copied onto the map from an east-oriented exemplar.

6 **Zephyrus**
Zephyrus (cm)

One of the four principal winds, which rises due west. This wind has been correctly placed.

7 **Nothus**
Nothus (cm)

One of the four principal winds, which rises due south. It has been wrongly placed in the west sinistral position. Its satellite winds are Euroauster (*sinister*) and Euronothus (*dexter*).

8 **boreas**
Boreas (cm)

This wind appears in the northern, principal position. Septentrio's dextral satellite Aquilo or Boreas, here shown as two separate winds.

9 **Aquilo**
Aquilo (cm)

This wind appears in the northern, sinistral position. Septentrio's dextral satellite Aquilo or Boreas, here shown as two separate winds.

10 **Favonius**
Favonius (cm)

One of the four principal winds, which rose due west. It has been misplaced in the map's southern, sinistral position.

The Twelve Signs of the Zodiac

11 **Gemini**
Gemini (cm)

The signs of the zodiac are the constellations through which the sun moves over the course of the year. They are here arranged in harmony with the map's other temporal schemes, the seasons and the months of the year.

12 **Cancer**
Cancer (cm)

13 **Leo**
Leo (cm)

14 **Virgo**
Virgo (cm)

15 ---
Libra (cm)

A stain on the manuscript folio renders this legend unreadable.

16 **Scorpius**
Scorpio (cm)

17 **Sagittari*us***
Sagittarius (cm)

18 **Capricorni*us***
Capricorn (cm)

19 **Aquari*us***
Aquarius (cm)

20 **Pisces**
Pisces (cm)

21 **– – –**
Aries (cm)

A stain on the manuscript folio renders this legend unreadable.

22 **Taurus**
Taurus (cm)

The Twelve Months

23 **Maius**
May (cm)

The twelve months, like the twelve winds, were a quaternary scheme: Isidore understood that there are three months in each season: the first when it is new, the second when it is mature, and the third when it is in decline (Isidore, *Etymologies* 5.35.2–3). These are placed in harmony with map's scheme of the four seasons. May is the first of the summer months, when the season is new.

24 **iunius**
June (cm)

June is the middle of the summer months, when the season is mature.

25 **iulius**
July (cm)

July is the last of the summer months, when the season is in decline.

26 **Augustus**
August (cm)

August is the first of the autumn months, when the season is new.

27 **Se– – –**
September (cm)

A stain on the manuscript folio obscures part of this legend. September is the middle of the autumn months, when the season is mature.

28 **October**
October (cm)

October is the last of the autumn months, when the season is in decline.

29 **Nouemb*er***
November (cm)

November is the first of the winter months, when the season is new.

30 **December**
December (cm)

December is the middle of the winter months, when the season is mature.

31 **Januari*us***
January (cm)

January is the last of the winter months, when the season is in decline.

32 **Februa– –**
ebruary (cm)

February is the first of the spring months, when the season is new.

33 **Marci*us***
March (cm)

March is the middle of the spring months, when the season is mature.

34 **Aprilis**
April (cm)

April is the last of the spring months, when the season is in decline.

The Four Seasons and their Qualities

35 **Estas Calida**
Hot Summer (cm)

Summer is associated with the south in accordance with theories about the Earth's climatic zones.

36 **Autumn*us* humidus**
Humid Autumn (cm)

Autumn is associated with the west.

37 **Hiemps frigida**
Cold Winter (cm)

Winter is associated with the north in accordance with theories about the Earth's climatic zones.

38 **Ver tepidu*m***
Warm Spring (cm)

Spring is associated with the east. The sun rises due east and sets due west on the spring and autumn equinoxes in the northern hemisphere.

The Four Ages of Man

39 **Iuuenta**
Youth (cm)

South is associated with summer and youth.

40 **Senecta**
Old Age (cm)

West is associated with autumn and old age.

41 **Decrepita**
Decrepitude (cm)

North is associated with winter and decrepitude.

42 **Infancia**
Infancy (cm)

East is associated with spring and infancy.

The Four Physical Qualities

43 **Calor**
Heat (cm)

44 **Humor**
Moisture (cm)

45 **Frigus**
Cold (cm)

46 **Tepor**
Warmth (cm)

The Human Microcosm

47 ***Spiritus* Ignis**
Breath Fire (cm)

This legend should complement others in describing the four component parts of the human microcosm, but the map-maker has written both *Spiritus* and *Ignis* in the southern quarter. It seems that the map-maker anticipated the scheme, which includes *Aqua,* to be the four classical elements.

48 **Aqua**
Water (cm)

49 **Corp*us***
Flesh (cm)

50 **Sangvi nis**
Blood (cm)

51 Asia
Asia (r)

The three continents are arranged on the T-O map at the centre of this diagram, with south at the top. Double lines, representing the T-O map's hydrographic framework, divide the *orbis terrarum* into its three parts.

52 **Aff rica**
Africa (r)

53 **Eu ropa**
Europe (r)

Bibliography

Primary Sources

Adam of Bremen, *Gesta Hammaburgensis ecclesiae pontificum*, MGH SRG 2, ed. B. Schmeidler (Leipzig: Hahnsche Buchhandlung, 1917)

——, *History of the Archbishops of HamburgBremen*, trans. Francis J. Tschan (New York: Columbia University Press, 2002)

Aethicus Ister, *The Cosmography of Aethicus Ister: Text, Translation, and Commentary*, ed. Michael W. Herren (Turnhout: Brepols, 2011)

Alexanders saga: Islandsk Oversættelse ved Brandr Jónsson, ed. Finnur Jónsson (Copenhagen: Kommissionen for det Arnamagnænske Legat, 1925)

Alfræði Íslenzk: Islandsk encyklopædisk litteratur. 3 vols., ed. Kristian Kålund (Copenhagen: Møllers bogtrykkeri, 1908–18)

Aristotle, *Meteorologica*, Loeb Classical Library, trans. H. D. P. Lee (Cambridge, MA: Harvard University Press, 1952)

Bede, *De natura rerum*, ed. J-P. Migne. *PL* 90 (Paris: J.-P. Migne and Garnier Frères, 1862), cols. 187–274

——, *On the Nature of Things and On Times*, trans. Calvin B. Kendall and Faith Wallis (Liverpool: Liverpool University Press, 2010)

——, *Bede, The Reckoning of Time*, trans. Faith Wallis (Liverpool: Liverpool University Press, 1999)

Biskupa sögur, vol. 1, ed. Sigurgeir Steingrímsson, Ólafur Halldórsson and Peter Foote, Íslenzk Fornrit 15.2 (Reykjavík: Hið Íslenska Fornrit Félag, 2003)

Biskupa sögur, vol. 2, ed. Ásdís Egilsdóttir, Íslenzk Fornrit 16:2 (Reykjavík: Hið Íslenska Fornrit Félag, 2002)

Danakonunga sögur, ed. Bjarni Guðnason, Íslenzk Fornrit 35 (Reykjavík: Hið Íslenska Fornrit Félag, 1968)

Derolez, Albert, *Lamberti S. Audomari Canonici Liber Floridus: Codex Autographus Bibliothecae Universitatis Gandavensis* (Ghent: Story-Scientia, 1968)

Dicuil, *Dicuili Liber de Mensura Orbis Terrae*, ed. C. A. Walckenaer (Paris: F. Didot, 1807)

Diplomatarium Islandicum, vol. 1, *834–1264* (Copenhagen: S. L. Møllers, 1857)

Diplomatarium Islandicum, vol. 6, *1245–1491* (Copenhagen: S. L. Møllers, 1904)

The Old Norse Elucidarius, ed. Evelyn Scherabon Firchow (Drawer: Camden House, 1992)

Eusebius, *Eusebii Pamphili Episcopi Caesariensis Onomasticon: Urbum et Locum Sacrae Scripturae. Graece cum Latina Hieronymi Interpretatione*, ed. F. Larsow and G. Parthey (London: Williams & Norgate, 1862)

Finnur Jónsson, ed., *Hauksbók, efter de Arnamagnæaske Håndskrifter no. 371, 544 og 675, 40* (Copenhagen: Thieles Bogtrykkeri, 1892–96)

—, ed., 'Lønskrift og lejligedsoptegnelser fra et par islandske håndskrifter', *Småstykker* 1–16 (Copenhagen: S. L. Møllers, 1886), pp. 185–94

Firchow, Evelyn Scherabon, ed., *The Old Norse Elucidarius* (Drawer: Camden House, 1992)

Geminus of Rhodes, *Geminos's Introduction to the Phenomena: A Translation and Study of a Hellenistic Survey of Astronomy*, trans. James Evans and J. Lennart Berggren (Oxford: Princeton University Press, 2006)

Geoffrey of Monmouth, *The History of the Kings of Britain: An Edition and Translation of De gestis Britonum (Historia regum Britanniae)*, ed. Michael D. Reeve and trans. Neil Wright (Woodbridge: Boydell, 2007)

Gerald of Wales, *Opera Omnia*, 8 vols., ed. J. S. Brewer, James F. Dimock and George F. Warner (London: Longman, 1861–91)

—, *The History and Topography of Ireland*, trans. John J. O'Meara (Mountrath: The Dolmen Press, 1982)

Grant, Edward, ed., *A Source Book in Medieval Science* (Cambridge, MA: Harvard University Press, 1974)

Grove, Jonathan, ed., 'Anonymous, stanzas addressed to fellow ecclesiastics 1 & 2 (*Anon Eccl 1-2*)', *Poetry on Christian Subjects, Part 2: The Fourteenth Century*, edited by Margaret Clunies Ross (Turnhout: Brepols, 2007), pp. 471–75

Guðbjörg Kristjánsdóttir, *Íslenska Teiknibókin* (Reykjavík: Crymogea, 2013)

Hákonar saga Hákonarsonar, ed. Þorleifur Hauksson, Sverrir Jakobsson, and Tor Ulset, Íslenzk Fornrit 31 (Reykjavík: Hið Íslenzka Fornritfélag, 2013)

Helpericus, *Liber de computo*, ed. J-P. Migne. *PL* 137, cols. 17–48

Hermann of Carinthia, *Introductorium in astronomiam Albumasaris Abalachi octo continens libros partials* (Venice: Sessa, 1506)

Herodotus, *The Landmark Herodotus: The Histories*, trans. Andrea L. Purvis, ed. Robert B. Strassler and introd. Rosalind Thomas (London: Quercus, 2008)

Historia Norwegie, ed. Inger Ekrem and Lars Boje Mortensen, trans. Peter Fisher (Copenhagen: Museum Tusculanum Press, 2003)

Hugh of Saint Victor, *Hugonis de Sancto Victore De archa Noe, Libellus de formatione arche*, ed. Patrice Sicard (Turnhout: Brepols, 2001)

Hugh of St Victor, *La "Descriptio mappe mundi" de Hugues de Saint-Victor: texte inédit avec introduction et commentaire*, ed. Patrick Gautier Dalché (Paris: Études Augustiniennes, 1988)

Isidore of Seville, *Etymologiarum sive originum libri xx*, 2 vols., ed. W. M. Lindsay (Oxford: Clarendon Press, 1911)

—, *The Etymologies of Isidore of Seville*, trans. Stephen A. Barney, W. J. Lewis, J. A. Beach, and Oliver Berghof (Cambridge: Cambridge University Press, 2006)

Íslendingabók Ara Fróða, AM. 113a and 113b, fol., ed. Jón Jóhannesson (Reykjavík: Háskóli Íslands, 1956)

Íslendingabók / Landnámabók, ed. Jakob Benediktsson, Íslensk Fornrit 1, 2 vols. (Reykjavík: Hið Íslenska Fornrit Félag, 1968)

Íslendingabók, The Book of the Icelanders; Kristni Saga, The Story of the Conversion, trans. Siân Grønlie (London: Viking Society for Northern Research, 2006)

Konungs Skuggsjá: speculum regale, ed. Magnús Már Lárusson (Reykjavík: H. F. Leiftur, 1955)

Kugler, Hartmut, *Die Ebstorfer Weltkarte: Kommentierte Neuausgabe in zwei Bänden*, 2. vols. (Berlin: Akademie Verlag, 2007)

Kålund, Kristian, ed., *Arne Magnusson. Brevveksling med Torfæus (Þormóður Torfason)*, (Copenhagen: Christiania, 1916)

Larsson, Ludvig, *Äldsta delen af Cod. 1812, 4to Gml. Kgl. Samling på Kgl. Biblioteket i København* (Copenhagen: S. L. Møllers, 1883)

Macrobius, Ambrosius Theodosius, *Commentarii in Somnium Scipionis*, ed. J. Willis (Leipzig: Teubner, 1963)

—, *Commentary on the Dream of Scipio*, trans. William Harris Stahl (New York: Columbia University Press, 1952)

Martianus Capella, *De nuptiis Philologiae et Mercurii*, ed. James Willis (Leipzig: Teubner, 1983)

, *Martianus Capella and the Seven Liberal Arts*, vol. 2: *The Marriage of Philology and Mercury*, trans. William Harris Stahl (New York: Columbia University Press, 1977)

McDonald, Sheryl, '*Nítíða saga*: A Normalised Icelandic Text and Translation', *Leeds Studies in English* 40 (2009), 119–45

Orosius, Paulus, *Pauli Orosii Historiarum adversum Paganos Librii VII*, ed. Carolus Zangemeister (Vienna: Apud C. Geroldi Filium Bibliopolam Academiae, 1882)

—, *Seven Books of History against the Pagans*, trans. A. T. Fear (Liverpool: Liverpool University Press, 2010)

Pliny the Elder, *Naturalis Historia*, vols. 2 and 4, ed. and trans. H. Rackham (Cambridge, MA: Harvard University Press, 1958, 1961)

Roger de Howden, *Du Yorkshire à l'Inde: une géographie urbaine et maritime de la fin du XIIe siècle (Roger de Howden?)*, ed. Patrick Gautier Dalché (Geneva: Droz, 2005)

Rómverjasaga, 2 vols., trans. Þorbjörg Helgadóttir (Reykjavík: Stofnun Árna Magnússonar í Íslenskum Fræðum, 2010)

Sacrobosco, Johannes de, *Libellus Iohannis de Sacrobosco, de Anni ratione, seu ut vocantur vulgo, Computus Ecclesiasticus* (Paris: Guillaume Cauellat, 1550)

—, *The Sphere of Sacrobosco and its Commentators*, ed. and trans. Lynn Thorndike (London: Cambridge University Press, 1949)

Saxo Grammaticus, *Saxonis Gesta Danorum*, ed. J. Olrik and H. Ræder (Copenhagen: Levin & Munksgaard, 1931)

Seneca, *Seneca VIII, Tragedies*, vol. 1, Loeb Classical Library, ed. John G. Fitch (Cambridge, MA: Harvard University Press, 2002)

Snorri Sturluson, *Heimskringla*, Íslensk Fornrit 26:1, ed. Bjarni Aðalbjarnarson (Reykjavík: Hið Íslenska Fornrit Félag, 1941)

—, *Snorra Edda: Prologue and Gylfaginning*, ed. Anthony Faulkes (London: Viking Society for Northern Research, 2005)

—, *Snorra Edda: Skaldskáparmál*, ed. Anthony Faulkes (London: Viking Society for Northern Research, 2005)

Solinus, Gaius Julius, *Collectanea rerum memorabilium*, ed. Th. Mommsen (Berlin: Weidmann, 1895)

Stjorn: Gammelnorsk Bibelhistorie fra verdens skabelse til det Babyloniske fangenskab, ed. C. R. Unger (Christiania: Feilberg & Landmarks Forlag, 1862)

Strabo, *Geographica*, Loeb Classical Library, ed. H. L. Jones (Cambridge, MA: Harvard University Press, 1924)

Theodoricus, 'Theodrici monachi Historia de antiquitate regum Norwagiensium', *Monumenta Historica Norvegiæ: Latinske Kildeskrifter til Norges Historie i Middelalderen*, ed. Gustav Storm (Oslo: A. W. Brøgger, 1880), pp. 1–68

—, *Theodoricus Monachus: The Ancient History of the Northern Kings*, trans. David McDougall and Ian McDougall (London: Viking Society for Northern Research, 1998)

Townsend, David, trans., *Alexandreis: A Twelfth Century Epic* (Philadelphia: University of Pennsylvania Press, 1996)

Virgil, *Georgics*, ed. and trans. Arthur S. Way (London: Macmillan, 1912)

Walter of Châtillon, *Alexandreis: A Twelfth Century Epic*, trans. David Townsend (Philadelphia: University of Pennsylvania Press, 1996)

Westrem, Scott D., *The Hereford Map: A Transcription and Translation of the Legends with Commentary* (Turnhout: Brepols, 2001)

William of Conches, *De Philosophia Mundi Libri Quatuor*, ed. J-P Migne *PL* 172

William of Newburgh, *The History of English Affairs*, book 1, ed. and trans. P. G. Walsh and M. J. Kennedy (Warminster: Aris and Phillips, 1988)

Secondary Sources

Akbari, Suzanne Conklin, *Idols in the East: European Representations of Islam and the Orient, 1100–1450* (Ithaca: Cornell University Press, 2009)

Andersson, Theodore M., 'The King of Iceland', *Speculum* 74 (1999), 923–34

Ashman Rowe, Elizabeth, 'Literary, Codicological, and Political Perspectives on *Hauksbók*', *Gripla* 19 (2008), 51–76

Ashurst, David, 'Journey to the Antipodes. Cosmological and Mythological Themes in *Alexanders saga*', in *Proceedings of the Eleventh International Saga Conference*, ed. Margaret Clunies Ross and Geraldine Barnes (Sydney: University of Sydney, 2000), 1–13

Ashurst, David, 'The Ironies in Cardinal William of Sabina's Supposed Pronouncement on Icelandic Independence', *Saga-Book* 31 (2007), 39–45

Anjac, Germaine, 'The Foundations of Theoretical Cartography in Archaic and Classical Greece', in *The History of Cartography*, vol. 1: *Cartography in Prehistoric, Ancient, and Medieval Europe and the Mediterranean*, ed. J. B. Harley and David Woodward (London: The University of Chicago Press, 1987), pp. 130–47

Aujac, Germaine, 'Greek Cartography in the Early Roman World', in *The History of Cartography* vol. 1: *Cartography in Prehistoric, Ancient, and Medieval Europe and the Mediterranean*, ed. J. B. Harley and David Woodward (London: University of Chicago Press, 1987), pp. 161–76

Bagge, Sverre, 'The Scandinavian Kingdoms', in *The New Cambridge Medieval History, vol. 5, c. 1198–c. 1300*, ed. David Abulafia (Cambridge: Cambridge University Press, 1999), pp. 720–42

Bagrow, Leo, 'Old Inventories of Maps', *Imago Mundi* 4 (1948), 18–20

Bang, Peter Fibiger and Dariusz Kołodziejczyk, ''Elephant of India': Universal Empire through Time and across Cultures', in *Universal Empire: A Comparative Approach to Imperial Culture and Representation in Eurasian History*, ed. Peter Fibiger Bang and Dariusz Kołodziejczyk (Cambridge: Cambridge University Press, 2012), pp. 1–40

Barnes, Geraldine, *The Bookish Riddarasögur: Writing Romance in Late Mediaeval Iceland* (Odense: University Press of Southern Denmark, 2014)

Birkholz, Daniel, *The King's Two Maps: Cartography and Culture in Thirteenth-Century England* (London: Routledge, 2004)

Blackburn, Bonnie and Leofranc Holford-Strevens, *The Oxford Companion to the Year* (Oxford: Oxford University Press, 1999)

Bober, Harry, 'An Illustrated Medieval School-Book of Bede's *De natura rerum*', *The Journal of the Walters Art Gallery* 20 (1957), 64–97

Boulhosa, Patricia Pires, *Icelanders and the Kings of Norway: Medieval Sagas and Legal Texts* (Leiden: Brill, 2005)

Bouloux, Nathalie, 'The Munich Map (c. 1130): Description, Meanings and Uses', in *A Critical Companion to English Mappae Mundi of the Twelfth and Thirteenth Centuries*, ed. Dan Terkla and Nick Millea (Woodbridge: Boydell, 2019), pp. 92–111 Breen, Katharine, 'Returning Home from Jerusalem: Matthew Paris's First Map of Britain in Its Manuscript Context', *Representations* 89 (2005), 59–93

Brégaint, David, 'Conquering Minds: *Konungs Skuggsjá* and the Annexation of Iceland in the Thirteenth Century', *Scandinavian Studies* 84 (2012), 439–66

Brink, Stefan, 'Naming the Land', in *The Viking World*, ed. Stefan Brink and Neil Price (London: Routledge, 2008), pp. 57–66

Brown, Michelle P., 'Making Manuscripts and Mappae Mundi', in *A Critical Companion to English Mappae Mundi of the Twelfth and Thirteenth Centuries*, ed. Dan Terkla and Nick Millea (Woodbridge: Boydell, 2019), pp. 20–43

Burnett, Charles, 'Does the Sea Breathe, Boil, or Bloat? A Textual Problem in Abū Maʿshar's Explanation of Tides', in *Mélanges offerts a Hossam Elkhadem par ses amis et ses élèves*, ed. Frank Daelemans and Hossam Elkhadem (Brussels: Archives et bibliothèques de Belgique, 2007), pp. 73–79

Burrow, J. A., *The Ages of Man: A Study in Medieval Writing and Thought* (Oxford: Clarendon Press, 1988)

Callow, Christopher, 'Putting Women in their Place? Gender, Landscape, and the Construction of *Landnámabók*', *Viking and Medieval Scandinavia* 7 (2011), 7–28

Campbell, Tony, 'Portolan Charts from the Late Thirteenth Century to 1500', in *The History of Cartography*, vol. 1: *Cartography in Prehistoric, Ancient, and Medieval Europe and the Mediterranean*, ed. J. B. Harley and David Woodward (London: The University of Chicago Press, 1987), pp. 371–462

Carrol, Bradley, and Dale Ostlie, *An Introduction to Modern Astrophysics* (San Francisco: Addison-Wesley, 2007)

Carruthers, Mary, and Jessica Weiss, 'A Little Book about Constructing Noah's Ark', in *The Medieval Craft of Memory: An Anthology of Texts and Pictures*, ed. Mary Carruthers and Jan M. Ziolkowski (Philadelphia: University of Pennsylvania Press, 2002), 41–70

Cassidy, Vincent H. de P., 'Voyage of an Island', *Speculum* 38 (1963), 595–602

Chekin, Leonid S., 'Mappae Mundi and Scandinavia', *Scandinavian Studies* 65 (1993), 487–520

——, 'Easter Tables and the Pseudo-Isidorean Vatican Map', *Imago Mundi* 51 (1999), 13–23

——, *Northern Eurasia in Medieval Cartography: Inventory, Text, Translation, and Commentary* (Turnhout: Brepols, 2006)

Cleasby, Richard, and Guðbrandur Vigfússon, *An Icelandic-English Dictionary* (Oxford: Clarendon Press, 1957)

Clunies Ross, Margaret, 'Land-taking and text-making in medieval Iceland', in *Text and Territory: Geographical Imagination in the European Middle Ages*, ed. Sylvia Tomasch and Sealy Gilles (Philadelphia: University of Pennsylvania Press, 1998), pp. 159–84

—, *The Cambridge Introduction to the Old Norse-Icelandic Saga* (Cambridge: Cambridge University Press, 2010)

Cole, Richard, 'Racial Thinking in Old Norse Literature: The Case of the *Blámaðr*', *SBVS* 39 (2015), 21–40

Conley, Tom, *The Self-Made Map: Cartographic Writing in Early Modern France* (Minneapolis: University of Minnesota Press, 1996)

Connolly, Daniel K., *The Maps of Matthew Paris: Medieval Journeys through Space, Time, and Liturgy* (Woodbridge: Boydell, 2009)

—, 'In the Company of Matthew Paris: Mapping the World at St. Albans Abbey', in *A Critical Companion to English Mappae Mundi of the Twelfth and Thirteenth Centuries*, ed. Dan Terkla and Nick Millea (Woodbridge: Boydell, 2019), pp. 147–78

Cosgrove, Denis, *Apollo's Eye: A Cartographic Genealogy of the Earth in the Western Imagination* (Baltimore: John Hopkins University Press, 2001)

—, 'Cultural cartography: Maps and Mapping in Cultural Geography / Les Cartes et la Cartographie en Géographie Culturelle', *Annales de Géographie* 660/661 (2008), 159–78

Delano-Smith, Catherine, and Roger J. P. Kain, *English Maps: A History* (London: British Library, 1999)

Derolez, Albert, *The Autograph Manuscript of the Liber Floridus: A Key to the Encyclopedia of Lambert of Saint-Omer* (Turnhout:Brepols, 1998)

Destombes, Marcel, *Mappemondes A.D. 1200–1500. Catalogue préparé par la commission de Cartes Anciennes de l'Union Géographique Internationale* (Amsterdam: N. Israel, 1964)

Diller, Aubrey, 'The Parallels on the Ptolemaic Maps', *Isis* 33 (1941), 4–7

Duhem, Pierre, *Le système du monde. Histoire des Doctrines Cosmologiques de Platon à Copernic*, 10 vols (Paris: Hermann et Fils, 1913–59)

Eastwood, Bruce S., 'Heraclides and Heliocentrism: Texts, Diagrams, and Interpretations', *Journal for the History of Astronomy* 23 (1992), 233–60

—, 'Invention and Reform in Latin Planetary Astrology', in *Latin Culture in the Eleventh Century. Proceedings of the Third International Conference on Medieval Latin Studies, Cambridge, September 9–12 1998*, vol. 1, ed. Michael W. Herren, C. J. McDonough, and Ross G. Arthur (Turnhout: Brepols, 2002), pp. 264–97

Eastwood Bruce S., and Gerd Grasshoff, *Planetary Diagrams for Roman Astronomy in Medieval Europe, ca. 800–1500* (Philadelphia: American Philosophical Society, 2004)

Edson, Evelyn, *Mapping Time and Space: How Medieval Mapmakers Viewed Their World* (London: British Library, 1998)

——, 'World Maps and Easter Tables: Medieval Maps in Context', *Imago Mundi* 48 (1996), 25–42

Ehrensvärd, Ulla, 'Colour in Cartography: A Historical Survey', in *Art and Cartography: Six Historical Essays*, ed. David Woodward (London: University of Chicago Press, 1987), pp. 123–46

Esmeijer, Anna C., *Divina Quaternitas: A Preliminary Study in the Method and Application of Visual Exegesis* (Assen: Gorcum, 1978)

Federico, Sylvia, *New Troy: Fantasies of Empire in the Late Middle Ages* (Minneapolis: University of Minnesota Press, 2003)

Fry, Donald K., 'Bede', in *Medieval Scandinavia: An Encyclopaedia*, ed. Philip Pulsiano (London: Garland, 1993), pp. 36–37

Gautier Dalché, Patrick, 'De la glose à la contemplation: place et fonction de la carte dans les manuscrits du haut Moyen Age', in *Testo E Immagine Nell'alto Medioevo* (Spoleto: Presso la sede del Centro, 1994), pp. 693–764

——, *Carte marine et portolan au XIIe siècle' Le 'Liber de existencia riveriarum et forma maris nostril Mediterranei' (Pise, circa 1200)* (Rome: École Française de Rome, 1995)

——, 'Maps in Words: The Descriptive Logic of Medieval Geography from the Eighth to the Twelfth Century', in *The Hereford World Map: Medieval World Maps and their Context*, ed. P. D. A. Harvey (London: British Library, 2006), pp. 223–42

——, 'Comment penser 'l'Océan? Modes de connaissances des fines orbis terrarum du nord-ouest (de l'antiquité au XIIIe siècle)', *Actes des congrès de la Société des historiens médiévistes de l'enseignement supérieur public* 17 (1986), 217–33

Gísli Sigurðsson, *The Medieval Icelandic Saga and Oral Tradition: A Discourse on Method* (Cambridge, MA: Harvard University Press, 2004)

Gjerløw, Lilli, *Libri Liturgici Provinciae Nidrosiensis Medii Aevi*, vol. 11: *Ordo Nidrosiensis Ecclesiae (Ordubók)* (Oslo: Universitetsforlaget, 1968)

Gjerløw, Lilli, *Liturgica Islandica I: Text* (Copenhagen: C.A. Reitzels Boghandel A/S, 1980)

Grant, Edward, 'Cosmology', in *Science in the Middle Ages*, ed. David C. Lindberg (London: University of Chicago Press, 1978), pp. 265–302

——, 'Eccentrics and Epicycles in Medieval Cosmology', in *Mathematics and its Application to Science and Natural Philosophy in the Middle Ages, Essays in Honor of Marshall Clagett*, ed. Edward Grant and John E. Murdoch (Cambridge: Cambridge University Press, 1987), pp. 189–214

——, *Planets, Stars, and Orbs: The Medieval Cosmos, 1200–1687* (Cambridge: Cambridge University Press, 1994)

——, *The Foundations of Modern Science in the Middle Ages: Their Religious, Institutional, and Intellectual Contexts* (Cambridge: Cambridge University Press, 1996)

Halldór Hermansson, *The Cartography of Iceland* (London: Milford, 1931)

Haraldur Sigurðsson, *Kortasaga Íslands: Frá öndverðu til loka 16. aldar* (Reykjavík: Bókaútgáfa Menningarsjóðs og Þjóðvinafélagsins, 1971)

Harley, J. B., *The New Nature of Maps: Essays in the History of Cartography*, ed. Paul Laxton (London: John Hopkins University Press, 2001)

Harvey, P. D. A., *Medieval Maps of the Holy Land* (London: British Library, 2012)

—, *Mappa Mundi: The Hereford World Map* (London: British Library, 1996)

Hermann, Pernille, 'Spatial and Temporal Perspectives in *Íslendingabók*: Historiography and Social Structures', *Viking and Medieval Scandinavia* 1 (2005), 73–89

Hiatt, Alfred, 'The Map of Macrobius before 1100', *Imago Mundi* 59:2 (2007), 149–76

—, *Terra incognita: Mapping the Antipodes before 1600* (Chicago: University of Chicago Press, 2008)

—, '*Beowulf* off the map', *Anglo-Saxon England* 38 (2009), 11–40

—, 'From Hulle to Carthage: Maps, England, and the Sea', in *The Sea and Englishness in the Middle Ages: Maritime Narratives, Identity, and Culture*, ed. Sebastian I. Sobecki (Cambridge: D.S. Brewer, 2011), pp. 133–58

—, 'The Sawley Map (c. 1190)', in *A Critical Companion to English Mappae Mundi of the Twelfth and Thirteenth Centuries*, ed. Dan Terkla and Nick Millea (Woodbridge: Boydell, 2019), pp. 112–26

Hill, Joyce, 'From Rome to Jerusalem: An Icelandic Itinerary of the Mid-Twelfth Century', *The Harvard Theological Review* 76:2 (1983), 175–203

Hodges, Kenneth, 'Introduction: Places of Romance', in *Mapping Malory: Regional Identities and National Geographies in Le Morte Darthur*, ed. Dorsey Armstrong and Kenneth Hodges (New York: Palgrave Macmillan, 2014), pp. 1–18

Jacoby, David, 'Latin Empire of Constantinople and Frankish States in Greece', in *The New Cambridge Medieval History: Volume V c. 1198–c. 1300*, ed. David Abulafia (Cambridge: Cambridge University Press, 1999), pp. 525–42

Jackson, Peter, 'The Mongols and Europe', in *The New Cambridge Medieval History: Volume V c. 1198–c. 1300*, ed. David Abulafia (Cambridge: Cambridge University Press, 1999), pp. 703–19

Jackson, Tatiana, 'Bjarmaland Revisited', *Acta Borealia* 19:2 (2002), 165–79

Jesch, Judith, 'Geography and Travel', in *A Companion to Old Norse-Icelandic Literature and Culture*, ed. Rory McTurk (London: Blackwell, 2008), pp. 119–35

Jesch, Judith, *The Viking Diaspora* (London: Routledge, 2015)

Jón Jóhannesson, *A History of the Old Icelandic Commonwealth*, trans. Haraldur Bessason (Winnipeg: University of Manitoba Press, 1974)

Jón Viðar Sigurðsson, *Chieftains and Power in the Icelandic Commonwealth*, trans. Jean Lundskær-Nielsen (Odense: Odense University Press, 1999)

——, 'Kings, Earls, and Chieftains. Rulers in Norway, Orkney and Iceland c.900–1300', in *Ideology and Power in the Viking and Middle Ages: Scandinavia, Iceland, Ireland, Orkney, and the Faeroes*, ed. Gro Steinsland, Jón Viðar Sigurðsson, Jan Erik Rekdal and Ian Beuermann (Leiden: Brill, 2011), pp. 69–108

Kalinke, Marianne, 'Arthur, King of Iceland', *Scandinavian Studies* 87:1 (2015), 8–32

Karamustafa, Ahmet T., 'Introduction to Islamic Maps', in *The History of Cartography* vol. 2, book 1: *Cartography in the Traditional Islamic and South Asian Societies*, ed. J. B. Harley and David Woodward (London: University of Chicago Press, 1992), pp. 1–9

Kedar, Benjamin Z., and Chr. Westergård-Nielsen, 'Icelanders in the Crusader Kingdom of Jerusalem: A Twelfth-Century Account', *Mediaeval Scandinavia* 11 (1978–9), 193–211

Kedwards, Dale, 'Wind Diagrams in Medieval Iceland', *Quaestio Insularis* 17 (2014), 92–107

——, 'Iceland, Thule, and the Tilensian Precedent in Medieval Historiography', *Arkiv för nordisk filologi* 130 (2015), 57–78

——, 'The World Image of the *Konungs skuggsjá*', in *Speculum Septentrionale: Konungs skuggsjá and the European Encyclopaedic Tradition*, ed. Karl Gunnar Johansson and Elise Kleivane (Oslo: Novus, 2018), pp. 71–92

——, 'Geography', in *A Critical Companion to Old Norse Literary Genre*, ed. Massimiliano Bampi, Carolyne Larrington, and Sif Rikharðsdóttir (Cambridge: D. S. Brewer, 2020), pp. 127–44

Kline, Naomi Reed, *Maps of Medieval Thought: The Hereford Paradigm* (Woodbridge: Boydell, 2001)

Knowles, Dom David, 'The Mappa Mundi of Gervase of Canterbury', *The Downside Review* 48:3 (1930), 237–47

Kristín Bjarnadóttir, 'Mathematical Education in Iceland in Historical Context: Socio-Economic Demands and Influences' (PhD diss., Roskilde University, 2006)

Kupfer, Marcia, 'Reflections in the Ebstorf Map: Cartography, theology and *dilectio speculationis*', in *Mapping Medieval Geographies: Geographical Encounters in the Latin West and Beyond, 300–1600*, ed. Keith D. Lilley (Cambridge: Cambridge University Press, 2013), pp. 100–26

——, 'The Hereford Map (c. 1300)', in *A Critical Companion to English Mappae Mundi of the Twelfth and Thirteenth Centuries*, ed. Dan Terkla and Nick Millea (Woodbridge: Boydell, 2019), pp. 227–52

Kålund, Kristian. 'En Islandsk vejviser for pilgrimme fra 12. århundrede', *Aarbøger for Nordisk Oldkyndighed og Historie*, 3rd series, 3 (1913), 51–105.

Lafferty, Maura K., 'Mapping Human Limitations: The Tomb Ecphrases Walter of Châtillon's Alexandreis', *The Journal of Medieval Latin* 4 (1994), 64–81

Laird, Edgar S., 'Robert Grosseteste, Albumasar, and Medieval Tidal Theory', *Isis* 81:4 (1990), 684–94

Lapidge, Michael, *The Anglo-Saxon Library* (Oxford: Oxford University Press, 2008)

Lassen, Annette, 'Indigenous and Latin Literature', in *The Routledge Research Companion to the Medieval Icelandic Sagas*, ed. Ármann Jakobsson and Sverrir Jakobsson (Abingdon: Routledge, 2017), pp. 74–87

Lavezzo, Kathy, *Angels on the Edge of the World: Geography, Literature, and English Community, 1000–1534* (Ithaca: Cornell University Press, 2006)

Leonard, Stephen Pax, 'Social Structures and Identity in Early Iceland', *Viking and Medieval Scandinavia* 6 (2010), 147–59

—, *Language, Society and Identity in Early Iceland* (Oxford: Wiley-Blackwell, 2012)

Lewis, C. S., *The Discarded Image: An Introduction to Medieval and Renaissance Literature* (Cambridge: Cambridge University Press, 1964)

Loredana, Teresi, 'Anglo-Saxon and Early Norman England *Mappaemundi*', in *The Foundations of Learning: The Transfer of Encyclopaedic Knowledge in the Early Middle Ages*, ed. R. Bremmer and C. Dekker (Paris: Peeters, 2007), pp. 341–77

Lynam, Edward, 'Early Maps of Scandinavia and Iceland (Synopsis of a Lecture)', *Saga-Book* 11 (1928–36), 1–4

Maqbul, S. Ahmad, 'Cartography of al-Sharif al-Idrisi', in *The History of Cartography* vol. 2, book 1: *Cartography in the Traditional Islamic and South Asian Societies*, ed. J. B. Harley and David Woodward (London: University of Chicago Press, 1992), pp. 156–72

Marani, Tomasso, 'Contextualising *Leiðarvísir*: Sant' Agnese, The Catacombs, The Pantheon and The Tiber', *Saga-Book* 33 (2009), 44–66

—, '*Leiðarvísir*. Its Genre and Sources, with Particular Reference to the Description of Rome' (PhD diss., University of Durham, 2012)

McCluskey, Stephen C., *Astronomies and Cultures in Early Medieval Europe* (Cambridge: Cambridge University Press, 1997)

Mel'nikova, Elena A., *Drevneskandinavskie Geograficheskie Sochineniia: Teksty, Perevod, Kommentarii* (Moscow: Nauka, 1986)

McDougall, Ian, 'Foreigners and Foreign Languages in Medieval Iceland', *Saga-Book* 22 (1986–89), 180–233

McGurk, Patrick, D. N. Dumville, M. R. Godden, and Ann Knock, ed., *An Eleventh-Century Anglo-Saxon Illustrated Miscellany (British Library Cotton Tiberius B. V. Part 1)* (Copenhagen: Rosenkilde and Bagger, 1983)

McNaughton, Douglas, 'A World in Transition: Early Cartography of the North Atlantic', in *Vikings: The North Atlantic Saga*, ed. William W. Fitzhugh and Elisabeth I. Ward (Washington: Smithsonian Institution Press, 2000), pp. 257–69

Mead, William R., 'Scandinavian Renaissance Cartography', in *The History of Cartography*, vol. 3, part 2, *Cartography in the European Renaissance*, ed.

David Woodward (London: University of Chicago Press, 2007), pp. 1781–1805

Melberg, Håkon, *Origin of the Scandinavian Nations and Languages. An Introduction, I–II* (Halden: Eget Forlag, 1953)

Merrills, Andrew H., *History and Geography in Late Antiquity* (Cambridge: Cambridge University Press, 2005)

Merrills, Andy, 'Geography and Memory in Isidore's *Etymologies*', in *Mapping Medieval Geographies: Geographical Encounters in the Latin West and Beyond, 300–1600*, ed. Keith D. Lilley (Cambridge: Cambridge University Press, 2013), pp. 45–65

Miller, Konrad, *Mappaemundi: die ältesten Weltkarten*, 6 vols (Stuttgart: Roth, 1895–98)

Mittman, Asa Simon, 'The Vercelli Map (c. 1217)', in *A Critical Companion to English Mappae Mundi of the Twelfth and Thirteenth Centuries*, ed. Dan Terkla and Nick Millea (Woodbridge: Boydell, 2019), pp. 127–46

Morstein-Marx, Robert, 'The Myth of Numidian Origins in Sallust's African Excursus (*Iugurtha* 17.7–18.12)', *The American Journal of Philology* 122 (2001), 179–200

Mund-Dopchie, Monique, *Ultima Thulé: Histoire d'un lieu et genèse d'un mythe* (Geneva: Droz, 2009)

Nansen, Fridtjof, *In Northern Mists: Arctic Exploration in Early Times* vol. 1, trans. Arthur G. Chater (London: William Heinemann, 1911)

Nordal, Gúðrun, *Tools of Literacy: The Role of Skaldic Verse in Icelandic Textual Culture of the Twelfth and Thirteenth Centuries* (London: University of Toronto Press, 2001)

North, John, *The Fontana History of Astronomy and Cosmology* (London: Fontana Press, 1994)

Obrist, Barbara, 'Wind Diagrams and Medieval Cosmography', *Speculum* 72 (1997), 33–84

——, 'William of Conches, Māshā'allāh, and Twelfth-Century Cosmology', *Archives d'Historie Doctrinale et Littérature du Moyen Âge* 76 (2009), 29–87

O'Loughlin, Thomas, 'An Early Thirteenth-Century Map in Dublin: A Window into the World of Giraldus Cambrensis', *Imago Mundi* 51:1 (1999), 24–39

——, 'Map Awareness in the Mid-Seventh Century: Jonas' *Vita Columbani*', *Imago Mundi* 62:1 (2010), 83–85

O'Reilly, Jennifer, 'Patristic and Insular Traditions of the Evangelists: Exegesis and Iconography', in *Le Isole Britanniche e Roma in età Romanobarbarica*, ed. A. M. Luiselli Fadda and É. Ó Carragáin (Rome: Herder Editrice e Libreria, 1998), pp. 49–94

Orri Vésteinsson, *The Christianisation of Iceland: Priests, Power, and Social Change 1000–1300* (Oxford: Oxford University Press, 2000)

Parkes, M. B., 'The Hereford Map: The Handwriting and Copying of the

Text', in *The Hereford World Map: Medieval World Maps and their Context*, ed. P. D. A. Harvey (London: British Library, 2006), pp. 107–17

Pedersen, Olaf, 'Astronomy', in *Science in the Middle Ages*, ed. David C. Lindberg (London: University of Chicago Press, 1978), pp. 303–37

—, 'In quest of Sacrobosco', *Journal for the History of Astronomy* 16 (1985), 175–220

Poole, Reginald L., *Medieval Reckonings of Time* (London: Macmillan, 1921)

Pringle, Denys, 'Architecture in the Latin East 1098–1571', in *The Oxford Illustrated History of the Crusades*, ed. Jonathan Riley-Smith (Oxford: Oxford University Press, 1995), pp. 160–83

Pritsak, Omeljan, *The Origin of Rus'*, vol. 1: *Old Scandinavian Sources other than the Sagas* (Cambridge MA: Harvard University Press, 1982)

Rafn, Carl Christian, *Antiquitates Americanae sive Scriptores Septentrionales rerum ante-Columbianarum* (Copenhagen: Schultz, 1837)

—, *Antiquités Russes d'après les monuments historiques des Islandais et des anciens Scandinaves*, vol. 2 (Copenhagen: De l'imprimerie des frères Berling, 1852)

Riant, P., *Expéditions et pèlerinages des Scandinaves en Terre Sainte au temps des Croisades* (Paris: Lainé et Harvard, 1865)

Ribémont, B., 'Encyclopaedias', in *Medieval Science Technology and Medicine: An Encyclopaedia*, ed. Thomas F. Glick, Steven John Livesey, and Faith Wallis (New York: Routledge, 2005), pp. 159–61

Richter, Michael, 'Medieval Ireland and Iceland – Worlds Apart?' in *Saluting Aron Gurevich: Essays in History, Literature, and Other Related Subjects*, ed. Yelena Mazour-Matusevich and Alexandra Korros (Liden: Brill, 2010), pp. 155–66

Romm, James S., *The Edges of the Earth in Ancient Thought: Geography, Exploration, and Fiction* (Princeton: Princeton University Press, 1992)

Ross, Alan S. C., *The 'Terfinnas' and 'Beormas' of Ohthere* (London: Viking Society for Northern Research, 1981)

Rudolph, Conrad, *The Mystic Ark: Hugh of Saint Victor, Art, and Thought in the Twelfth Century* (New York: Cambridge University Press, 2014)

Scafi, Alessandro, 'Defining *mappaemundi*', in *The Hereford World Map: Medieval World Maps and their Context*, ed. P. D. A. Harvey (London: British Library, 2006), pp. 345–54

Scardigli, Piergiuseppe, and Fabrizio D. Raschellà, 'A Latin-Icelandic Glossary and Some Remarks on Latin in Medieval Scandinavia', in *Idee, Gestalt, Geschichte: Festschrift Klaus von See: Studien zur europäischen Kulturtradition*, ed. Gerd Wolfgang Weber (Odense: Odense University Press, 1988), pp. 299–323

Schöller, Bettina, *Wissen speichern, Wissen ordnen, Wissen übertragen: schriftliche und bildliche Aufzeichnungen der Welt im Umfeld der Londoner Psalterkarte* (Zürich: Chronos, 2015)

Scully, Diarmuid, 'Augustus, Rome, Britain and Ireland on the Hereford mappa mundi: Imperium and Salvation', *Peregrinations: Journal of Medieval Art and Architecture* 4:1 (2013), 107–133

Seaver, Kirsten, *Maps, Myths, and Men: The Story of the Vinland Map* (Stanford: Stanford University Press, 2004)

Sharpe, Richard, *Titulus: Identifying Medieval Latin Texts: An Evidence-Based Approach* (Turnhout: Brepols, 2003)

Sigurdson, Erika, 'The Church in Fourteenth-Century Iceland: Ecclesiastical Administration, Literacy, and the Formation of an Elite Clerical Identity' (PhD diss., University of Leeds, 2011)

Simek, Rudolf, *Altnordische Kosmographie: Studien und Quellen zu Weltbild und Weltbeschreibung in Norwegen und Island vom 12. Bis zum 14. Jahrhundert* (Berlin: Walter de Gruyter, 1990)

—, 'Skandinavische Mappae Mundi in der europäischen Tradition', in *Ein Weltbild vor Columbus: Die Ebstorfer Weltkarte*, ed. Hartmut Kugler, Eckhard Michael, and Horst Appuhn (Weinheim: VCH, 1991), pp. 167–84

—, '*Hierusalem civitas famosissima*: Die erhalten Fassungen des hochmittelalterlichen *Situs Jerusalem* (mit Abbildungen zur gesamten handschriflichen Überlieferung)', *Codices Manuscripti* 12 (1992), 121–51

—, 'Cosmology', in *Medieval Scandinavia: An Encyclopaedia*, ed. Philip Pulsiano (New York: Garland, 1993), pp. 110–11

—, 'Völundarhús - Domus Daedali: Labyrinths in Old Norse Manuscripts', *Nowele* 21 (1993), 323–68

—, *Heaven and Earth in the Middle Ages: The Physical World before Columbus*, trans. Angela Hall (Woodbridge: Boydell, 1996)

—, 'Snorri als Kosmograph', in *Snorri Sturluson: Beiträge zu Werk und Rezeption*, ed. Hans Fix (Berlin: Walter de Gruyter, 1998), pp. 255–66

—, 'Scandinavian World Maps', in *Trade, Travel, and Exploration in the Middle Ages: An Encyclopedia*, ed. John Block Friedman and Kristen Mossler Figg (London: Routledge, 2000), pp. 537–38

Springborg, Peter, 'Weltbild mit Löwe Die *Imago Mundi* von Honorius Augustodunensis in der altwestnordischen Textüberlieferung', in *Cultura classica e cultura germanica septentrionale*, ed. Pietro Janni, Diego Poli, and Carlo Santini (Rome: Herder, 1988), pp. 167–219

Stefán Karlsson, *The Icelandic Language* (London: Viking Society for Northern Research, 2004)

—, 'Resenshandrit', *Opuscula* 4 (1970), 269–78

Svanhildur Óskarsdóttir, 'Universal History in Fourteenth-Century Iceland: Studies in AM 764 4to' (PhD diss.,University of London, 2001)

—, 'Writing Universal History in Ultima Thule: The Case of AM 764 4to', *Mediaeval Scandinavia* 14 (2004), 185–94

—, 'The World and its Ages: The Organisation of an 'encyclopaedic' Narrative in MS AM 764 4to', in *Sagas, saints and settlements*, ed. Gareth Williams and Paul Bibire (Leiden: Brill, 2004), pp. 1–11

Sverrir Jakobsson, *Við og veröldin: Heimsmynd Íslendinga 1100–1400* (Reykjavík: Háskólaútgáfan, 2005)

—, 'Hauksbók and the Construction of an Icelandic World View', *Saga-Book* 31 (2007), 22–38

Syme, Ronald, 'Military Geography at Rome', *Classical Antiquity* 7 (1988), 227–51

Tenney, Frank, 'Classical Scholarship in Medieval Iceland', *The American Journal of Philology* 30:2 (1909), 139–52

Terkla, Dan, 'Introduction: Where to Fix Cadiz?', in *A Critical Companion to English Mappae Mundi of the Twelfth and Thirteenth Centuries*, ed. Dan Terkla and Nick Millea (Woodbridge: Boydell, 2019), pp. 1–20

—, 'Books and Maps: Anglo-Saxon Glastonbury and Geospatial Awareness', in *A Critical Companion to English Mappae Mundi of the Twelfth and Thirteenth Centuries*, ed. Dan Terkla and Nick Millea (Woodbridge: Boydell, 2019), pp. 44–67

—, 'Books and Maps: Anglo-Norman Durham and Geospatial Awareness', in *A Critical Companion to English Mappae Mundi of the Twelfth and Thirteenth Centuries*, ed. Dan Terkla and Nick Millea (Woodbridge: Boydell, 2019), pp. 68–91

—, 'The Duchy of Cornwall Map Fragment (c. 1286)', in *A Critical Companion to English Mappae Mundi of the Twelfth and Thirteenth Centuries*, ed. Dan Terkla and Nick Millea (Woodbridge: Boydell, 2019), pp. 197–226

Thoroddsen, Þorvald, *Landfræðissaga Íslands. Hugmyndir manna um Ísland, náttúruskoðun þess og rannsóknir, fyrr og síðar*, 4 vols (Reykjavík: Ísafoldatprentsmiðja, 1892–1904)

Tolmacheva, Marina, 'Geography, Chorography', in *Medieval Science Technology and Medicine: An Encyclopaedia*, ed. Thomas F. Glick, Steven Livesey, and Faith Wallis (London: Routledge, 2005), pp. 186–91

Toulmin Smith, Joshua, *The Discovery of America by the Northmen in the Tenth Century* (London: Charles Tilt, 1839)

van Duzer, Chet, 'The Psalter Map (c. 1262)', in *A Critical Companion to English Mappae Mundi of the Twelfth and Thirteenth Centuries*, ed. Dan Terkla and Nick Millea (Woodbridge: Boydell, 2019), pp. 179–96

Vasaru, Mervi Koskela, *Bjarmaland* (Oulu: University of Oulu, 2016)

Vídalín, Arngrímur, 'Óláfr Ormsson's *Leiðarvísir* and its Context: The Fourteenth-Century Text of a Supposed Twelfth-Century Itinerary', *JEGP* 117:2 (2018), 212–34

Villands Jensen, Kurt, 'The Blue Baltic Border of Denmark in the High Middle Ages', in *Medieval Frontiers: Concepts and Practices*, ed. David Abulafia and Nora Berend (Aldershot: Ashgate, 2002), pp. 173–83

von den Brincken, Anna-Dorothee, *Finnes Terrae: Die Enden der Erde und der vierte Kontunent auf mittelalterlichen Weltkarten* (Hanover: Hahnsche Buchhandlung, 1992)

Wallis, Faith, 'Chronology and Systems of Dating', in *Medieval Latin: An Introduction and Bibliographical Guide*, ed. Frank Anthony Carl Mantello and A. G. Rigg (Washington DC: The Catholic University of America Press, 1996), pp. 383–87

——, 'Bede', in *Medieval Science Technology and Medicine: An Encyclopaedia*, ed. Thomas F. Glick, Steven Livesey, and Faith Wallis (London: Routledge, 2005), pp. 81–83

——, *The Calendar and the Cloister: Oxford, St John's College MS17*, McGill University Library, Digital Collections Program, 2007, http://digital.library.mcgill.ca/ms-17 (accessed 29th July 2019)

Wellendorf, Jonas, 'Zoroaster, Saturn and Óðinn: The Loss of Language and the Rise of Idolatry', in *The Performance of Christian and Pagan Storyworlds*, ed. Lars Boje Mortensen and Thomas M. S. Lehtonen with Alexandra Bergholm (Turnhout: Brepols, 2013), pp. 143–70

Werlauff, Erich Christian, *Symbolas ad Geographiam Medii Ævi* (Copenhagen: Shultz, 1821)

Whaley, Diana, 'A Useful Past: Historical Writing in Medieval Iceland', in *Old Icelandic Literature and Society*, ed. Margaret Clunies Ross (Cambridge: Cambridge University Press, 2000), pp. 161–202

Whitfield, Peter, *The Charting of the Oceans: Ten Centuries of Maritime Maps* (Rohnert Park: Pomegranate Artbooks, 1996)

Whitaker, Ian, 'The Problem of Pytheas' Thule', *The Classical Journal* 77:2 (1982), 148–64

Wiedemann, Thomas, 'Sallust's 'Jugurtha': Concord, Discord, and the Digressions', *Greece & Rome* 40 (1993), 48–57

Wingfield, Emily, *The Trojan Legend in Medieval Scottish Literature* (Cambridge: D. S. Brewer, 2014)

Woodward, David, 'Medieval *Mappaemundi*', in *The History of Cartography*, vol. 1: *Cartography in Prehistoric, Ancient, and Medieval Europe and the Mediterranean*, ed. J. B. Harley and David Woodward (London: The University of Chicago Press, 1987), pp. 286–370

Þorbjörg Helgadóttir, 'On the Sallust Translation in *Rómverja saga*', *Saga-Book* 22:3–4 (1987–88), 263–77

INDEX

Page numbers in **bold** refer to figures.

Abraham 111, 148, 212, 217
Abū Maʿshar (Albumasar) 39–40
Adam 34, 35, 111, 114, 156–57, 172, **173**, 217
Adam of Bremen, *Deeds of the Bishops of Hamburg-Bremen* 120, 121, 132–34, 145, 165, 228
Aethicus Isther, *Cosmographia* 208, 212, 229
Africa
  continent 2, 15, 16, 17, 18, 122
  depiction on the Viðey Maps 109–10, 218–21
ages of man (*aetates hominum*) 103, 124–25, 150–51, 153, 155, 205–06, 238–39
Akbari, Suzanne Conklin 114
Alexanders saga (*Saga of Alexander the Great*) 13–14, 34
Alexandria 45, 215
Algorismus 64, 68, 105
Angevin Empire 112, 221
Antioch 45–47, 53
antipodes 33–34
Apostles 35, 45, 47, 53, 102n2, 157, 161
Apulia 47, 224
Arabia 111, 143, 217
Aristotle 39, 76–77, 85
Ari Þorgilsson the wise 69, 161–64, 166, 169
Armenia 209
Ascalon 214, 215
Ashman Rowe, Elizabeth 50
Asia
  continent 2, 15, 16, 18, 113
  depiction on the Viðey Maps 110–11, 148, 207–17
  and fantasies of Scandinavian origin 17, 138–41, 146, 180
Asia Minor 45, 111, 116, 140, 212
Assyria 213
Augustine (of Hippo), Saint 35
Árni Magnússon 9–10, 105–06
Ásgeir Jónsson (scribe for Árni Magnússon) 96n84, 105

Babylon 45, 47, 111, 213–14
  see also Cairo
Bactria 208
Beatus of Liebana 102n2
Bede, the Venerable 35, 39, 53, 55–56, 68, 88–89, 91, 127–28, 133, 136, 152, 153
Bible 107, 111
Biblical Kings *see* Three Magi
Birkholz, Daniel 177
Bjarms (*Bjarmar* / *Beormas*) 111–12, 123, 228
Bjarni Bergþórsson (computist) 56
Boulhosa, Patricia Pires 164, 177
Breen, Katharine 123
Breta sögur 136–37
Brittany 112, 113, 221–22
Brynjólfur Sveinsson (bishop of Skálholt) 70, 106
Byrhtferth 153, 156
Byzacium 219

Cairo 213, 218
Capella, Martianus, *De nuptiis Philologiae et Mercurii* 28, 82

Cappadocia 213, 216
cardinal directions 115, 123, 149–50, 153, 156–57, 174–75, 203, 234
  see also orientation
Caria 148, 210, 212
Carmania 148, 210
Carrhae 111, 114, 207, 212
Caspian Gates 207
Catalan Atlas 48, **49**
Cesarea 216
Chaldea 212, 213
Chekin, Leonid 104, 107, 116, 216
chieftains *see goðar*
Christ in Majesty 172, **173**
Church of the Holy Sepulchre 50–51, 53
Cicero, *Somnium Scipionis* 25–26, 33, 60–61, 75
Cicilia 212
Clavus, Claudius 3, 144
climatic zones 16, 18, 23, 26–28, 35, 61-62, 63, 99, 115, 116, 117, 195
Cnut the Great 119, 125
Colchis 208
Colour 33, 55, 63, 64, 90–91
Columbanus, Saint 12–13
Commagene 213
Commonwealth (Icelandic)
  organisation 160, 167–70, 174, 176
  end of 176–78, 227
computus 43–45, 53, 68, 70–71, 80, 87–89, 181
Conley, Tom 12
Constantinople 47, 61, 113, 115, 122, 182, 223–24
Cotton Map (Anglo-Saxon Cotton Map) 119–20, 123n15, 128, 141–42, 144
Crates of Mallos 26, 39, 91
Creation 57, 69, 172, **173**
Cyrene 219

Daniel of Morley 78
dansk tunga (*Dacisca lingua*) 125–26
De Lisle Psalter 151
Denmark
  depiction on medieval maps 120–21, 122, 123, 125–26, 225
  as border territory 121–22, 126
deserts 35n30, 110, 208, 221
Destombes, Marcel 24, 104
Dicuil, *De mensura orbis terrae* 130–31
Duchy of Cornwall Map 124–25, 151
Dudo of St Quentin 125n22

Earth, depiction on diagrams **41**, 44–45, **58**–59, 66, **74**–75, 82–**83**, 84, 89, **95**, 98
Easter 43, 70, 88
Ebstorf Map 4, 14, 152, 175, 207, 226
eclipses 31, 66, 68, 94–98, **95**, 105
Egypt 102, 113, 114, 215, 218
ekphrasis 12–14
Elucidarius 48, 151–52, 157, 217
England 47, 93, 119, 123, 222, 223
Ethiopia 110, 221
ethnonyms 112
Euphrates, River 111, 114, 211, 213–14
Europe
  continent 2, 15–18, 23, 28, 48, 64, 140
  depiction on the Viðey Maps 101, 102, 111–113, 115, 120–21
  alternative name *Enea* 15–17, 138–39
  see also Iceland
Eusebius of Caesarea 107, 143, 217
Evangelists 11, 171–74, **173**

fornaldarsögur ('sagas of ancient times') 112, 181
France 112, 113, 122, 222, 225
Frisia 112n32, 113, 225

Galatia 212, 214
Galicia 222
Galilee 215
Garamannia 220
Gascony 112, 221–22
Gaul 225–26
Gaulo 220
Gautier Dalché, Patrick 12, 108
Geoffrey of Monmouth, *History of the Kings of Britain* 136–37, 139, 140
geography
  as analytical category 5–8, 180–81
  and periodisation 113–14

relation to cartography 8, 12
Gerald of Wales 142, 165
Germania 113, 224–25
Gervase of Canterbury 4, 176
Getulia 110, 112, 220
Gizurr Ísleifsson (bishop of Skálholt) 163, 165, 166, 176
goðar (Icelandic chieftains) 160, 161, 164–65
as priests 164–66, 167
Gough Map of Great Britain 140
Grágás 176
Greece 112n32, 113, 122, 223
Grosseteste, Robert (bishop) 40
Grønlie, Siân 163
Guðmundur Arason (bishop of Hólar) 67, 80

habitability *see* climatic zones
Halldór Hermansson 2, 7
Haraldur Sigurðsson 2, 104
Harley, J. B. 9
Haukr Erlendsson *see* Hauksbók
Hauksbók 49, 136
Hákon IV Hákonarson of Norway 176, 177
Hákon Ormsson (*ráðsmaðr*) 69–70
Hebron 111, 217
Heimskringla 14–16, 61, 126, 139, 164, 170–71, 174
Helpericus (computist) 56, 68, **86**, 88–89, 94
Henry III of England 172
Hereford Map 102, 117, 175, 207, 208, 209, 211, 212, 214, 215, 216, 217, 219, 220, 221, 223, 224, 225, 227, 228
depiction of Iceland 128, 140, 141, 142, 143, 144
depiction of Scandinavia 121–22, 125
donor inscription 14, 124
as *estoire* 4, 114
languages 123–25
Hermann, Pernille 163, 168
Herodotus 101, 107
Hiatt, Alfred 123
Historia Norwegiae 134–35, 137–38
Hjalti Þorsteinsson (scribe) 10
Holy Roman Empire 112–13, 122
Homer, *Iliad* 12
Horeb, Mount *see* Sinai, Mount
Hugh of Saint Victor
Descriptio mappa mundi 107–108
Libellus de formatione arche 153–56, 174, 219
Hungrvaka 165, 168, 175, 176, 182
Hyrcania 208

Iberia (kingdom in the Caucasus) 211
Iceland
name origins 119–20,
association with Thule 127–28, 130–31, 132–38
appearance on medieval maps 141–43
appearance on the Viðey Maps 125, 126, 128, 142–46, 178, 181, 226–27
see also Thule
India 47, 61, 209
Ireland 47, 142, 222–23
Isauria 212
Isidore of Seville
Etymologies 2, 28, 42n50, 57, 73, 75, 107, 109–10, 111, 114, 129, 140, 172
On the nature of things 12–13
Italy 224
Íslendingabók 69, 131, 162–63, 168, 169–70, 174, 176
Íslendingasögur ('sagas of Icelanders') 168, 169, 177, 182, 227

Jacobus Angelus 3
Jaffa 111, 215
Jericho 215
Jerusalem 15–16, 42, 48–53, **52**, 60, 111, 180, 217–18
John the Baptist 43–44
Jonas of Bobbio 12
Jón Jóhannesson 168
Jón Sigurðsson 162
Judah 217
Julian calendar 23, 24, **41**, 42–44, 55, 56, 57, 88, 89

Kalinke, Marianne 136–37
Ketill Þorsteinsson (bishop of Hólar) 162, 163, 165, 166, 168, 176

Kiev *see* Rus'
Kline, Naomi Reed 109
Knýtlinga saga 125
Konnungs skuggsjá 16, 61
kringla heimsins 2, 15–16, 61, 101, 109, 146, 147, 181, 218
Kristín Bjarnadóttir 70
Kristni saga 161, 164–65, 176, 182
Kålund, Kristian 24, 64, 65, 66, 91, 104, 126, 228

Laird, Edgar S. 40
Lambert of Saint Omer *see* Liber Floridus
Landnámabók 119, 127–28, 131, 133, 135–37, 142, 144, 145, 146, 167, 182, 226
landvættir ('guardian-wights') 170–72, 174
languages on medieval maps 123–25
  English 123–24
  Old English 120
  Old Norse 28–31, 34, 42, 120, 122–23, 175
  Insular French (Anglo Norman) 4, 123–25
  Latin 23, 28, 122–23, 175
  macaronic (mixed-language) inscriptions 123, 216, 228
  see also dansk tunga, translation
Latin *see* languages on medieval maps
Lavezzo, Kathy 177
lawspeaker *see lögsögumaðr*
Lebanon, Mount 214
Leiðarvísir 50–51, 121, 126
Lewis, C. S. 11–12
Liber Floridus 27–**32**, 34, 36, 39, 50, 60, 63
Libya 109, 219
Lombardy 47, 224
Louis IX of France, Saint 13
Ludvig Larsson 166
lögsögumaðr (lawspeaker) 162, 167–68

Macrobius, Ambrosius Theodosius, *Commentary on the Dream of Scipio* 25–35, 39, 59, 60, 61, 67, 68, 92, 93
Magnús Einarsson (bishop of Skálhólt) 162, 164, 166, 175
Magnús Erlendsson, Earl of Orkney, Saint 43, 44, 80
Magnús Hákonarson VI of Norway 16, 75, 80
mappa mundi
  definition 1, 179
  medieval usage 4, 13, 92–93, 176
  other words meaning map 4, 10–11, 36
Māshā'allāh 78
Massagetes 114, 148, 207
Master Walter (unknown computist) 24, 42–43, 57
Matthew Paris 102, 112, 121, 123, 140
Mauritania 109, 220
Media 210
Mediterranean Sea 2, 3, 13–14, 16, 17–18, 39, 47, 92–93, 102, 103, 112, 226
Mesopotamia 111, 211
microcosm 103, 109, 148, 151–52, 156, 205–06, 239
Midianites 217
Moesia 113, 223, 229
moon
  phases 35–36, 59, 63, **90**, 91, 94–98, 190
  saltus lune 56, 88
  see also tides
multilingualism *see* languages on medieval maps

Nazareth 214–15
Nile, River 2, 47, 102, 103, 109, 113, 114, 218
Nítíða saga 61–62, 112n32
Noah's Ark 153, 209
  see also Hugh of Saint Victor
Normandy 112, 113, 221
Norway, depiction on medieval maps 111, 120–21, 125, 127, 227
Numidia 109, 220

orbis terrarum see kringla heimsins
O'Reilly, Jennifer 171
orientation 55, 99–100, 115–17, 152, 174, 180, 203
  see also cardinal directions

Orosius, Paulus 2, 107, 111–12, 207, 208, 209, 210, 212, 219, 220, 221, 223, 228, 229
Orri Vésteinsson 164, 166, 168
Ólafr II of Norway, Saint 11
Ólafr Tryggvason 126, 170

Palestine 111, 180, 214, 215, 216
Pamphylia 211, 212
papar (hermits) 131
parallels 23, 26–30, 33, 61, 122, 190–91
  equator 2, 23, 26–29, 31, 33–34, 50, 61–63, 99, 190–91
  tropics 23, 27–29, 31, 33, 38, 132–33, 190–91
  polar circles 2, 23, 27–29, 61–62, 99, 129, 116, 183, 190–91
Parthia 111, 209
Pentapolis 219
Persia 210
Philistis 214, 215
Phrygia 111, 116–17, 140, 210, 211, 212, 228–29
Pierre d'Ailly (theologian) 85, 117
Planets
  domiciles 94, **95**, 96–97
  heliocentrism **74**, 77–78
  inferior planets (Mercury and Venus) 75–79, 85, 98
  orbits **41**, 42, 44, 59, 72–73, 75–79, 81–82
  retrograde motion 73, 76, 78–79, 82, 84, 99
  superior planets (Mars, Jupiter, and Saturn) 82–85, 98
  see also moon, sun
Pliny the Elder 39, 75, 82, 107, 120, 129, 208
portolan charts (sea charts) 3, 10, 13, 18, 48
priesthood in Iceland
  register of forty Icelandic priests 148, 157, 158–69, 174–75, 176–78
  social status 164–67
  see also goðar, as priests
Psalter Maps 102
Ptolemy, Claudius
  Geography 3, 144
  Hypotheses of the Planets 76–77
  Almagest 77, 78n47
Ptolomais 214
Pyrenees 113, 226
Pytheas Masilliensis (of Marseille) 127, 129–30, 131, 132–33, 141, 142, 146

Quarters of Iceland 149, 157, 158–61, 162, 166, 168–71, 174–75, 177, 203
  primary colonists 169–70, 174
  see also landvættir

Rafn, Carl Christian 6, 23–24, 64, 104, 116, 143–44
retrograde motion *see* planets
Revelation 143, 171–72
riddarasögur ('knights' sagas') 42n50, 181–82
Robertus Anglicus 87
Roger of Howden, *Expositio mappa mundi* 108
Rome 25, 51, 115, 121, 126, 140, 181, 224
Rómverja saga 109–10, 220
Rus' 6, 104, 111, 120, 121, 122, 180, 227–28, 229
Russia *see* Rus'

Sacrobosco, Johannes de 68, 75, 77, 78, 79, 80–82, 87, 89, 99, 105, 153
Sallust 107, 109–10, 220
Sawley Map 141, 142, 207
Saxo Grammaticus, *Gesta Danorum* 128, 134–35, 145
Saxony 113, 120, 121–22, 126, 224–25
Scotland 47, 112n32, 222–23
Scythia 111, 114, 116–17, 130n36, 228
seasons 44, 103, 109, 147, 148, 150–51, 152, 153, 155–56, 171, 205, 236–38
Seneca 75, 132
Seres 208
Sidon 216, 217
Simek, Rudolf 7, 10, 16, 24, 34–35, 64, 88, 91, 104
Sinai, Mount 111, 143, 217
Snorra Edda 16–18, 61, 138–41
Snorri Sturluson 14, 16, 17, 61, 69, 126, 138–140, 166, 176, 211

Solinus, Gaius Julius, *Collection of Remarkable Facts* 129
Spain 10, 47, 112n32, 113, 222
Sparta 113, 223, 229
stars 71, 73, 81, 172
star of Bethlehem 48–**49**, 68
Strabo 129–130, 142
Sturlungaöld ('Age of the Sturlungs') 176
sun
  on diagrams 23, **29**, 31–33, 38, 44–45, **54**, **74**–75, 82–85, **90**–91, 94–98, **188**–190
  ecliptic 26, 28, 31, 33, 37, 38, 43–44, 59, 33
  equinox 28, 31, 36, 38n41, 39, 44, 87, 153, 205, 238
  heat of 16, 17, 18, 35, 61, 87
  influence on tides 23, 36–40, 55, 92, 96–98
  and microcosm 152
  midnight sun 29, 127–30, 133, 142, 183
  size of 59, 87
  solstice 28, 31, 36–39, 43–44, 55, 79, 107n18, 130, 132–33, 153
  see also planets, heliocentrism; tides
Sweden
  depiction on medieval maps 6, 15–16, 111, 121, 145, 227
  Götaland 6, 111, 120, 123, 227
  Svealand 6, 111, 120, 123, 227
Syria 215
Sæmundr Sigfússon the wise 165–66

Tanais, River 2, 15, 17, 103, 110, 113, 138, 139–40, 146, 183, 216
Tanakvísl *see* Tanais, River
Tarsus 212
Teiknibók 11, 172–**73**
Terkla, Dan 124-25
Theodoricus Monachus, *The Ancient History of the Northern Kings* 128, 131, 135, 143
Thrace 113, 223
Three Magi 42, **46**, 48–**49**
Thule 226
  name origins 127, 129–30
  in Roman literature 131–32
  see also Iceland
tides 23, 24, 35–40, 42, 45, 55–57, 59, 60, 68, **90**, 92, 94–98, 183, 190
Tigris, River 114, 211
T-O maps
  form 2, 12–14, 16, 18, 101–03, 106, 109, 115, 158
  in manuscript illumination 11, 172, **173**
Translation 29–31, 34, 42, 47, 60, 79, 81–82, 88, 99, 122, 181
Troglodytes 110, 112, 219
Troy 16-17, 111, 114, 139, 140, 146, 210–11
  see also Asia, and fantasies of Scandinavian origin
Tyre 216

Vanir (Norse divinities) 17, 139, 146, 216
Vatican Map 116
vernacular *see* language
Viðey (monastery) 65, 69–70, 140
Virgil 19, 25, 26, 60, 61, 131–32, 134, 137
Vínland 6, 113, 143–44

Wallis, Faith 71, 88
Walter of Châtillon, *Alexandries* 13, 42n50
Westrem, Scott D. 212, 227
William of Conches 16, 27, 39, 78, **86**, 88, 91–94, 99, 103
William of Newburgh 128, 137, 143, 145
winds 91, 92, 152, 155
  on diagrams 57, 109, 153
  on *mappae mundi* 103, 104, 148, 149–50, 153, 204–05, 234–35
Woodward, David 174

zodiac 23, 26, 27, **29**, 31–33, 37–38, 45, 59, 67, 68, 72, 82, 94–97, 148, 150, 154, 156, 190–91, 235–36

Þorlákr Runólfsson (bishop of Skálholt) 162, 176
Þormóður Torfason 10

Æsir (Norse divinities) 17, 138, 140

Studies in Old Norse Literature

1 EMOTION IN OLD NORSE LITERATURE
Translations, Voices, Contexts
Sif Rikhardsdottir

2 THE SAINT AND THE SAGA HERO
Hagiography and Early Icelandic Literature
Siân E. Grønlie

3 DAMNATION AND SALVATION IN OLD NORSE LITERATURE
Haki Antonsson

4 MASCULINITIES IN OLD NORSE LITERATURE
Edited by Gareth Lloyd Evans and Jessica Clare Hancock

5 A CRITICAL COMPANION TO OLD NORSE LITERARY GENRE
Edited by Massimiliano Bampi, Carolyne Larrington and Sif Rikhardsdottir